The IMA Volumes
in Mathematics
and Its Applications

Volume 11

Series Editors
George R. Sell   Hans Weinberger

Institute for Mathematics and Its Applications
IMA

The **Institute for Mathematics and Its Applications** was established by a grant
from the National Science Foundation to the University of Minnesota in 1982. The
IMA seeks to encourage the development and study of fresh mathematical concepts
and questions of concern to the other sciences by bringing together mathematicians and
scientists from diverse fields in an atmosphere that will stimulate discussion and collaboration.

The IMA Volumes are intended to involve the broader scientific community in this
process.

Hans Weinberger, Director
George R. Sell, Associate Director

IMA Programs

1982–1983 Statistical and Continuum Approaches to Phase Transition

1983–1984 Mathematical Models for the Economics of Decentralized Resource
Allocation

1984–1985 Continuum Physics and Partial Differential Equations

1985–1986 Stochastic Differential Equations and Their Applications

1986–1987 Scientific Computation

1987–1988 Applied Combinatorics

1988–1989 Nonlinear Waves

1989–1990 Dynamical Systems and Their Applications

Springer Lecture Notes from the IMA

The Mathematics and Physics of Disordered Media
Editors: Barry Hughes and Barry Ninham
(Lecture Notes in Mathematics, Volume 1035, 1983)

Orienting Polymers
Editor: J.L. Ericksen
(Lecture Notes in Mathematics, Volume 1063, 1984)

New Perspectives in Thermodynamics
Editor: James Serrin
(Springer-Verlag, 1986)

Models of Economic Dynamics
Editor: Hugo Sonnenschein
(Lecture Notes in Economics, Volume 264, 1986)

Mary F. Wheeler
Editor

Numerical Simulation in Oil Recovery

With 84 Illustrations

Springer-Verlag
New York Berlin Heidelberg
London Paris Tokyo

Mary F. Wheeler
Department of Mathematics
Rice University
Houston, Texas 77251, USA

ISBN-13:978-1-4684-6354-5   e-ISBN-13:978-1-4684-6352-1
DOI: 10.1007/978-1-4684-6352-1

AMS Classifications: 35A40, 35K55, 35L65, 35J70, 39A10

Library of Congress Cataloging-in-Publication Data
Numerical simulation in oil recovery.
    (The IMA volumes in mathematics and its applications ; v. 11)
    In part the proceedings of the Symposium on Numerical Simulation in Oil Recovery, held Dec.
1-12, 1986, at the Institute for Mathematics and Its Applications, University of Minnesota.
    Includes bibliographies.
    1. Oil fields–Production methods–Computer simulation–Congresses. I. Wheeler, Mary F. (Mary
Fanett) II. Symposium on Numerical Simulation in Oil Recovery (1986 : Institute for Mathematics
and Its Applications, University of Minnesota) III. Series.
TN871.N88   1988   622′.3382′0724   87–28708

Camera-ready copy provided by the author.
Printed and bound by R.R. Donnelley & Sons, Harrisonburg, Virginia.

The IMA Volumes in Mathematics and Its Applications

Current Volumes:

Forthcoming Volumes:

CONTENTS

FOREWORD

This IMA Volume in Mathematics and its Applications

NUMERICAL SIMULATION IN OIL RECOVERY

is in part the proceedings of a workshop which was an integral part of the 1986-87 IMA program on SCIENTIFIC COMPUTATION. We are grateful to the Scientific Committee: Bjorn Engquist (Chairman), Roland Glowinski, Mitchell Luskin and Andrew Majda for planning and implementing an exciting and stimulating year-long program. We especially thank the Workshop Organizer, Mary Fanett Wheeler for organizing a workshop which brought together many of the major figures in a variety of research fields connected with reservoir modelling for a fruitful exchange of ideas.

George R. Sell

Hans Weinberger

PREFACE

During 1 December through 12 December 1986 a Symposium on Numerical Simulation in Oil Recovery was held at the Institute for Mathematics and its Applications, University of Minnesota. The major research emphasis of this meeting was the modeling of fractures, heterogeneities, viscous fingering, and diffusion-dispersion effects in the flow in porous media. The participants included well-known applied mathematicians, chemical engineers, physicists and hydrologists from universities, national laboratories, and industrial companies. The objective of this meeting was to develop interaction and cooperation between researchers with similar interests in fluid flow but with somewhat diverse research backgrounds. The papers in this volume were part of the many excellent presentations made during this symposium.

We wish to thank the contributers to this volume as well as those who because of prior commitments were unable to provide a manuscript. Special thanks to Dr. Robert Jerry Blackwell. As one of the pioneers in the study of viscous fingering his experience and enthusiasm did much to make this workshop a success. In addition we gratefully acknowledge the help and support of the staff of the IMA, Professors Hans Weinberger and George Sell, Mrs. Pat Kurth, and Mr. Robert Copeland. Their assistance in arranging the workshop and their warm hospitality made this a most enjoyable and stimulating meeting.

Mary Fanett Wheeler

EFFECTIVE BEHAVIOR OF TWO-PHASE
FLOW IN HETEROGENEOUS RESERVOIR

Brahim AMAZIANE
Centre de Mathématiques, I.N.S.A. 403
20, Avenue A. Einstein, 69621 Villeurbanne, France

Alain BOURGEAT
Faculté des Sciences et Techniques
Université de Saint-Etienne
23, rue du Dr Paul Michelon, 42023 Saint-Etienne, France

A multiscale analysis based on two widely different length scales that exist in a reservoir, the characteristic length of the unit porous cell and the global scale of the reservoir, is carried out to investigate two-phase flow model in periodic system. This paper aim is to rigorously derive reservoir homogenized or global equations from exact local two-phase flow equations in the porous medium, not only for justifying the homogenized equations but also for delineating the meaning of the effective parameters in the global equations.

1. Introduction.

The need to describe diffusion / conduction or transport phenomena in heterogeneous media with complex heterogeneities geometry, in term of homogenized equations with effective or bulk parameters have inspired a large amount of litterature. But in fluid flow there are essentially two problems which are adressed. The first one is to get the Darcy law (i.e. homogenized equation) in the porous media from the Stokes equation (exact local equation) in each pore. The second one is to get an effective or bulk conductivity / diffusivity or permeability from the exact local conductivity/diffusivity or permeability. But whatever the approaches to this problem, periodic (Sanchez - Palencia 1980) or random model (Papanicolaou 1980) using multiscale method (Bensoussan et al. 1978) or volume averaging (Ryan 1980) or moment analysis (Brenner 1980) or generalised function method (Perrins et al. 1979), they are all dealing with linear phenomena like linear diffusion / conduction or permeability in one-phase flow.

We are using multiscale technique in order to give rigorous proofs of convergence by means of functional analysis and H-convergence (Murat 1978). In this technique, the assumption of periodicity looks like rather restrictive but we must admit it is by now the only technique which allow to adress non linear phenomena as the two-phase flows system of equations. We must also recall that this assumption of periodicity could be released; a non uniform periodicity (i.e. periodic coefficients depending on $\underline{x}$ the space variable) is sufficient moreover to get analogous result, and to account for a wide range of physical heterogeneities. We must also recognize that in case of diffusion / conduction or transport equation in heterogeneous media numerical results of the multiscale technique with accordance with periodicity assumptions are in good accordance with experimental data for random media (Chang 1982).

Two-phase flow equations are used to modeling oil reservoir exploitation by fluid drive. An immiscible wetting phase (water) or a miscible fluid phase (hydrocarbon mixture) is injected and displace a second phase (oil) in a porous medium (reservoir). Reservoirs are made up of various porous media and fluids displacement process are then described by exact local equations.

In the section that folllows we derive homogenized equations, i.e. simplified equations describing two-phase flows in equivalent homogeneous media, from local equations which account for heterogeneities, gravity and capillary effects. Homogenized model is rigorously derived by multiscale technique and justified by functional analysis. Numerical studies have been performed and results from homogenized equations are compared to those one obtained from exact local equations.

In the last section we are investigating a more complicated exact local model which account for saturation jumps at the interfaces due to capillary pressure and relative permeability curves.

2. Formulation of Two-Phase Flow in Heterogeneous Reservoirs.

Standing governing equations describing two immiscible fluid phases with no mass transfer between the fluids (Ewing 1983) are made up of equations describing conservation of mass in each phase coupled via Capillary Pressure law $\mathcal{P}_c(x,S)$, figure 2-1, and relative permeability curves $k_{ri}(S)$ in each phase, S_i and P_i, $i = 1,2$, where subcript 1 is for the wetting phase.

But in the sequel we will use a formulation obtained after transformation suggested by Chavent (1976) and Kruzkov et al. (1977).

Defining the total and phase mobilities as

$$d(S) := \frac{k_{r1}(S)}{\mu_1} + \frac{k_{r2}(S)}{\mu_2} \qquad (2\text{-}1)$$

$$\lambda_i(S) := \frac{k_{ri}(S)}{d(S)\mu_i}, \quad i = 1,2 \qquad (2\text{-}2)$$

defining a reservoir "reduced" pressure P and a "reduced" wetting saturation S as

$$P := \frac{P_1 + P_2}{2} + P_{cm}(x) \int_{S_c}^{S} (\lambda_1(S) - \frac{1}{2}) \frac{dP_c(s)}{ds} \, ds, \qquad (2\text{-}3)$$

$$S := \frac{S_1 - S_{1m}}{S_{1M} - S_{1m}}$$

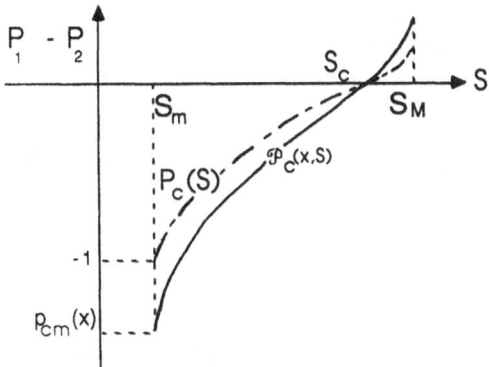

figure 2-1 . Capillary pressure curve at fixed x, $P_1 - P_2 := \mathcal{P}_2(S) := p_{cm}(x) \times P_c(S)$

are defining the total volumetric flow, the capillary flow and gravitational flow as

$$q_0 := -K(x)d(S)\ (\nabla P + \gamma_1(S)\nabla P_{cm} + \gamma_2(S)\nabla P_g) \qquad (2\text{--}4)$$

$$q_1 := -K(x)\ \nabla p_{cm}(x), \qquad (2\text{-}5)$$
$$q_2 := -K(x)\ \nabla P_g(x), \qquad (2\text{-}6)$$

where P_g is the gravitational potential , $P_g := -\rho_m g Z(x)$, K is the absolute permeability tensor and

$$\gamma(S) := \int_{S_c}^{S} (\lambda_1(S) - 1/2)\frac{dP_c}{dS}(s)ds \qquad \gamma_1(S) := \int_{S_c}^{S} \frac{d\lambda_1}{dS}(s)\ P_c(s)ds$$

$$\gamma_2(S) := (\lambda_1\rho_1 + \lambda_2\rho_2)/\rho_m$$

with ρ_i the phase density of fluid phase i, ρ_m the mean density $1/2(\rho_1 + \rho_2)$.
we obtain

for $x \in \Omega$, $t \in [t_0, t_1]$, $\qquad (2\text{-}7)$
$\nabla . q_0(S,P) = $ sources,

$$\Phi(x)\frac{\partial S}{\partial t} + \nabla . (-K(x)p_{cm}(x)a(S)\nabla S + \lambda_1(S)q_0 + b_1(S)q_1 + b_2(S)q_2) = \text{sources} \qquad (2\text{-}8)$$

where

$$a(S) := d(S)\lambda_1(S)\lambda_2(S) \times -\frac{dP_c(S)}{dS} \qquad (2\text{-}9)$$

$$b_1(S) := d(S)\ \lambda_1(S)\ \lambda_2\ (S) \times P_c(S) \qquad (2\text{-}10)$$

$$b_2(S) := d(S)\ \lambda_1(S)\ \lambda_2(S) \times (-\frac{\rho_1 - \rho_2}{\rho_m}) \qquad (2\text{-}11)$$

The equation for the saturation (2-8) is a non linear diffusion-convection equation. The coefficient of the capillary diffusion therms, the three last terms in the left side of (2-8), consist of a gravity free transport term and of capillary and gravitational flow modifying terms.

In the sequel we will denote the diffusion flux by

$$r := -K(x)p_{cm}(x) \, a(S)\nabla S \tag{2-12}$$

and the flow

$$\psi := r + \lambda_1(S) \, q_o + b_1(S) \, q_1 + b_2(S)q_2. \tag{2-13}$$

This system (2-7), (2-8) appear to be very similar to the governing equation for miscible displacement :

$$\nabla.q := \nabla.[- \frac{K(x)}{\mu(C)}(\nabla P + \rho(C)\nabla P_g)] = \text{sources} \tag{2-14}$$

$$\Phi \frac{dC}{dt} + \nabla. \, [D(x,C)\nabla C + C.q] = \text{sources} \tag{2-15}$$

where C is the concentration of the invading fluid and D is the diffusion/dispersion tensor.

But in the equation (2-15), the coefficient of the diffusion term does'nt degenerate and the convection term is "linear" in C.

3. Homogenization without Interface Phenomena.

In this section and in the next ones, we consider a network of uniformly spaced cells which is part of a reservoir, fig.3-1; this reservoir area, Ω, under consideration, is assumed far enough from wells and boundaries, in order to avoid boundary layers. Heterogeneous cells are equally spaced and of size 1 which is small comparatively to the reservoir area size L; then $\epsilon := $ 1/L is a small parameter.

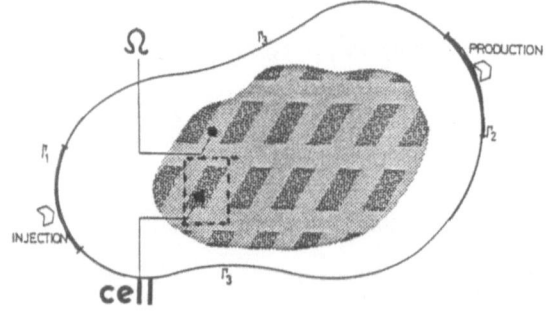

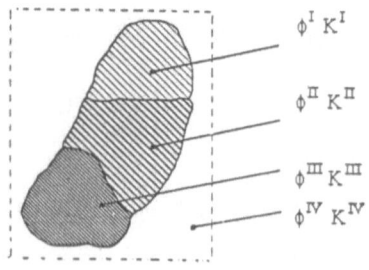

figure 3-1 .Schematic heterogeneous periodic porous media area .

figure 3-2 .A typical cell in the Reservoir area under consideration,Ω

Porosity Φ and permeability K, are rapidly oscillating functions on the space variable and are ε-periodic uniformly (i.e. are functions of x/ε only) or non uniformly (i.e. are functions of both x and x/ε).

Assuming continuity of saturation at the contact surface of two porous media, as in Bourgeat (1985) and in the next sections , is equivalent to getting the same relative permeability curve in both porous media. But to take account for possible saturation jumps at the interface between the two porous media we need to consider unlike curves of capillary and unlike curves of relative permeability in each porous medium, as in the last section 5.

Under appropriate conditions on the coefficients as

$q_j \in [L^2(\Omega)]^n$ $j = 1,2,$

$d, \gamma, a, b_j, \gamma_j, \lambda_j, \; j = 1,2,$ continuous and bounded applications of
$$\mathbb{R} \quad \text{into} \quad \mathbb{R}.$$

$d(S) \geq m > o,$ $a(s) \geq o,$ $\lambda_1(S) \in [o,1]$ $\forall S \in \mathbb{R}.$

β^{-1}, the inverse of

$$\beta(S) := \int_o^S (a(s))^{1/2} ds$$

is a Hölder function of order $\theta, o < \theta < 1$ and $\underset{o<S<1}{\text{Sup}} \; [b_j(S)/(a(S))^{1/2}] < + \infty \; j=1,2$

with initial conditions

$S(x,o) = S_o(x),$ $x \in \Omega$

and with appropriate boundary conditions, we may prove rigorousely the following result (Bourgeat 1985)

Theorem 3.1

The global equations, i.e. the limit of (2-7), (2-8) when $\varepsilon \to o$ is :

$$\nabla.q_o(S,P) = \text{sources}$$
$$(3.1)$$
$$\phi^\sim \frac{\partial S}{\partial t} + \nabla.(-K^\# a(S)p_{cm}(x)\nabla S + \lambda_1(S)q_o + b_1(S)q_1 + b_2(S)q_2) = \text{sources} \qquad (3-2)$$

where $\phi^\sim$ and $K^\#$ are effective parameters and are no longer depending on x. Equations (3.1), (3.2) are equations of an equivalent homogeneous porous media.

In the more general case, i.e. K non isotropic and non diagonal we get

$K^\#_{ij} = $ average on a cell Y of $\{K_{ij}(y) + \alpha_{ij}(y)\}$ (3.3)

where Y is the magnified cell (variable y) obtained from the initial cell (variable x) by doing the change of variable $y := x/\varepsilon$

and where the correction term α_{ij} is

6

$$\alpha_{ij}(y) := K_{ik} \frac{\partial w^k}{\partial y_j}(y) \qquad (3\text{-}4)$$

with $w^k(y)$ is Y-periodic solution of the p.d.e.

$$- \nabla.(K(y)\nabla w^k(y)) = \nabla.(K(y)e_k) \qquad \text{on } Y \qquad (3\text{-}5)$$

$e_k := $ the k-th vector in the canonical base of $\mathbb{R}^3$.

However to get this resultl some additional assumptions are required, like

$$d(S) = \text{constant} \qquad (3\text{-}6)$$

and

$$q_o^{\varepsilon} \text{ bounded in } L^{\infty}(\Omega) \qquad (3\text{-}7)$$

In some special cases we get some well known results. For instance in the case where Ω is one-dimensional, (3-6) and (3-7) are no more necessary and we get

$$K^{\#} = \text{ the arithmetic mean, on cell, of } \{K(y)\} := K^{\sim} \qquad (3\text{-}8)$$

In the case where K is diagonal non isotropic, when the flow q_o^{ε} is orthogonal to the x_2-axis and with heterogeneous media made up of stratas orthogonal to the x_1-axis we get

$$K^{\#}_{11} = K^{\sim}_{11} \text{ and } K^{\#}_{22} := \text{the harmonic mean on a cell of } \{K(y)\} \qquad (3\text{-}9)$$

4. Numerical Studies.

Numerical simulations have been done on layered porous sample core, 1-D simulations, fig.4-1 to 4-8, and on a layered quarter five-spot, 2-D Simulations fig.4-9 to 4-16. Numerical Simulations have been performed at INRIA-France with the code BIDIMIX, INRIA-IFP-SNEA(P), Chavent et al. (1987), Chavent et al. (1985).

1-D. Simulations

Simulations 1 to 5 are on a porous sample core, fig.4-1, of length 100 meters and width 10 meters. Viscosities are $\mu_1 = 1.05$ cp and $\mu_2 = 1.52$ cp, mobility ratio is approximately 0.1. In both simulations we consider a Peclet number $Pe := (L\phi^{\sim}q_o)/K^{\#}$.

In Simulations 3 to 5 we are comparing the effect of the Porous media I / Porous media II ratio, i.e. l_1/l_2, on homogenized results accuracy.

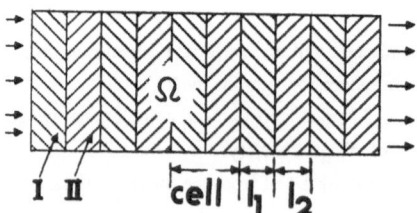

figure 4-1 .Porous sample core,with five cells and 10 stratas.

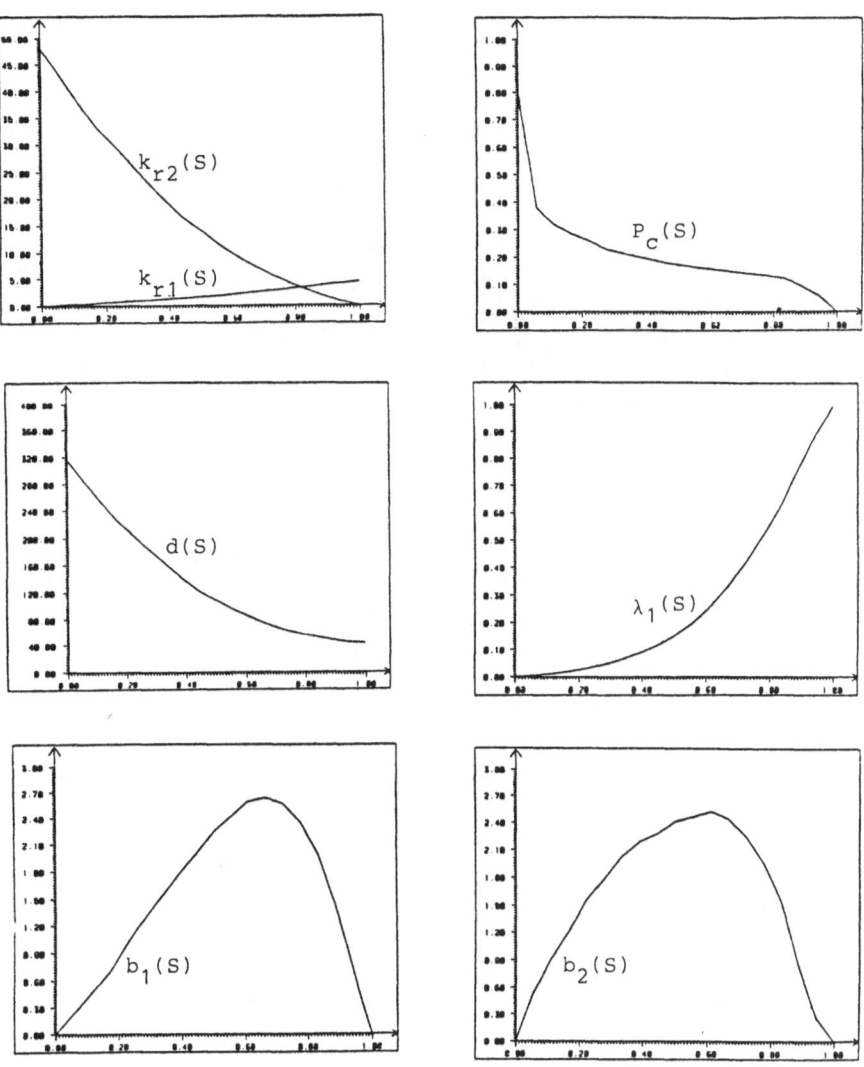

figure 4-2   Typical curves used for numerical simulations

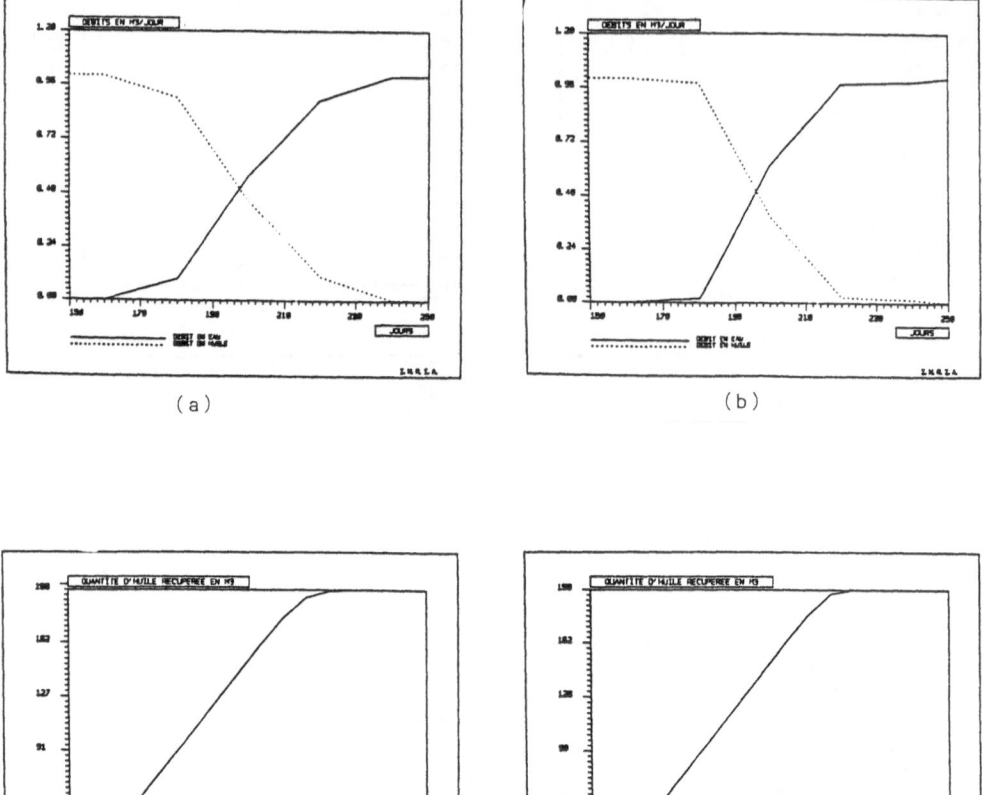

(a)   (b)

(c)   (d)

fig 4-3. Simulation 1, amont of Recovered Oil (a) heterogeneous
(b) homogenized. Flow Rates on Right Side (c) heterogeneous,
(d) homogenized.

In simulation 1, datas are : Total volumetric injected flow $q_o = 1 m^3$/day, 5 cells, with media I : $\phi_1 = 0.1$, $K_1 = 60$ md and media II : $\phi_2 = 0.2$, $K_2 = 300$ md; homogenized medium : $\phi^{\sim} = 0.2$, $K^\# = 100$ md, Pe = 0.2.

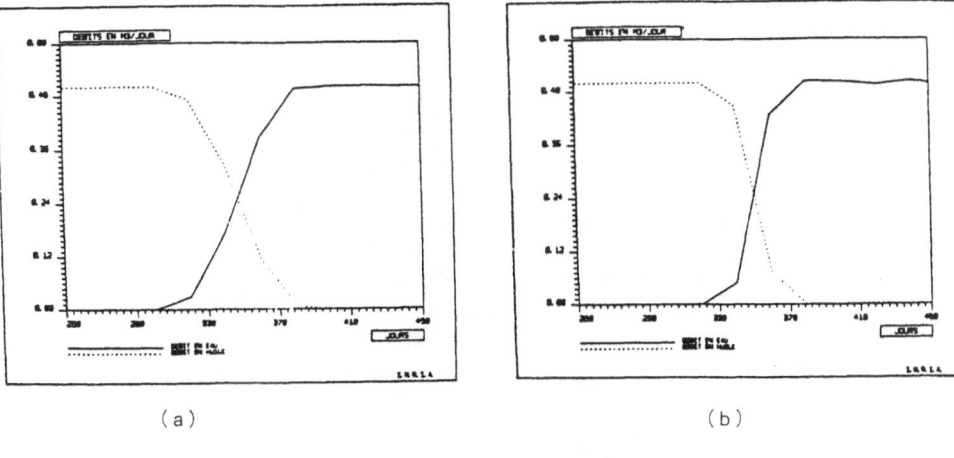

(a)   (b)

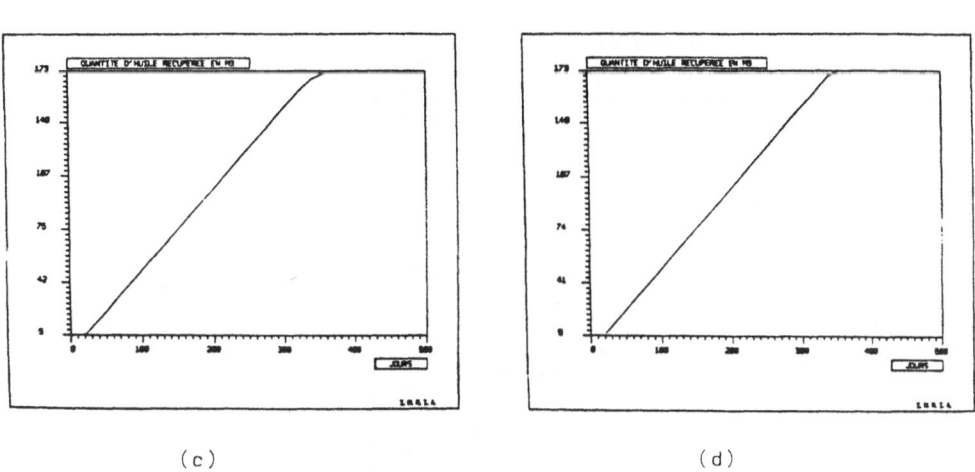

(c)   (d)

fig 4-4. Simulation 2, amont of recovered Oil (a) heterogeneous
(b) homogenized   Flow Rates on Right Side (c) heterogeneous,
(d) homogenized

In simulation 2, datas are : Total volumetric injected flow $q_o = 0.5$ m^3/day, 10 cells, with media I : $\phi_1 = 0.05$, $K_1 = 3$ md and media II : $\phi2 = 0.3$, $K_2 = 300$ md; homogenized medium : $\phi^\sim = 0.175$, $K^\# = 5.94$ md, Pe = 1.47.

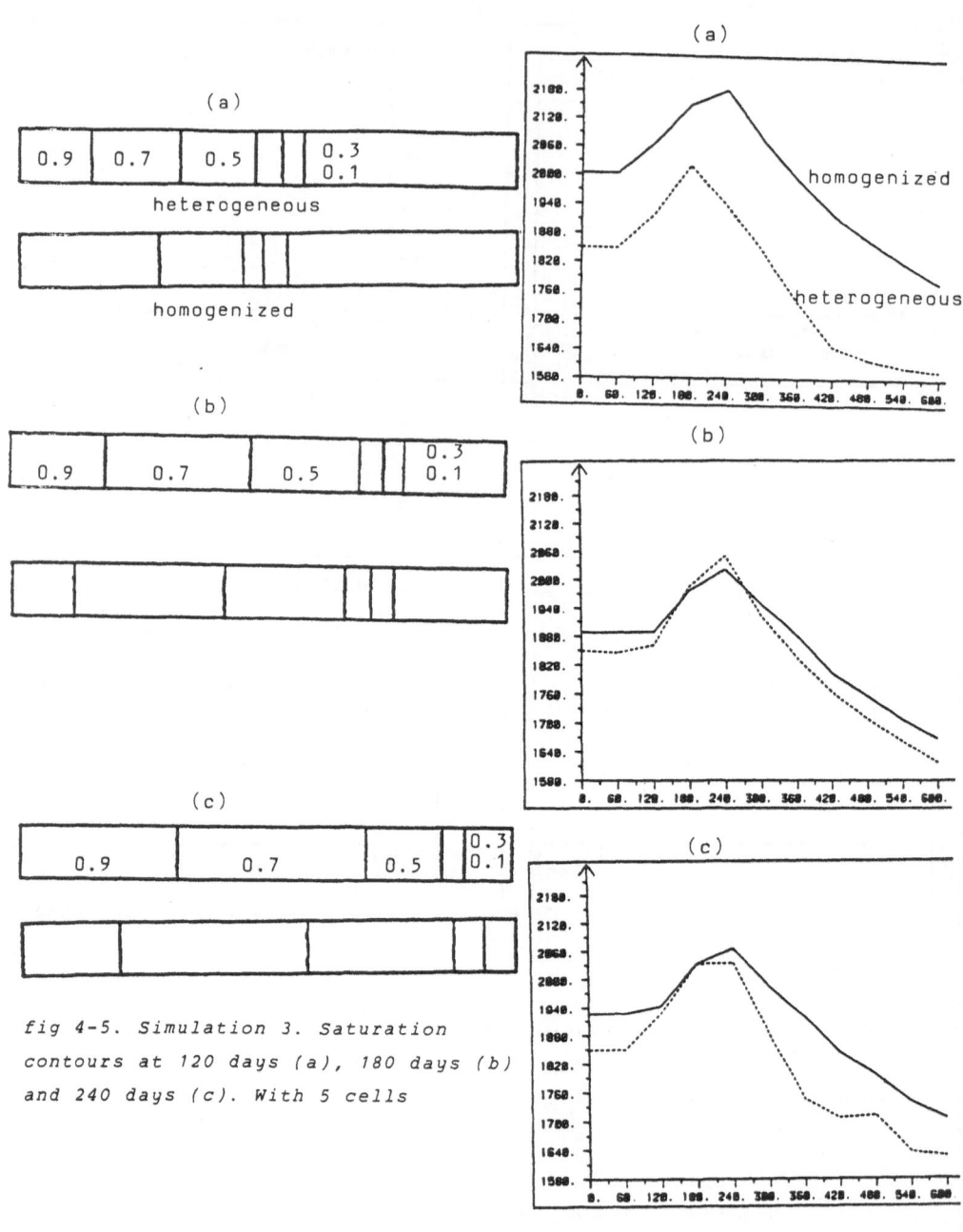

(a)

| 0.9 | 0.7 | 0.5 | | 0.3
0.1 |

heterogeneous

homogenized

(b)

| 0.9 | 0.7 | 0.5 | | 0.3
0.1 |

(c)

| 0.9 | 0.7 | 0.5 | 0.3
0.1 |

fig 4-5. Simulation 3. Saturation contours at 120 days (a), 180 days (b) and 240 days (c). With 5 cells

(a) homogenized / heterogeneous

(b)

(c)

fig 4-6. Simulation 3. Pressure Curves (bar) vs. time, at the middle of the core sample. Core with 5 cells (a), 10 cells (b), 20 cells (c)

In simulation 3, datas are : Total volumetric injected flow $q_o = 0.5$ m^3/day, media I : $\phi_1 = 0.03$, $K_1 = 0.3$ md and media II : $\phi_2 = 0.3$, $K_2 = 300$ md; homogenized medium : $\phi^{\sim} = 0.165$ $K^{\#} = 0.6$ md, Pe = 13.76.

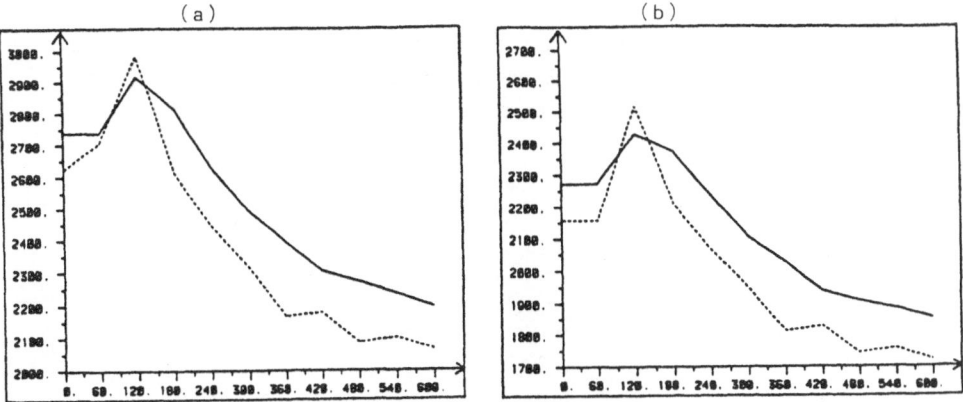

fig 4-7. Simulation 4, ratio porous media I / porous media II:=l1/l2=
1/4. Pressure curves (bar) vs. time at the middle of the core in the
media I (a), in the media II (b). $\tilde{\Phi}=0.097$, K =0.4, Pe=12.1.

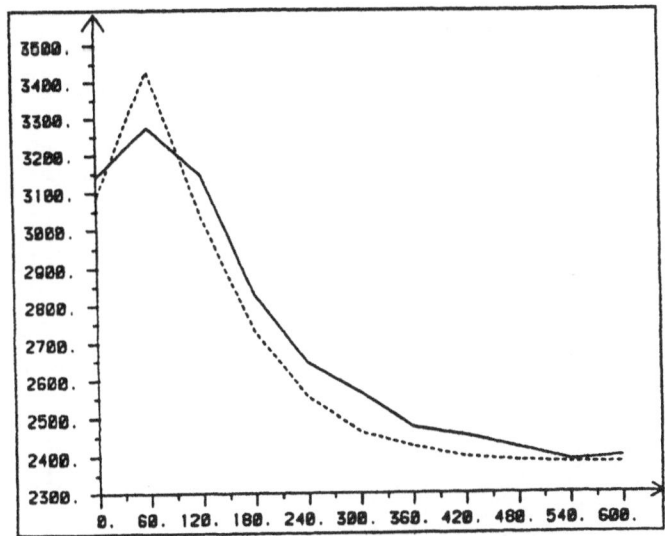

fig 4-8. Simulation 5, ratio porous media I / porous media II
:=l1/l2=1/10 pressure curves (bar) vs. time, at the middle of the
core in the media I $\tilde{\Phi}=0.057$, K =0.333 and Pe=8.6.

In simulations 4 and 5, above, heterogeneous medium datas are the same as in
simulation 3, with 5 cells.

2-D Simulations.

Simulations have been done on a quarter five-spot, fig.4-9, of length 100 meters, with a 10×10 grid made up of 10 stratas and five cells. Viscosities are $\mu_1 = 1.05$ Cp and $\mu_2 = 1.52$ cp, mobility ratio is 0.1. We consider two Peclet numbers Pex : $= (L\phi^\sim qo)/K^\#_{11}$ and Pey : $= (L\phi^\sim qo)/K^\#_{22}$.

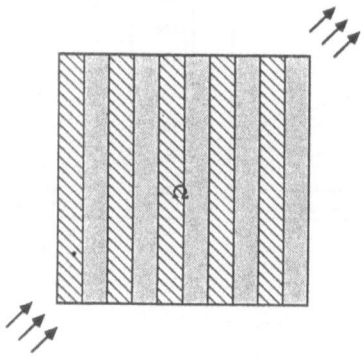

figure 4-9 . Quarter five-spot, 5 cells, 10 stratas.

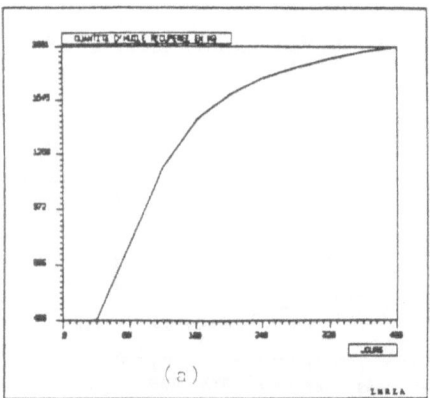

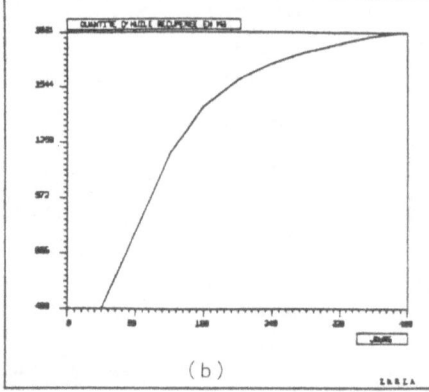

fig 4-10 Simulation 6. Recovery curves, (a) heterogeneous, (b) homogenized. $q_0 = 10 m^3/day$, Media I: $\Phi_1 = 0.1$ $K_{11} = 60$ md $K_{22} = 100md$, Media II: $\Phi_2 = 0.3$ $K_{11} = 300$ md $K_{22} = 300$ md, Homogenized medium: $\Phi^\sim = 0.2$ $k_{11} = 100$ md $K_{22} = 200$ md Pex=2 Pey=1

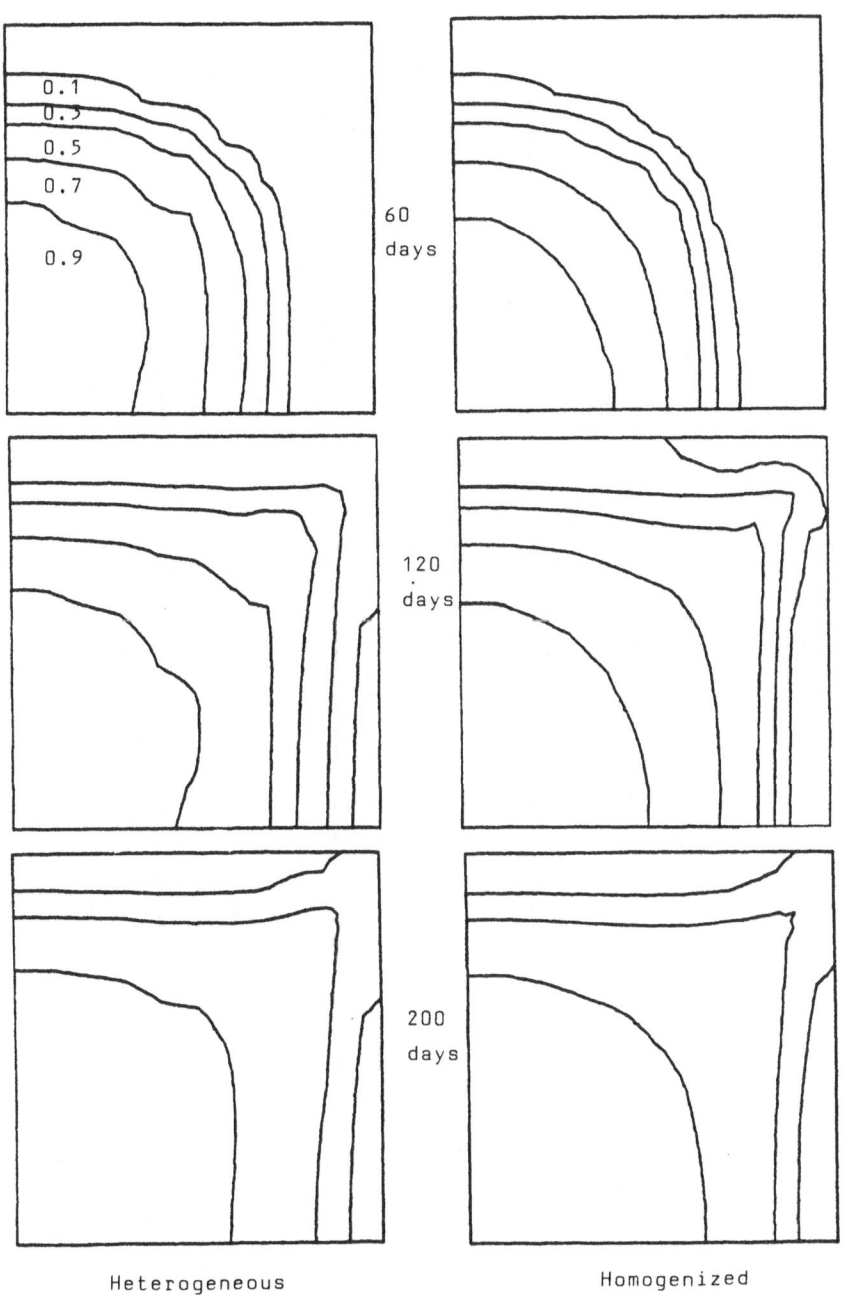

fig 4-11. Simulation 6, Evolution of Saturation contours

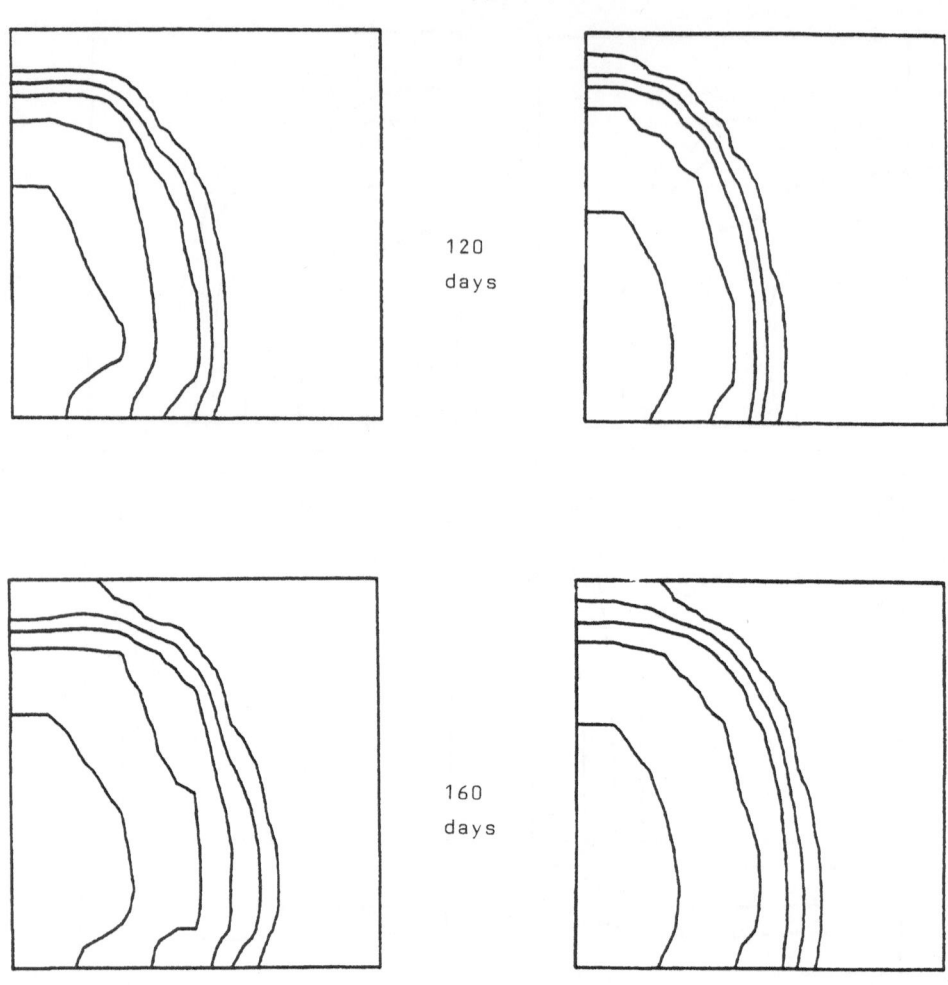

120
days

160
days

Heterogeneous   Homogenized

fig 4-12. Simulation 7, evolution of Saturation contours
$q_0 = 5m^3/day$, *Media* $I : \Phi_1 = 0.1$ $K_{11} = 6$ *md* $K_{22} = 30$ *md,Media* $II : \Phi_2 = 0.3$
$K_{11} = 300$ *md* $K_{22} = 300$ *md, Homogenized medium* $: \tilde{\Phi} = 0.2$ $K_{11} = 11.76$ *md*
$K_{22} = 165$ *md* $Pex = 8.5$ $Pey = 0.6$.

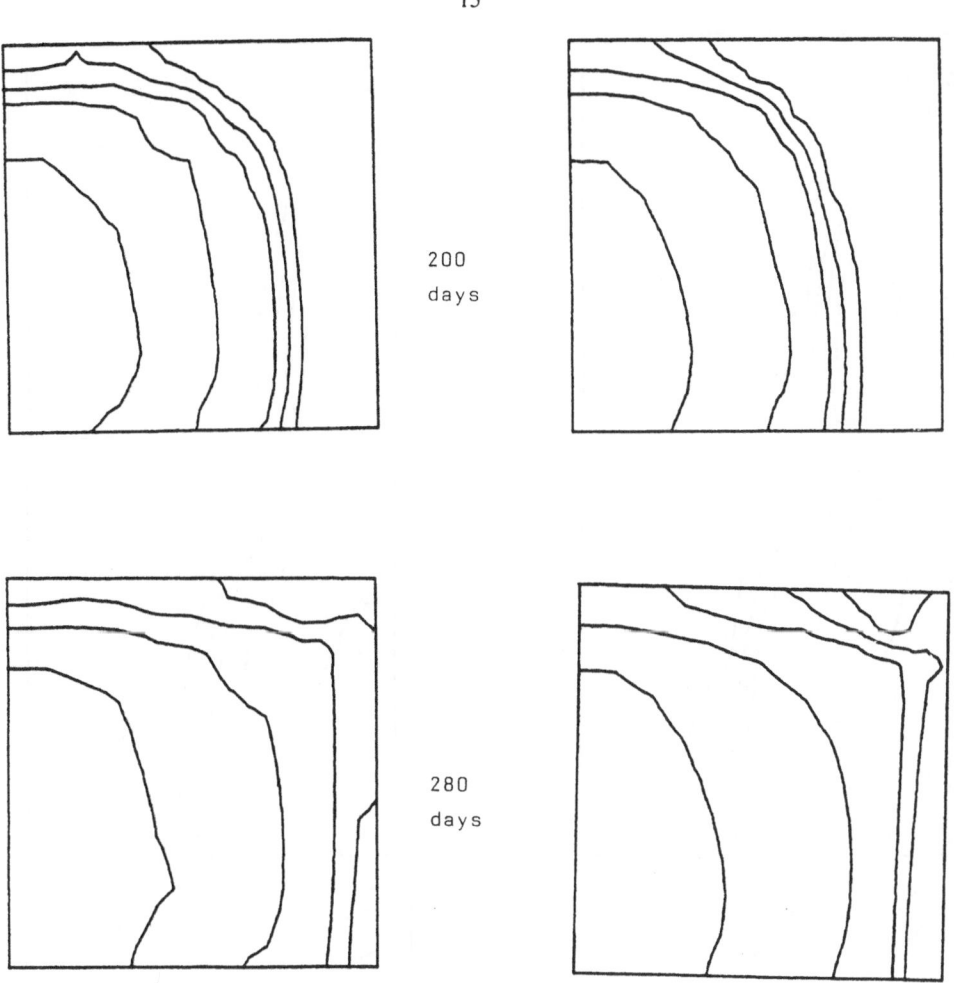

200
days

280
days

fig 4-13. Simulation 7, evolution of saturation contours

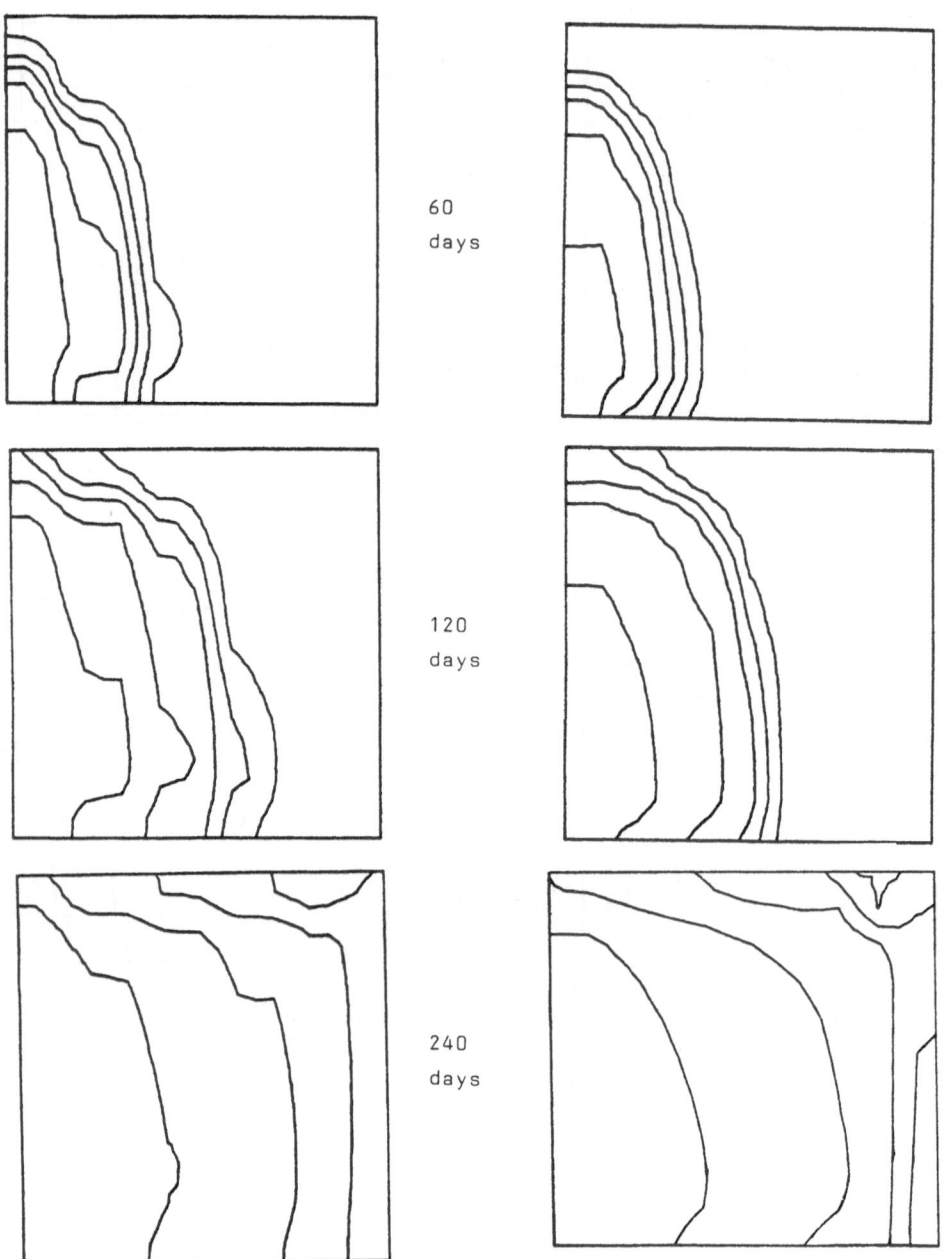

fig 4-14. Simulation 8, evolution of saturation curves. $q_0 = 5m^3/day$,
Media I : $\Phi_1 = 0.03$ $K_{11} = 3$ md $K_{22} = 30$ md, Media II : $\Phi_2 = 0.3$ $K_{11} = 300$ md
$K_{22} = 300$ md, Homogenized medium : $\tilde{\Phi} = 0.165$ $K_{11} = 5.94$ $K_{22} = 165$ md
Pex = 1.41 Pey = 0.6.

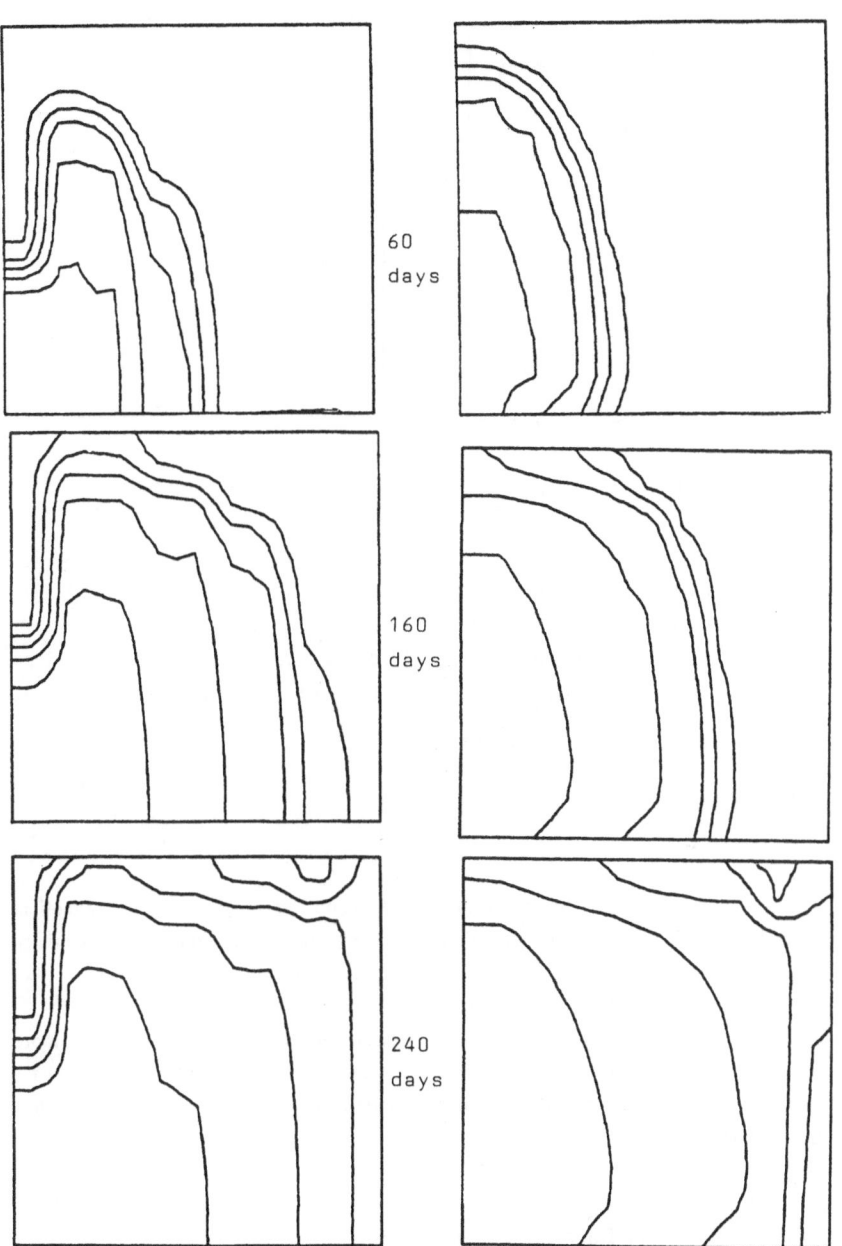

60
days

160
days

240
days

fig 4-15. Simulation 9, evolution of saturation curves

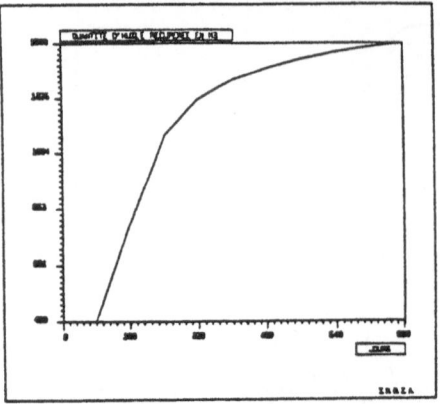 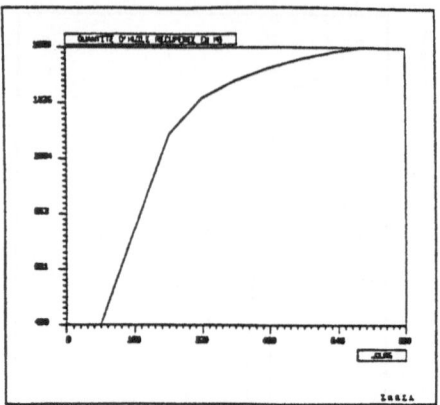

Heterogeneous   Homogenized

fig 4-16. Simulation 9, Recovery curves.
$q_0 = 5m^3/day$, *Media I* : $\Phi_1 = 0.03$ $K_{11} = 3$ md $K_{22} = 3$ md,
Media II : $\Phi_2 = 0.3$ $K_{11} = 300$ md $K_{22} = 300$ md, *Homogenized medium* :
$\tilde{\Phi} = 0.165$ $K_{11} = 5.94$ md $K_{22} = 151.5$ md, $Pex = 13.9$ $Pey = 0.55$

Results obtained with homogenization method show good agreementwith numerical simulations in the case of no interface saturation jump. Simulation presented here are essentially with stratas, but with 2-D cells, i.e. periodicity in both x_1 and x_2 directions, boundary layers effect are much more important and to get like good result a minimum of 10×10 cells required.

In case of saturation jumps at the interface, as in section 5, the situation is much more complicated. To our knowledge it is for the first time that this problem is adressed in this way. Though results presented are valid for simplified equations, they are good premices to further theoretical and numerical investigations.

5. Homogenization with Interface Phenomena.

In this section we are investigating an exact local model taking account Capillary Pressure and Relative Permeability within each media in a cell, figure 5-1 and 5-2.

$$\mathcal{P}^j_c(x,S) := c^jS + \alpha^j$$

$$P^j_c(S) := -\frac{c^j}{\alpha^j}S - 1$$

$$p^j_{cm}(x) = -\alpha^j \qquad , \quad j = I,II \qquad (5\text{-}1)$$

and in order to decouple equations, the diffusion flow which is with (5-1)

$$r^j := K^j(x)(d^j(S)\lambda_1^j(S)\lambda_2^j(S)c^j\nabla S^j, \qquad (5\text{-}2)$$

is linearized by assuming β^j constant in each media

$$\beta^j := (d^j(S)\lambda_1^j(S)\lambda_2^j(S)), \qquad (5\text{-}3)$$

and then

$$r^j = K^j(x)\beta^jc^j\nabla S^j. \qquad (5\text{-}4)$$

Furthermore, to study the saturation we will have a first simplified and totally Pressure decoupled, model

in Ω_ε^j, $\qquad \Omega_\varepsilon^j$ denoting the open set filled with porous media j, $j = 1,2$

$$\phi^\varepsilon(x)\frac{\partial S_\varepsilon^j}{\partial t} + \nabla.(r_\varepsilon^j + \lambda_1^j(S_\varepsilon^j)q_o^\varepsilon) = \text{sources} \qquad (5\text{-}5)$$

with q_o^ε, the transport flow, given with enough strong convergence property as ε, the size cell, tends to zero.

To study the Pressure we will consider a simplified pressure equation decoupled from the saturation

$$q_o^\varepsilon := -K^\varepsilon(x)\nabla P_\varepsilon^j \quad \text{in } \Omega_\varepsilon^j \qquad (5\text{-}6)$$

$$\nabla.q_o^\varepsilon = \text{sources} \qquad (5\text{-}7)$$

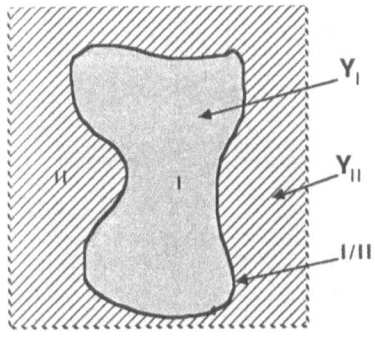

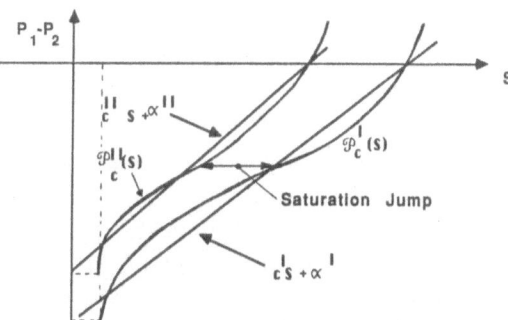

figure 5-1 .Y-cell with porous medium I,
porous medium II with interface I/II

figure 5-2 .Capillary pressure curves, $\mathcal{P}c$, and
linearized capillary pressure in medium I and II

In addition to usual boundary and initial conditions, equation (5-5) has saturation jump condition

$$0 = [[cS_\varepsilon - \alpha]]_{I/II} = (c^J S_\varepsilon^I - \alpha^I) - (c^{II} S_\varepsilon^{II} - \alpha^{II}), \quad \text{on the interface I/II} \qquad (5\text{-}8)$$

and a flux continuity condition

$$[[r_\varepsilon + \lambda_1 (S_\varepsilon) q_o^\varepsilon). \nu]]_{I/II} = 0, \qquad (5\text{-}9)$$

with ν the outward normal to the interface I/II

In the same way, equation (5-6) has a pressure jump condition

$$[[P_\varepsilon]]_{I/II} = \varepsilon g(x,t) \qquad (5\text{-}10)$$

and a flux continuity condition

$$[[q_o^\varepsilon. \nu]]_{I/II} = 0 \qquad (5\text{-}11)$$

In case of $\alpha^j = 0$ we get a rigourous mathematical result, B. Amaziane 1987, when $\alpha^j \neq 0$ we could get the same result only in a heuristic way.

Theorem 5.1

There is a global equation, corresponding to the local exact equation (5-5)

$$\Phi^* \frac{\partial S}{\partial t} + \nabla.(r^\# + \Lambda_1^\#(S).q^\#) = \text{sources} \qquad (5\text{-}12)$$

where the effective diffusion flux is

$$r^\# := H^\# (\frac{\theta^I}{c^I} + \frac{\theta^{II}}{c^{II}})^{-1} \nabla S \quad , \quad \theta^j := \frac{\text{meas}|Y^j|}{\text{meas}|Y^I \cup Y^{II}|} \qquad (5\text{-}13)$$

and the effective porosity,

$$\Phi^* = \frac{1}{\text{meas}|Y|} [\frac{1}{c^I} \int_{Y^I} \Phi^I(y) dy + \frac{1}{c^{II}} \int_{Y^{II}} \Phi^{II}(y) dy] \times (\frac{\theta^I}{c^I} + \frac{\theta^{II}}{c^{II}})^{-1} \qquad (5\text{-}14)$$

with $H^\#$ the usual homogenized coefficient, as in (3-3), associated to the local coefficient
$H^\varepsilon(x) := K^{j,\varepsilon}(x) \beta^{j,\varepsilon}(x)$;
moreover $H^\# = K^\# \beta$ only when $\beta^{j,\varepsilon}$ tends to β a-e.
In the general case we know only that $\lambda_1^\varepsilon q_o^\varepsilon$ has a limit but it is not possible to identify this limit.
With enough strong assumption on the convergence of q_o^ε, for instance with

$$q_o^\varepsilon := K^\varepsilon(x) Q^\varepsilon \ , \quad Q^\varepsilon \text{ parallel to the } x_1\text{-axis} \qquad (5\text{-}15)$$
$$Q^\varepsilon \to Q^o \quad \text{strongly in } L^2(\Omega) \qquad (5\text{-}16)$$

and assuming heterogeneous media made up of stratas orthogonal to the x_1-axis
then

$$\Lambda_1^\#(S) q^\# = K^\# [\theta_1^I \lambda_1^I (\frac{w + \alpha^I}{c^I}) + \theta_1^{II} \lambda_1^{II} (\frac{w + \alpha^{II}}{c^{II}})] Q^o \qquad (5\text{-}17)$$

21

where w is defined by

$$S := (\frac{\theta^I}{c^I} + \frac{\theta^{II}}{c^{II}})w + (\frac{\theta^I}{c^I}\alpha^I + \frac{\theta^{II}}{c^{II}}\alpha^{II}) \qquad (5\text{-}18)$$

To study the pressure equation (5-6), (5-7), we use two length scales expansion as in B.Amaziane (1987), and we get a result akin to the standard result, i.e. a global pressure equation

$$q_o^{\#} := K^{\#}\nabla P, \qquad (5\text{-}19)$$

$$\nabla.q_o^{\#} = sources, \qquad in \ all \ \Omega \qquad (5\text{-}20)$$

Acknowledgments.

We gratefully acknowledge G. Chavent and J. Jaffre for useful discussions and for providing the BIDIMIX Code which was used to simulate heterogeneous and homogenized reservoirs in section 4.

References

B. Amaziane, Thesis University of Lyon, To appear, (1987).

A. Bensoussan, J.L. Lions, G. Papanicolaou, *Asymptotic Analysis For Periodic Structures*, North-Holland, Amsterdam, (1978).

A. Bourgeat, "Homogenization of Two Phase Flow equations", Proc. of Symposia in Pures Mathematics, Vol.**45**, part.1, pp.157-163; A.M.S., (1986).

H. Brenner, "Dispersion Resulting From Flow Through Spatially Periodic Porous Media", Phil.Trans.Roy.Soc.London, **297**, 81 (1980).

H.C. Chang, "Effective Diffusion and Conduction in Two-Phase Media : A Unified Approach", AIChE Journal, vol.**29**, N°5, pp.846-853, (1983).

G.Chavent, "A New Formulation of Diphasic Incompressible Flows in Porous Media", Lect.Notes in Math., Vol.**503**, Springer-Verlag Berlin, pp.258-270, (1976).

G. Chavent, B. Cockburn, G. Cohen, J. Jaffre, "Une méthode d'éléments finis pour la simulation dans un réservoir de déplacements bidimensionnels d'huile par de l'eau", INRIA report 353, (1985).

G. Chavent, J. Jaffre, *Mathematical Models and Finite Elements for Reservoir Simulation*, North-Holland, (1987).

R.E.Ewing, *The mathematics of Reservoir Simulation*, Frontiers in Applied Mathematics, SIAM, pp.3-35, (1983).

S.N.Kruzkov, Sukorjanskii S.M., "Boundary Value Problems for Systems of Equations of Two-Phase porous Flow type", Maths.USSR-Sb, **33** (1977).

F. Murat, "Compacité par Compensation", An.Sc.Normale Pisa, **5**, pp. 489-508, (1978).

G.Papanicolaou, S.Varahan,"Diffusion in Region with Many Small Holes", Lectures Notes in Control and Information, Vol.**75**, pp.190-206, Springer Verlag (1980).

W.T.Perrins , Me Kenzie, Mc Phedran R.C., "Transport Properties of Regular Arrays of Cylinders", Proc.R. Soc.Lond., A **369**, 207, (1979).

D.Ryan, R.G.Carbonell, S.Whitaker, "Effective Diffusivities for Catalyst Pellets Under Reactive Conditions", Chem.Eng.Sci., **35**, 10, (1980).

E.Sanchez-Palencia, *Non homogeneous Media and Vibration Theory,*, Lectures Notes in Physics, vol.**127**, Springer Verlag, Berlin (1980).

THE DOUBLE POROSITY MODEL FOR SINGLE PHASE FLOW IN NATURALLY FRACTURED RESERVOIRS

Todd Arbogast

Department of Mathematics
University of Chicago
Chicago, Illinois 60637

Institute for Mathematics
and its Applications
University of Minnesota
Minneapolis, Minnesota 55455

1. Introduction

As early as 1953, Pirson [6] gave a qualitative description of a fractured reservoir as a reservoir with two porous structures—matrix (porous rock) and fracture. It was not until the early 1960's, however, that a quantitative description of this double porosity concept appeared [3], [9]. Briefly, a fractured reservoir is a region of space $\Omega \subset \mathbb{R}^d$ (d=2 or 3) which is partitioned by the thin fractures into disjoint, simply connected matrix blocks $\Omega_i \subset \Omega$. The idea is to consider at each point of space Ω not one but two porosities (actually, two sets of reservoir properties), one associated with the matrix, and the other associated with the fracture system. The matrix porosity, denoted $\phi_i(\underline{x})$ for $\underline{x} \epsilon \Omega_i$, is the macroscopic matrix pore void space per unit bulk volume. The fracture porosity, $\phi(\underline{x})$ for $\underline{x} \epsilon \Omega$, is the macroscopic fracture void space per unit bulk volume.

Some of the fluid flows in the matrix, while the rest flows in the fracture system. The matrix flow is certainly of Darcy type. We also assume a macroscopic Darcy flow over the reservoir Ω for the fracture system flow, even though it is literally constrained to the physical fractures. This is a reasonable assumption for reservoirs with thin, interconnected fractures and low fluid velocities. Consequently, at each point of space we have tensors for both the matrix permeability $k_i(\underline{x})$, $\underline{x} \epsilon \Omega_i$, and the fracture permeability $k(\underline{x})$, $\underline{x} \epsilon \Omega$.

Several authors have used the double porosity concept to model single phase flow in a fractured reservoir, most notably Barenblatt, et al. [3], Warren and Root [9], kazemi [5], and de Swaan O. [8]. Many other authors have considered double porosity models for multi-phase, multi-component systems (for examples, see [7] and its references).

The most important and difficult thing to model is the exchange of fluid between the matrix and the fracture system. Generally, this exchange is assumed to be in a quasi-steady state; that is, an ad hoc transfer function is introduced which does not depend explicitly on time but rather depends on the difference between the matrix fluid pressure and the fracture system fluid pressure. This assumption effectively ignores the pattern of flow in an individual matrix block. Fluid exchange actually takes place at the surfaces of the blocks. The rate of exchange depends upon the fluid pressures, the geometry of the matrix blocks, and the past history of the matrix flow.

Two exceptions to the quasi-steady state approach are to be found in the papers of Kazemi [5] and de Swaan O. [8]. In both of these papers, the fluid exchange is related to the flow through the matrix blocks' surfaces. Unfortunatly, both papers also constrain the fracture flow to the physical fractures, rather than considering it macroscopically spread out over the entire reservoir. This places restrictions on the geometry of the reservoir and otherwise severely complicates any field-scale numerical solution procedure. These two models are not genuinely of the double porosity type, at least not as it has been described above, as at each point of space there is only a single porosity.

While the quasi-steady state assumption is often reasonable for single phase flow when the matrix blocks are small [3], [5], [9], it is still valuable to properly model the unsteady state nature of the fluid exchange without unnecessarily restricting the geometry nor complicating the approximation process. Obviously, such a model is needed for simulations where accuracy is quite important or where the quasi-steady state assumption is invalid; moreover, this model should give us greater understanding of the quasi-steady state assumption itself, and it should provide a stepping-stone to properly modeling multi-phase, multi-component fractured reservoir flow.

Such a model has been described and analyzed by the author in [1]. The approach taken there is being used by the author in modeling a completely miscible displacement [2], and by Douglas, et al. [4] in modeling an immiscible waterflood. Here, one of the assumptions of the modeling process in [1] shall be relaxed, as described below.

Pressure must be a continuous variable; hence, the matrix pressure at the surface of each block and the pressure of the fluid in the physical fractures around the block must be equal. On each block, this fact gives rise to a boundary condition for the matrix flow. Let $p(\underline{x},t)$ denote the fracture system pressure. The pressure in a physical fracture is not necessarily the value of p evaluated at the fracture location, as p is a macroscopic variable with respect to the system of fractures. A reasonable value for the pressure in a physical fracture is given by taking some local average of p. It is appropriate to take the physical fracture pressure around $\partial\Omega_i$ as some average in the vicinity of the entire block, since the diameters of the matrix blocks are small compared to the smallest dimension of the reservoir itself (i.e., the reservoir is *naturally* fractured).

In [1], it was assumed that the variation of p over each Ω_i was so small as to be negligible. A simple local average of p was taken, giving a constant (in space) pressure (and density) over $\partial\Omega_i$. Consequently, fluid could only be absorbed or emitted uniformly by each matrix block; fluid could not flow from one side to another through any block.

In the model to be presented here, the variation of p near each Ω_i will influence the matrix flow. At each time, the physical fracture fluid density around $\partial\Omega_i$ shall be assumed to be the restriction of a globally linear function. More generally, one could take a higher degree polynomial; the model will generalize. However, if the spatial variation over the blocks is so large as to require this generality, one can hardly claim to have a double porosity reservoir.

An outline of the paper follows. In section 2 we will derive our model from physical considerations by applying the double porosity concept to a naturally fractured reservoir. The matrix/fracture fluid exchange will influence the matrix flow through a boundary condition on each block as described above, and it will influence the fracture system flow through a macroscopically distributed source. In the last three sections, we will extend most of the results of [1] to the present model. We will again make the (unphysical) assumption that f_e is in $L^2(\Omega \times J) = L^2(J; L^2(\Omega))$; that is, f_e will not be concentrated at points. This will ensure a smooth solution to the differential problem. We will analyze the model in section 3 by showing that it is mathematically well posed. In section 4, we will describe a finite element method that is easy to implement in field-scale simulation, and we will prove its convergence at the optimal rate in section 5.

2. Derivation of the Model

We will now apply the double porosity concept to a single phase fluid in a naturally fractured reservoir. We will assume that the fluid is an ideal liquid; that is, a fluid of constant viscosity $\mu > 0$ and compressibility $c > 0$:

$$\rho^{-1}d\rho = c\,dp, \qquad \sigma^{-1}d\sigma = c\,dq, \tag{2.1}$$

where $\rho(\underline{x},t)$ and $p(\underline{x},t)$ are the fracture system fluid density and pressure, respectively, and $\sigma(\underline{x},t)$ and $q(\underline{x},t)$ are the corresponding quantities for the matrix fluid.

First consider fluid flow in the matrix blocks. Since the blocks are small and typically the fracture system has a higher flow capacity (permeability) than the matrix, we make the following two assumptions:

1. An individual matrix block will interact only with the fractures surrounding it; hence, matrix blocks do not interact directly with each other, nor with external sources or sinks. (Kazemi's numerical results [5] indicate that this is a good assumption.)

2. At any given time, the spatial variation of the fluid density in the physical fractures around a block is sufficiently well taken into account by a linear function.

Over the i *th* matrix block, assumption 1 and conservation of mass combined with Darcy's law and (2.1) give us

$$\phi_i \sigma_t - \nabla \cdot (\kappa_i \nabla \sigma - \sigma^2 \Gamma_i) = 0, \qquad (\underline{x}, t) \epsilon \Omega_i \times J, \qquad (2.2)$$

where the t subscript denotes partial differentiation in time, $\kappa_i(\underline{x}) = k_i(\underline{x})/\mu c$ (a tensor), $\Gamma_i(\underline{x})$ is the gravitational constant times $k_i(\underline{x})/\mu$ applied to the gradient of the vertical coordinate, and $J = (0, T]$ is the time interval of interest. It is sufficient in reservoir simulation to linearize the quadratic term; hence, let $\sigma_0(\underline{x})$ be some reference density function and approximate

$$\sigma^2 = [(\sigma - \sigma_0) + \sigma_0]^2 \approx 2\sigma_0\sigma - \sigma_0^2. \qquad (2.3)$$

The i *th* matrix block flow equation becomes

$$\phi_i \sigma_t - \nabla \cdot [\kappa_i \nabla \sigma - (2\sigma - \sigma_0)\underline{\gamma}_i] = 0, \qquad (\underline{x}, t) \epsilon \Omega_i \times J, \qquad (2.4)$$

where $\underline{\gamma}_i(\underline{x}) = \sigma_0(\underline{x})\Gamma_i(\underline{x})$.

By assumption 1, the boundary condition for (2.4) is given entirely by continuity of pressure (equivalently, of density, by (2.1)) with the fracture flow. As explained in the introduction, we should locally average the macroscopic fracture density to obtain the fluid density in the physical fractures. For simplicity, let us assume that the fractures are infinitely thin so that $\overline{\Omega} = \cup_i \overline{\Omega}_i$. Now, to make this local averaging process explicit, let $\{\chi_i(\underline{x})\}$ be some partition of unity over Ω such that each χ_i is or is approximately the characteristic function of Ω_i; that is, $\sum_i \chi_i \equiv 1$, $\chi_i \geq 0$, the support of $\chi_i \approx \Omega_i$, and $\int \chi_i dx = |\Omega_i| = $ the measure of Ω_i. Later, the χ_i will be described further.

There are many ways to linearly approximate ρ near Ω_i. Perhaps the best way to approximate ρ by its local averages is to express it in terms of an orthonormal expansion. Let $\{1, \Lambda_{i,1}(\underline{x}), \dots, \Lambda_{i,d}(\underline{x})\}$ be an orthonormal basis for the linear functions with respect to the inner product given by integration against the weight $|\Omega_i|^{-1}\chi_i$. (For example, perform Gram-Schmidt orthogonalization and normalization to $\{1, x_1, \dots, x_d\}$. If Ω_i is a rectangular parallelepiped and χ_i is its characteristic function, then the $\Lambda_{i,j}$ are just scaled Legendre polynomials.) Let $\Delta_i(\underline{x}) = (\Lambda_{i,1}(\underline{x}), \dots, \Lambda_{i,d}(\underline{x}))$, and let $(\cdot, \cdot)$ be the $L^2(\Omega)$ or $(L^2(\Omega))^d$ inner product. Then we have that

$$\rho = \frac{1}{|\Omega_i|}[(\rho, \chi_i) + (\rho, \Delta_i \chi_i) \cdot \Delta_i] + O([\text{diam}(\Omega_i)]^2), \qquad (\underline{x}, t) \epsilon \Omega_i \times J, \qquad (2.5)$$

and (by assumption 2) our boundary condition shall be

$$\sigma = \frac{1}{|\Omega_i|}[(\rho, \chi_i) + (\rho, \Delta_i \chi_i) \cdot \Delta_i], \qquad (\underline{x}, t) \epsilon \partial\Omega_i \times J. \qquad (2.6)$$

The initial condition for (2.4) must be given:

$$\sigma(\underline{x},0) = \sigma^0(\underline{x}), \qquad \underline{x} \in \Omega_i. \tag{2.7}$$

The fracture flow is governed by an equation analogous to (2.4), except that two source terms appear. The effect of external sources (sinks) $f_e(\underline{x},t)$ has been reserved for the fracture system (assumption 1). In addition, the fluid produced through the matrix blocks' surfaces is another source. Denote by $f_i(\underline{x},t)$ the source from the ith block. Then the entire matrix source is $\sum_i f_i$. Hence,

$$\phi \rho_t - \underline{\nabla} \cdot [\, \kappa \underline{\nabla} \rho - (2\rho - \rho_0)\underline{\gamma}\,] = f_e + \sum_i f_i, \qquad (\underline{x},t) \in \Omega \times J, \tag{2.8}$$

where $\kappa(\underline{x}) = k(\underline{x})/\mu c$ (a tensor), $\rho_0(\underline{x})$ is the fracture system's reference density function, $\underline{\gamma}(\underline{x}) = \rho_0(\underline{x})\underline{\Gamma}(\underline{x})$, and $\underline{\Gamma}(\underline{x})$ is the gravitational constant times $k(\underline{x})/\mu$ applied to the gradient of the vertical coordinate.

We will now define the matrix source term. We must be careful to model the fluid exchange from the point of view of the fracture system in a manner that is consistent with the boundary condition (2.6).

At each point $\underline{x} \in \partial\Omega_i$, the ith block loses (or gains) an amount of fluid equal to

$$-[\, \kappa_i \underline{\nabla}\sigma - (2\sigma - \sigma_0)\underline{\gamma}_i\,] \cdot \underline{\nu}_i. \tag{2.9}$$

where $\underline{\nu}_i(\underline{x})$ is the outer unit normal to $\partial\Omega_i$. This fluid should be macroscopically spread out as f_i over the domain Ω near Ω_i in such a way that f_i has the same macroscopic effect. Since the block only detects the linear variation of the fracture flow, f_i and (2.9) should have the same effect on a linear fracture flow.

Since (2.9) is a distribution (supported on $\partial\Omega_i$), its effect is determined by considering its action on a test function $\varphi \in C^\infty(\Omega)$:

$$\int_{\partial\Omega_i} [\, \kappa_i \underline{\nabla}\sigma - (2\sigma - \sigma_0)\underline{\gamma}_i\,] \cdot \underline{\nu}_i \, \varphi \, ds$$

$$= \int_{\Omega_i} \underline{\nabla} \cdot \{[\, \kappa_i \underline{\nabla}\sigma - (2\sigma - \sigma_0)\underline{\gamma}_i\,]\varphi\} \, dx$$

$$= (\underline{\nabla} \cdot [\, \kappa_i \underline{\nabla}\sigma - (2\sigma - \sigma_0)\underline{\gamma}_i\,], \varphi)_i + (\kappa_i \underline{\nabla}\sigma - (2\sigma - \sigma_0)\underline{\gamma}_i, \underline{\nabla}\varphi)_i$$

$$= (\phi_i \sigma_t, \varphi)_i + (\kappa_i \underline{\nabla}\sigma - (2\sigma - \sigma_0)\underline{\gamma}_i, \underline{\nabla}\varphi)_i, \tag{2.10}$$

by the divergence theorem and (2.4). (Here, $(\cdot,\cdot)_i$ is the inner product on $L^2(\Omega_i)$ or $(L^2(\Omega_i))^d$.) If we now restrict φ to lie in the set of linear functions, then

$$\varphi = \frac{1}{|\Omega_i|}[(\varphi, \chi_i) + (\varphi, \Delta_i \chi_i) \cdot \Delta_i] \tag{2.11}$$

and

$$\int_{\partial\Omega_i} [\, \mathbf{x}_i \underline{\nabla}\sigma - (2\sigma - \sigma_0)\underline{\mathbf{x}}_i \,] \cdot \underline{\nu}_i \, \varphi \, ds$$

$$= \frac{1}{|\Omega_i|} \Big[(\phi_i \sigma_t, 1)_i \, (\varphi, \chi_i) + (\phi_i \sigma_t, \Delta_i)_i \cdot (\varphi, \Delta_i \chi_i)$$

$$+ (\mathbf{x}_i \underline{\nabla}\sigma - (2\sigma - \sigma_0)\underline{\mathbf{x}}_i, \, \underline{\nabla}\Delta_i)_i \cdot (\varphi, \Delta_i \chi_i) \Big]$$

$$= \Big[\frac{1}{|\Omega_i|} \Big\{ (\phi_i \sigma_t, 1)_i + [(\phi_i \sigma_t, \Delta_i)_i + (\mathbf{x}_i \underline{\nabla}\sigma - (2\sigma - \sigma_0)\underline{\mathbf{x}}_i, \, \underline{\nabla}\Delta_i)_i] \cdot \Delta_i \Big\} \chi_i, \, \varphi \Big].$$

$$(2.12)$$

(Note that $\underline{\nabla}\Delta_i$ is a tensor in the above expression.) Consequently,

$$f_i = -\frac{1}{|\Omega_i|} \Big\{ (\phi_i \sigma_t, 1)_i + [(\phi_i \sigma_t, \Delta_i)_i + (\mathbf{x}_i \underline{\nabla}\sigma - (2\sigma - \sigma_0)\underline{\mathbf{x}}_i, \, \underline{\nabla}\Delta_i)_i] \cdot \Delta_i \Big\} \chi_i,$$

$$(\underline{x}, t) \epsilon \Omega \times J. \quad (2.13)$$

Higher order variations in φ were ignored in defining f_i, just as higher order variations in p were ignored in defining the boundary condition (2.6). To put it another way, the function f_i is the best localized (by χ_i) linear approximation to the distribution (2.9). Mass is conserved by f_i. Moreover, f_i takes into account the spatial variation of the source (2.9), up to the linear order.

Note that χ_i defines and weights the region of space over which the i th block influences the fracture flow. Effectively, χ_i has related the scale of the matrix macroscopic averaging (which is on the order of the size of the rock grains) to the scale of the fracture system macroscopic averaging (which is on the order of the size of the blocks).

As a boundary condition for (2.8), let us simply take the "no flow" Neumann condition:

$$[\, \mathbf{x} \underline{\nabla} p - (2p - p_0)\underline{\mathbf{x}} \,] \cdot \underline{\nu} = 0, \qquad (\underline{x}, t) \epsilon \partial\Omega \times J, \qquad (2.14)$$

where $\underline{\nu}(\underline{x})$ is the outer unit normal to $\partial\Omega$. Finally, give the initial condition as

$$p(\underline{x}, 0) = p^0(\underline{x}), \qquad \underline{x} \epsilon \Omega. \qquad (2.15)$$

In summary, (2.4), (2.6), and (2.7) define the flow in the matrix blocks, while (2.8) with (2.13), (2.14), and (2.15) define the fracture system flow.

3. Analysis of the Model

For the analysis below, we shall tacitly assume the following hypotheses. Ω is a simply connected, bounded domain with smooth boundary, and each Ω_i is a convex domain. Each of the functions ϕ, ϕ_i, κ, κ_i, Γ, Γ_i, and χ_i is bounded and sufficiently smooth. Additionally, ϕ and the ϕ_i are bounded below by positive constants, and the tensors κ and κ_i, for all i, are symmetric and uniformly positive-definite.

Let $W^{r,p}(\Omega)$ denote the usual Sobolev space of $r \geq 0$ times differential functions in $L^p(\Omega)$, $1 \leq p \leq \infty$. Let $H^r(\Omega) = W^{r,2}(\Omega)$, and let $H^{-r}(\Omega)$ be its dual. $H^1_0(\Omega) = \{u \in H^1(\Omega) | u = 0$ on $\partial\Omega\}$. $H^s(J;H^r(\Omega))$, s a nonnegative integer, is the usual space having the norm

$$\|u\|_{H^s(J;H^r(\Omega))} = \left\{ \int_J \sum_{\ell=0}^{s} \left\| \frac{\partial^\ell u}{\partial t^\ell} \right\|^2_{H^r(\Omega)} dt \right\}^{1/2} .$$

Similarily, we have the space $L^\infty(J;H^r(\Omega))$. It will be convenient to define the space

$$H^1_\Lambda(\Omega) = \left\{ v \in H^1(\Omega) \middle| \begin{array}{l} \text{the trace of v on } \partial\Omega \text{ is identical to the} \\ \text{restriction of a linear function to } \partial\Omega \end{array} \right\} .$$

Of course, in each of these spaces, the domain Ω may be changed to Ω_i, $\partial\Omega$, $\partial\Omega_i$, or even $\Omega_m = \cup_i \Omega_i$.

C and ϵ will always denote generic positive constants.

Following [1], we will use the method of continuity to show that the model is well posed. We will present the argument in a more direct form here. This form of the argument is completely analogous to that used in the error analysis below.

For each $\lambda \epsilon [0,1]$, consider the problem defined by

$$\phi u_t - \nabla \cdot (\kappa \nabla u - 2 \underline{\chi} u) = G(\underline{x},t) + \lambda \sum_i F_i, \qquad (\underline{x},t) \epsilon \Omega \times J, \qquad (3.1a)$$

$$(\kappa \nabla u - 2 \underline{\chi} u) \cdot \underline{\nu} = u_0(\underline{x}), \qquad (\underline{x},t) \epsilon \partial\Omega \times J, \qquad (3.1b)$$

$$u(\underline{x},0) = u^0(\underline{x}), \qquad \underline{x} \epsilon \Omega, \qquad (3.1c)$$

and, for each i,

$$\phi_i v_t - \nabla \cdot (\kappa_i \nabla v - 2 \underline{\chi}_i v) = G_i(\underline{x},t), \qquad (\underline{x},t) \epsilon \Omega_i \times J, \qquad (3.2a)$$

$$v = \frac{1}{|\Omega_i|} [(u, \chi_i) + (u, \Delta_i \chi_i) \cdot \Delta_i], \qquad (\underline{x},t) \epsilon \partial\Omega_i \times J, \qquad (3.2b)$$

$$v(\underline{x},0) = v^0(\underline{x}), \qquad \underline{x} \epsilon \Omega_i. \qquad (3.2c)$$

and

$$F_i = -\frac{1}{|\Omega_i|}\Big\{(\phi_i v_t, 1)_i + [(\phi_i v_t, \Delta_i)_i + (\mathbf{x}_i \underline{\nabla} v - 2\underline{\mathbf{X}}_i v, \underline{\nabla} \Delta_i)_i]\cdot\Delta_i\Big\}\chi_i,$$
$$(\underline{x}, t)\epsilon\,\Omega\times J. \qquad (3.3)$$

Note that u_0 is a function of $\underline{x}$ only.

The weak form of (3.2a) for a test function $\psi\epsilon H^1_{\Lambda}(\Omega_i)$ (where ψ is identical to the linear function $\overline{\psi}_i$ on $\partial\Omega_i$) is,

$$(\phi_i v_t, \psi)_i + (\mathbf{x}_i \underline{\nabla} v, \underline{\nabla}\psi)_i - (2\underline{\mathbf{X}}_i v, \underline{\nabla}\psi)_i$$
$$= (G_i, \psi)_i + \int_{\partial\Omega_i}(\mathbf{x}_i \underline{\nabla} v - 2\underline{\mathbf{X}}_i v)\cdot\underline{\nu}_i\,\overline{\psi}_i\,ds$$
$$= (G_i, \psi - \overline{\psi}_i)_i - (F_i, \overline{\psi}_i). \qquad (3.4)$$

A similar weak form of (3.1a,b) for a test function $\varphi\epsilon H^1(\Omega)$ can be found. If we add this weak form to λ times the sum on i of (3.4), we get

$$(\phi u_t, \varphi) + \lambda\sum_i(\phi_i v_t, \psi)_i + (\mathbf{x}\underline{\nabla}u, \underline{\nabla}\varphi) + \lambda\sum_i(\mathbf{x}_i\underline{\nabla}v, \underline{\nabla}\psi)_i$$
$$= (2\underline{\mathbf{X}}u, \underline{\nabla}\varphi) + \lambda\sum_i(2\underline{\mathbf{X}}_i v, \underline{\nabla}\psi)_i + (G, \varphi) + \lambda\sum_i(G_i, \psi - \overline{\psi}_i)_i$$
$$+ \int_{\partial\Omega}u_0\,\varphi\,ds + \lambda\sum_i(F_i, \varphi - \overline{\psi}_i). \qquad (3.5)$$

A priori energy estimates can be derived from (3.5). It is necessary to take $\overline{\psi}_i = |\Omega_i|^{-1}[(\varphi, \chi_i) + (\varphi, \Delta_i\chi_i)\cdot\Delta_i]$ so that the last term above vanishes. This condition is met for the usual choices $\varphi=u$, $\psi=v$ and $\varphi=u_t$, $\psi=v_t$ because of the boundary condition (3.2b). We obtain

$$\|u_t\|^2_{L^2(J;L^2(\Omega))} + \|u\|^2_{L^\infty(J;H^1(\Omega))} + \lambda\{\|v_t\|^2_{L^2(J;L^2(\Omega_m))} + \|v\|^2_{L^\infty(J;H^1(\Omega_m))}\}$$
$$\leq C\{\|G\|^2_{L^2(J;L^2(\Omega))} + \|u_0\|^2_{H^{1/2}(\partial\Omega)} + \|u^0\|^2_{H^1(\Omega)}$$
$$+ \lambda[\sum_i\|G_i\|^2_{L^2(J;L^2(\Omega_i))} + \|v^0\|^2_{H^1(\Omega_m)}]\}$$
$$+ \epsilon\{\|u\|^2_{L^2(J;H^2(\Omega))} + \lambda\|v\|^2_{L^2(J;H^2(\Omega_m))}\}, \qquad (3.6)$$

where C is independent of λ, and ϵ is as small as we like. The last two terms on the right side above arise from bounding the terms containing $\underline{\mathbf{X}}$ and $\underline{\mathbf{X}}_i$ in such a way that only the H^1-norms of these quantities appear (see [1]). Consequently, C depends on $\underline{\mathbf{X}}$ and $\underline{\mathbf{X}}_i$ only through their H^1-norms. This is important to notice, since $\underline{\mathbf{X}}$ and $\underline{\mathbf{X}}_i$ are proportional to ρ_0 and σ_0, respectively.

Energy estimates of (3.4) and (3.2b) with $\psi=v$ and with $\psi=v_t$ can be made directly.

Because v_t appears explicitly in F_i, an integration by parts in time argument is needed when $\psi=v$ (the same argument is needed to treat $(2\underline{x}_i v, \underline{\nabla}\psi)_i$ when $\psi=v_t$). The final result shows that

$$\|v_t\|^2_{L^2(J;L^2(\Omega_m))} + \|v\|^2_{L^\infty(J;H^1(\Omega_m))}$$

$$\leq C\{\sum_i \|G_i\|^2_{L^2(J;L^2(\Omega_i))} + \|v^0\|^2_{H^1(\Omega_m)} + \|u_t\|^2_{L^2(J;L^2(\Omega))} + \|u\|^2_{L^\infty(J;L^2(\Omega))}\}$$

$$+ \epsilon\|v\|^2_{L^2(J;H^2(\Omega_m))}, \tag{3.7}$$

where C and ϵ are as above. Hence, we can omit the λ's in (3.6).

Elliptic regularity and direct estimates of (3.1a) and (3.2a) finally show that

$$\|u\|_{H^1(J;L^2(\Omega))} + \|u\|_{L^2(J;H^2(\Omega))} + \|v\|_{H^1(J;L^2(\Omega_m))} + \|v\|_{L^2(J;H^2(\Omega_m))}$$

$$\leq C\{\|G\|_{L^2(J;L^2(\Omega))} + \|u_0\|_{H^{1/2}(\partial\Omega)} + \|u^0\|_{H^1(\Omega)}$$

$$+ \sum_i \|G_i\|_{L^2(J;L^2(\Omega_i))} + \|v^0\|_{H^1(\Omega_m)}\}, \tag{3.8}$$

where C is as above.

This result and the method of continuity give us the

Lemma:

If $G \in L^2(J;L^2(\Omega))$, $u_0 \in H^{1/2}(\partial\Omega)$, $u^0 \in H^1(\Omega)$, $G_i \in L^2(J;L^2(\Omega_i))$, and $v^0 \in H^1(\Omega_i)$ (for each i) satisfy the compatibility relations

$$\frac{1}{|\Omega_i|}[(u^0, \chi_i) + (u^0, \Delta_i\chi_i)\cdot\Delta_i] = v^0, \qquad \underline{x}\in\partial\Omega_i, \text{ for each i,} \tag{3.9}$$

then the problem (3.1)-(3.3) with $\lambda=1$ has a unique solution $u\in H^1(J;L^2(\Omega))\cap L^2(J;H^2(\Omega))$ and $v\in H^1(J;L^2(\Omega_m))\cap L^2(J;H^2(\Omega_m))$ that satisfies (3.8) with the constant C depending on $\underline{x}$ and $\underline{x}_i$ only through their H^1-norms.

We can now easily derive the theorem below as a corollary to the lemma. We need only set

$$G = \underline{\nabla}\cdot(\rho_0\underline{x}) + f_e - \sum_i \frac{1}{|\Omega_i|}(\sigma_0\underline{x}_i, \underline{\nabla}\Delta_i)_i\cdot\Delta_i\chi_i$$

$$= \underline{\nabla}\cdot(\rho_0^2\underline{\Gamma}) + f_e - \sum_i \frac{1}{|\Omega_i|}(\sigma_0^2\underline{\Gamma}_i, \underline{\nabla}\Delta_i)_i\cdot\Delta_i\chi_i, \tag{3.10a}$$

$$u_0 = -\rho_0\underline{x}\cdot\underline{\nu} = -\rho_0^2\underline{\Gamma}\cdot\underline{\nu}. \tag{3.10b}$$

$$u^0 = \rho^0, \tag{3.10c}$$

$$\underline{G}_i = \underline{\nabla}\cdot(\sigma_0\underline{\chi}_i) = \underline{\nabla}\cdot(\sigma_0^2\underline{\Gamma}_i), \tag{3.10d}$$

and

$$v^0 = \sigma^0 \tag{3.10e}$$

in (3.1)-(3.3) to obtain our model (2.4), (2.6)-(2.8), (2.13)-(2.15), where $u=\rho$ and $v=\sigma$.

Theorem:

If $f_e \epsilon L^2(J;L^2(\Omega))$, $\rho_0^2 \epsilon H^1(\Omega)$, $\sigma_0^2 \epsilon H^1(\Omega_m)$, $\rho^0 \epsilon H^1(\Omega)$, and $\sigma^0 \epsilon H^1(\Omega_m)$ are such that

$$\frac{1}{|\Omega_i|}[(\rho^0, \chi_i) + (\rho^0, \Delta_i\chi_i)\cdot\Delta_i] = \sigma^0, \qquad \underline{x} \epsilon \partial\Omega_i, \text{ for each } i, \tag{3.11}$$

then the double porosity model has a unique solution $\rho \epsilon H^1(J;L^2(\Omega)) \cap L^2(J;H^2(\Omega))$ and $\sigma \epsilon H^1(J;L^2(\Omega_m)) \cap L^2(J;H^2(\Omega_m))$ which varies continuously with the data:

$$\|\rho\|_{H^1(J;L^2(\Omega))} + \|\rho\|_{L^2(J;H^2(\Omega))} + \|\sigma\|_{H^1(J;L^2(\Omega_m))} + \|\sigma\|_{L^2(J;H^2(\Omega_m))}$$
$$\leq C\{\|f_e\|_{L^2(J;L^2(\Omega))} + \|\rho_0^2\|_{H^1(\Omega)} + \|\sigma_0^2\|_{H^1(\Omega_m)} + \|\rho^0\|_{H^1(\Omega)} + \|\sigma^0\|_{H^1(\Omega_m)}\}, \tag{3.12}$$

where C depends on the H^1-norms of ρ_0 and σ_0, and, for solutions (ρ_j, σ_j) arising from data $(f_{e,j}, \rho_{0,j}, \sigma_{0,j}, \rho^0_j, \sigma^0_j)$, $j=1$ and 2,

$$\|\rho_1 - \rho_2\|_{H^1(J;L^2(\Omega))} + \|\rho_1 - \rho_2\|_{L^2(J;H^2(\Omega))}$$
$$+ \|\sigma_1 - \sigma_2\|_{H^1(J;L^2(\Omega_m))} + \|\sigma_1 - \sigma_2\|_{L^2(J;H^2(\Omega_m))}$$
$$\leq C\{\|f_{e,1} - f_{e,2}\|_{L^2(J;L^2(\Omega))} + \|\rho_{0,1}^2 - \rho_{0,2}^2\|_{H^1(\Omega)} + \|\sigma_{0,1}^2 - \sigma_{0,2}^2\|_{H^1(\Omega_m)}$$
$$+ \|\rho_{0,1} - \rho_{0,2}\|_{H^1(\Omega)} + \|\sigma_{0,1} - \sigma_{0,2}\|_{H^1(\Omega_m)}$$
$$+ \|\rho^0_1 - \rho^0_2\|_{H^1(\Omega)} + \|\sigma^0_1 - \sigma^0_2\|_{H^1(\Omega_m)}\}, \tag{3.13}$$

where C depends on the H^1-norms of $\rho_{0,2}$ and $\sigma_{0,2}$ as well as on the $L^2(H^2)$-norms of ρ_1 and σ_1.

While higher order regularity for the solution has not been demonstrated, it is trivial to at least see that the solution is smooth on the interior of its domain when the model's coefficients (including the χ_i) and data are smooth.

4. A Finite Element Method

For parameters h and h_i, all i, in $(0,1]$, let $M_h \subset H^1(\Omega)$ and, for each i, $N_{i,h_i} \subset H^1_0(\Omega_i)$ be standard Galerkin finite dimensional H^1-approximation spaces of order $R \geq 2$ in h and $S_i \geq 2$ in h_i, respectively. Specifically, we need to be able to approximate $u \in H^r(\Omega)$, $1 \leq r \leq R$, such that

$$\inf_{\varphi \in M} \|u - \varphi\|_{H^1(\Omega)} \leq C \|u\|_{H^r(\Omega)} h^{r-1}, \tag{4.1}$$

and, for each i, $v \in H^{s_i}(\Omega_i) \cap H^1_0(\Omega_i)$, $1 \leq s_i \leq S_i$, such that

$$\inf_{\psi \in N_i} \|v - \psi\|_{H^1(\Omega_i)} \leq C \|v\|_{H^{s_i}(\Omega_i)} h_i^{s_i-1}. \tag{4.2}$$

(We have and will continue to suppress the h and h_i parameters in the notation.) For convenience, we will assume that M contains the constant functions. We will also need the space

$$N_i^* = N_i + \text{span}\{1, \Lambda_{i,1}, \ldots, \Lambda_{i,d}\}.$$

For any positive integer N, let $\Delta t = T/N$. For any function u, we will use the following notation:

$$t_n = n \Delta t,$$

$$u^n = u(t_n),$$

$$u^{n-1/2} = \frac{u^n + u^{n-1}}{2},$$

and

$$\partial u^n = \frac{u^n - u^{n-1}}{\Delta t},$$

where $n = 0, 1, \ldots, N$ (except that $n \geq 1$ in the last two expressions).

It will be useful to define the following elliptic projections. For $u \in H^1(\Omega)$, let $\tilde{u}$ denote the unique function in M for which

$$(\kappa \underline{\nabla}(u - \tilde{u}), \underline{\nabla}\varphi) = 0, \qquad \text{for all } \varphi \in M, \tag{4.3a}$$

and

$$\int_{\Omega} (u - \tilde{u}) \, dx = 0. \tag{4.3b}$$

For $v \in H^1_\Lambda(\Omega_i)$, let $\hat{v}$ denote the unique function in $\mathbf{N}_i^*$ for which

$$(\mathbf{x}_i \underline{\nabla}(v - \hat{v}), \underline{\nabla}\psi)_i = 0, \qquad \text{for all } \psi \in \mathbf{N}_i, \tag{4.4a}$$

and

$$\hat{v}(\underline{x}) = v(\underline{x}), \qquad \underline{x} \in \partial\Omega_i. \tag{4.4b}$$

If $v \in H^1(\Omega_m)$ is such that $v \in H^1_\Lambda(\Omega_i)$, for all i (σ is such a function), then $\hat{v}$ is defined on all of Ω_m, and it is doubly valued on the boundaries of the matrix blocks. Since we will always consider $\hat{v}$ only on an Ω_i, its boundary values can always be considered to be those of its trace from inside Ω_i, and, hence, no confusion should arise.

We shall now describe the finite element method. It will be a straightforward modification of the method in [1]. At each time t_n, we will approximate p^n by $U^n \in \mathfrak{M}$ and, on Ω_i, σ^n by $V^n \in \mathbf{N}_i^*$. For convenience, start the method with

$$U^0 = \tilde{p}^0 \tag{4.5}$$

and

$$V^0 = \hat{\sigma}^0. \tag{4.6}$$

The equations for V^n, $n \geq 1$, will amount to the following:

$$(\phi_i \partial V^n, \psi)_i + (\mathbf{x}_i \underline{\nabla} V^{n-1/2}, \underline{\nabla}\psi)_i - (2\underline{x}_i V^{n-1/2}, \underline{\nabla}\psi)_i$$
$$= -(\sigma_0 \underline{x}_i, \underline{\nabla}\psi)_i, \qquad \text{for all } \psi \in \mathbf{N}_i, \tag{4.7a}$$

$$V^n = \frac{1}{|\Omega_i|}[(U^n, \chi_i) + (U^n, \Delta_i \chi_i) \cdot \Delta_i], \qquad \underline{x} \in \partial\Omega_i. \tag{4.7b}$$

This calculation depends on U^n. In turn, the calculation for U^n will depend on V^n through the matrix source term. We can decouple the two calculations by splitting the boundary condition (4.7b) as

$$V^n = \frac{1}{|\Omega_i|}[(U^{n-1}, \chi_i) + (U^{n-1}, \Delta_i \chi_i) \cdot \Delta_i] + \frac{1}{|\Omega_i|}[(\partial U^n, \chi_i) + (\partial U^n, \Delta_i \chi_i) \cdot \Delta_i]\Delta t,$$
$$\underline{x} \in \partial\Omega_i. \tag{4.8}$$

Then, for $n = 1, \ldots, N$,

$$V^n = W^n + \frac{1}{|\Omega_i|}[(\partial U^n, \chi_i) Z_0 + (\partial U^n, \Delta_i \chi_i) \cdot \underline{Z}]\Delta t, \tag{4.9}$$

where $W^n \in \mathbf{N}_i^*$ satisfies

$$\left[\phi_i \frac{W^n - V^{n-1}}{\Delta t}, \psi\right]_i + \left[\kappa_i \nabla \frac{W^n + V^{n-1}}{2}, \nabla\psi\right]_i - \left[2\underline{\chi}_i \frac{W^n + V^{n-1}}{2}, \nabla\psi\right]_i$$
$$= -(\sigma_0 \underline{\chi}_i, \nabla\psi)_i, \qquad\qquad \text{for all } \psi \in \mathbf{N}_i. \qquad (4.10a)$$

$$W^n = \frac{1}{|\Omega_i|}[(U^{n-1}, \chi_i) + (U^{n-1}, \Delta_i \chi_i)\cdot\Delta_i], \qquad \underline{x}\in\partial\Omega_i. \qquad (4.10b)$$

and where $\underline{Z}=(Z_1,\ldots,Z_d)$ and the $Z_j\in\mathbf{N}_i^*$, $j=0,1,\ldots,d$, satisfy

$$\left[\phi_i \frac{Z_j}{\Delta t}, \psi\right]_i + \left[\kappa_i \nabla \frac{Z_j}{2}, \nabla\psi\right]_i - \left[2\underline{\chi}_i \frac{Z_j}{2}, \nabla\psi\right]_i$$
$$= 0, \qquad\qquad \text{for all } \psi \in \mathbf{N}_i. \qquad (4.11a)$$

$$Z_0 = 1, \qquad \underline{x}\in\partial\Omega_i, \qquad\qquad (4.11bi)$$

$$Z_j = \Lambda_{i,j}, \qquad \underline{x}\in\partial\Omega_i, \ j=1,\ldots,d. \qquad (4.11bii)$$

Over the time interval (t_{n-1}, t_n), W^n accounts for the flow in the block arising from the fluid present there at time t_{n-1} (which is V^{n-1}) with no change in the boundary condition, while each Z_j accounts for the flow arising in an empty block experiencing some unit change over its boundary. The combination (4.9) is precisely the solution to (4.7).

Let

$$Q_{i,j} = \frac{1}{|\Omega_i|}\left\{\left[\left(\phi_i \frac{Z_j}{\Delta t}, 1\right)\right]_i + \left[\left(\phi_i \frac{Z_j}{\Delta t}, \Delta_i\right)_i + \left[\kappa_i \nabla \frac{Z_j}{2} - Z_j\underline{\chi}_i, \nabla\Delta_i\right]_i\right]\cdot\Delta_i\right\}\chi_i,$$
$$j = 0,1,\ldots,d, \qquad (4.12)$$

and $\underline{Q}_i=(Q_{i,1},\ldots,Q_{i,d})$. The matrix source term should be approximated as follows:

$$f_i^{n-1/2} \approx -\frac{1}{|\Omega_i|}\left\{ (\phi_i\partial V^n, 1)_i + [(\phi_i\partial V^n, \Delta_i)_i\right.$$
$$\left. + (\kappa_i\nabla V^{n-1/2} - (2V^{n-1/2} - \sigma_0)\underline{\chi}_i, \nabla\Delta_i)_i]\cdot\Delta_i\right\}\chi_i$$

$$= -\frac{1}{|\Omega_i|}\left\{\left[\left(\phi_i \frac{W^n - V^{n-1}}{\Delta t}, 1\right)\right]_i + \left[\left(\phi_i \frac{W^n - V^{n-1}}{\Delta t}, \Delta_i\right)_i\right.\right.$$
$$\left.\left. + \left[\kappa_i\nabla \frac{W^n + V^{n-1}}{2} - (W^n + V^{n-1} - \sigma_0)\underline{\chi}_i, \nabla\Delta_i\right]_i\right]\cdot\Delta_i\right\}\chi_i$$

$$- \frac{1}{|\Omega_i|}\left\{(\partial U^n, \chi_i)Q_{i,0} + (\partial U^n, \Delta_i\chi_i)\cdot\underline{Q}_i\right\}\Delta t. \qquad (4.13)$$

The equations for U^n, $n \geq 1$, can now be expressed as

$$(\phi \partial U^n, \varphi) + \sum_i \frac{1}{|\Omega_i|} \left\{ (\partial U^n, \chi_i)(Q_{i,0}, \varphi) + (\partial U^n, \Delta_i \chi_i) \cdot (\underline{Q}_i, \varphi) \right\} \Delta t$$

$$+ (\kappa \underline{\nabla} U^{n-1/2}, \underline{\nabla} \varphi) - (2 \underline{\gamma} U^{n-1/2}, \underline{\nabla} \varphi)$$

$$= -(\rho_0 \underline{\gamma}, \underline{\nabla} \varphi) + (f_e^{n-1/2}, \varphi)$$

$$- \sum_i \frac{1}{|\Omega_i|} \left\{ \left[\left(\phi_i \frac{W^n - V^{n-1}}{\Delta t}, 1 \right) \right] (\chi_i, \varphi) + \left[\left(\phi_i \frac{W^n - V^{n-1}}{\Delta t}, \Delta_i \right) \right]_i \right.$$

$$\left. + \left[\kappa_i \underline{\nabla} \frac{W^n + V^{n-1}}{2} - (W^n + V^{n-1} - \sigma_0) \underline{\gamma}_i, \underline{\nabla} \Delta_i \right]_i \right] \cdot (\Delta_i \chi_i, \varphi) \right\},$$

$$\text{for all } \varphi \in \mathcal{M}. \quad (4.14)$$

This has completed the method. In summary, (4.5) and (4.6) give the values of U^0 and V^0. A single factorization (for each i) can be used to obtain the Z_j from (4.11), and then the $Q_{i,0}$ and $\underline{Q}_i$ can be calculated from (4.12). Now, successively for $n=1,\ldots,N$, solve (4.10) for W^n, (4.14) for U^n, and then define V^n by (4.9).

Note that only the block problems (4.10), (4.11) that sit over the quadrature points of the fracture calculation (4.14) need be computed. The block problems are independent of each other, so they can be solved in parallel. Hence, as in [1], this is a field-scale method.

Before going on to an analysis of the convergence of the method, we should remark that the linear systems that arise in (4.3), (4.4), (4.10), (4.11), and (4.14) are not singular. This is known for the first four systems, provided only that Δt is not too large. Since uniqueness implies existence, for (4.14) it is sufficient to verify that

$$\left[\phi \frac{U}{\Delta t}, \varphi \right] + \sum_i \frac{1}{|\Omega_i|} \left\{ (U, \chi_i)(Q_{i,0}, \varphi) + (U, \Delta_i \chi_i) \cdot (\underline{Q}_i, \varphi) \right\}$$

$$+ \left[\kappa \underline{\nabla} \frac{U}{2}, \underline{\nabla} \varphi \right] - (\underline{\gamma} U, \underline{\nabla} \varphi)$$

$$= 0, \qquad \text{for all } \varphi \in \mathcal{M}, \qquad (4.15)$$

has $U=0$ as its only solution in $\mathcal{M}$. If we set $\varphi = U$ in (4.15), then for Δt not too large, the first and last two terms on the left side above taken together are positive-definite. For similar reasons, the other term is at least positive-semidefinite. (This reflects the stabilizing effect of the matrix which is more fully explored in [1] for its simpler model.) We can easily see this fact by writing the term out via (4.12). With $\Lambda_{i,0} = 1$,

$$\frac{1}{|\Omega_i|}\left\{(U, \chi_i)(Q_{i,0}, U) + (U, \Delta_i\chi_i)\cdot(\mathbf{Q}_i, U)\right\}$$

$$= \frac{1}{|\Omega_i|^2}\sum_{j=0}^{d}(U, \Lambda_{i,j}\chi_i)\left\{\sum_{k=0}^{d}\left[\left[\phi_i\frac{Z_j}{\Delta t}, \Lambda_{i,k}\right]_i + \left[\kappa_i\underline{\nabla}\frac{Z_j}{2} - z_j\underline{\varnothing}_i, \nabla\Lambda_{i,k}\right]_i\right]\right.$$

$$\times (\Lambda_{i,k}\chi_i, U)\Big\}$$

$$= \frac{1}{|\Omega_i|^2}\sum_{j=0}^{d}\sum_{k=0}^{d}(U, \Lambda_{i,j}\chi_i)\left[\left[\phi_i\frac{Z_j}{\Delta t}, Z_k\right]_i + \left[\kappa_i\underline{\nabla}\frac{Z_j}{2} - z_j\underline{\varnothing}_i, \nabla Z_k\right]_i\right](U, \Lambda_{i,k}\chi_i),$$

(4.16)

by (4.11). Now, the expression in square brackets on the far right side of (4.16) is a $(d+1)\times(d+1)$-tensor which is the sum of $1+2d$ terms. $1+d$ of these terms are outer products of $(d+1)$-vectors (and so are positive-semidefinite), while the other d terms are dominated by these, so long as Δt is not too large. Hence, (4.16) is nonnegative, and $U=0$ is the only solution to (4.15).

5. Convergence of the Method

Following [10], for $n=0,1,\ldots,N$, consider each approximation error as the sum of two pieces: $\rho^n - U^n = (\rho^n - \tilde{\rho}^n) + (\tilde{\rho}^n - U^n)$ and $\sigma^n - V^n = (\sigma^n - \hat{\sigma}^n) + (\hat{\sigma}^n - V^n)$. It is known that

$$\left\|\frac{\partial^k(\rho - \tilde{\rho})}{\partial t^k}\right\|_{H^\ell(\Omega)} \leq \left\|\frac{\partial^k\rho}{\partial t^k}\right\|_{H^r(\Omega)}h^{r-\ell}, \qquad k=0,1\ ;\ \ell=0 \text{ and } \ell=-1 \text{ if } R\geq 3\ ;\ 1\leq r\leq R, \qquad (5.1a)$$

$$\|\underline{\nabla}(\rho - \tilde{\rho})\|_{L^2(\Omega)} \leq \|\rho\|_{H^r(\Omega)}h^{r-1}, \qquad 1\leq r\leq R, \qquad (5.1b)$$

$$\left\|\frac{\partial^k(\sigma - \hat{\sigma})}{\partial t^k}\right\|_{L^2(\Omega_m)} \leq \sum_i\left\|\frac{\partial^k\sigma}{\partial t^k}\right\|_{H^{s_i}(\Omega_i)}h_i^{s_i}, \qquad k=0,1\ ;\ 1\leq s_i\leq S_i \text{ (all } i), \qquad (5.2a)$$

and

$$\|\underline{\nabla}(\sigma - \hat{\sigma})\|_{L^2(\Omega_m)} \leq \sum_i\|\sigma\|_{H^{s_i}(\Omega_i)}h_i^{s_i-1}, \qquad 1\leq s_i\leq S_i \text{ (all } i). \qquad (5.2b)$$

At first glance, it may appear that the linear functions $\bar{\sigma}_i$ where $\sigma - \bar{\sigma}_i\epsilon H^1_0(\Omega_i)$ must be included in the bounding norms of (5.2). However, $\|\underline{\nabla}\bar{\sigma}_i\|^2_{L^2(\Omega_i)} \leq \|\underline{\nabla}\sigma\|^2_{L^2(\Omega_i)}$, so Poincaré's inequality shows that this is not necessary.

It remains to estimate the errors

$$\zeta^n = \tilde{\rho}^n - U^n \epsilon \mathbf{M}$$

and

$$\xi^n = \hat{\sigma}^n - V^n \epsilon \mathbf{N}_i^* \text{ (all } i),$$

which satisfy an equation that is similar to (3.5). We will derive it in stages below.

For $n = 1, \ldots, N$, let us first combine the weak form of the average of the equations for ρ at times t_n and t_{n-1} ((2.8), (2.13), and (2.14)) with the defining relation for $\tilde{\rho}$ (4.3a). After some manipulation, this equation is

$$
(\phi \partial \tilde{\rho}^n, \varphi) + (\kappa \nabla \tilde{\rho}^{n-1/2}, \nabla \varphi)
$$

$$
\begin{aligned}
= \, & ((2\tilde{\rho}^{n-1/2} - \rho_0)\underline{x}, \nabla\varphi) + (f_e^{n-1/2}, \varphi) \\
& - \sum_i \frac{1}{|\Omega_i|} \left[\left\{ (\phi_i \partial\hat{\sigma}^n, 1)_i + [(\phi_i \partial\hat{\sigma}^n, \Delta_i)_i \right. \right. \\
& \qquad\qquad \left. \left. + (\kappa_i \nabla\hat{\sigma}^{n-1/2} - (2\hat{\sigma}^{n-1/2} - \sigma_0)\underline{x}_i, \nabla\Delta_i)_i] \cdot \Delta_i \right\} \chi_i, \varphi \right] \\
& - (\phi(\rho_t^{n-1/2} - \partial\tilde{\rho}^n), \varphi) + (2(\rho^{n-1/2} - \tilde{\rho}^{n-1/2})\underline{x}, \nabla\varphi) \\
& - \sum_i \frac{1}{|\Omega_i|} \left[\left\{ (\phi_i(\sigma_t^{n-1/2} - \partial\hat{\sigma}^n), 1)_i + [(\phi_i(\sigma_t^{n-1/2} - \partial\hat{\sigma}^n), \Delta_i)_i \right. \right. \\
& \qquad\qquad - (\sigma^{n-1/2} - \hat{\sigma}^{n-1/2}, \nabla\cdot\kappa_i\nabla\Delta_i)_i \\
& \qquad\qquad \left. \left. - (2(\sigma^{n-1/2} - \hat{\sigma}^{n-1/2})\underline{x}_i, \nabla\Delta_i)_i] \cdot \Delta_i \right\} \chi_i, \varphi \right],
\end{aligned}
$$

$$
\text{for all } \varphi \in \mathcal{M}, \quad (5.3)
$$

where (4.4b) has been used in integrating $(\kappa_i\nabla(\sigma^{n-1/2} - \hat{\sigma}^{n-1/2}), \nabla\Delta_i)_i$ by parts.

Next, we can find a weak form for the average of the equation for σ (2.4) at times t_n and t_{n-1}. We will take a test function $\psi \in \mathbf{N}_i^*$ for which $\overline{\psi}_i$ is the linear function associated to it (i.e., integrate against $\psi - \overline{\psi}_i \in \mathbf{N}_i \subset H^1_0(\Omega_i)$). Combined with the defining relation (4.4a), after some manipulation, we see that

$$
(\phi_i \partial\hat{\sigma}^n, \psi)_i + (\kappa_i \nabla\hat{\sigma}^{n-1/2}, \nabla\psi)_i
$$

$$
\begin{aligned}
= \, & (2(\hat{\sigma}^{n-1/2} - \sigma_0)\underline{x}_i, \nabla\psi)_i \\
& + \frac{1}{|\Omega_i|} \left[\left\{ (\phi_i \partial\hat{\sigma}^n, 1)_i + [(\phi_i \partial\hat{\sigma}^n, \Delta_i)_i \right. \right. \\
& \qquad\qquad \left. \left. + (\kappa_i \nabla\hat{\sigma}^{n-1/2} - (2\hat{\sigma}^{n-1/2} - \sigma_0)\underline{x}_i, \nabla\Delta_i)_i] \cdot \Delta_i \right\} \chi_i, \overline{\psi}_i \right] \\
& - (\phi_i(\sigma_t^{n-1/2} - \partial\hat{\sigma}^n), \psi)_i + (2(\sigma^{n-1/2} - \hat{\sigma}^{n-1/2})\underline{x}_i, \nabla\psi)_i \\
& + \frac{1}{|\Omega_i|} \left[\left\{ (\phi_i(\sigma_t^{n-1/2} - \partial\hat{\sigma}^n), 1)_i + [(\phi_i(\sigma_t^{n-1/2} - \partial\hat{\sigma}^n), \Delta_i)_i \right. \right. \\
& \qquad\qquad - (\sigma^{n-1/2} - \hat{\sigma}^{n-1/2}, \nabla\cdot\kappa_i\nabla\Delta_i)_i \\
& \qquad\qquad \left. \left. - (2(\sigma^{n-1/2} - \hat{\sigma}^{n-1/2})\underline{x}_i, \nabla\Delta_i)_i] \cdot \Delta_i \right\} \chi_i, \overline{\psi}_i \right],
\end{aligned}
$$

$$
\text{for all } \psi \in \mathbf{N}_i^*, \quad (5.4)
$$

where we recall that

$$\overline{\Psi}_i = \frac{1}{|\Omega_i|}[(\overline{\Psi}_i, \chi_i) + (\overline{\Psi}_i, \Delta_i \chi_i) \cdot \Delta_i].\tag{5.5}$$

We can write the defining equation for V^n (4.7a) in a similar manner:

$$(\phi_i \partial V^n, \psi)_i + (\kappa_i \underline{\nabla} V^{n-1/2}, \underline{\nabla}\psi)_i$$

$$= ((2V^{n-1/2} - \sigma_0)\underline{x}_i, \underline{\nabla}\psi)_i$$

$$+ \frac{1}{|\Omega_i|}\left[\left\{(\phi_i \partial V^n, 1)_i + [(\phi_i \partial V^n, \Delta_i)_i \right.\right.$$

$$\left.\left. + (\kappa_i \underline{\nabla} V^{n-1/2} - (2V^{n-1/2} - \sigma_0)\underline{x}_i, \underline{\nabla}\Delta_i)_i] \cdot \Delta_i\right\}\chi_i, \overline{\psi}_i\right],$$

$$\text{for all } \psi \in \mathbf{N}_i^*.\tag{5.6}$$

We are now ready to derive the equation for ζ^n and ξ^n, $n=1,\dots,N$. Add the difference of (5.3) and (4.14) (with (4.13)) to the sum on i of the difference of (5.4) and (5.6). The result is

$$(\phi \partial \zeta^n, \varphi) + \sum_i (\phi_i \partial \xi^n, \psi)_i + (\kappa \underline{\nabla} \zeta^{n-1/2}, \underline{\nabla}\varphi) + \sum_i (\kappa_i \underline{\nabla} \xi^{n-1/2}, \underline{\nabla}\psi)_i$$

$$= (2\zeta^{n-1/2}\underline{x}, \underline{\nabla}\varphi) + \sum_i (2\xi^{n-1/2}\underline{x}_i, \underline{\nabla}\psi)_i$$

$$- \sum_i \frac{1}{|\Omega_i|}\left[\left\{(\phi_i \partial \xi^n, 1)_i + [(\phi_i \partial \xi^n, \Delta_i)_i \right.\right.$$

$$\left.\left. + (\kappa_i \underline{\nabla} \xi^{n-1/2} - 2\xi^{n-1/2}\underline{x}_i, \underline{\nabla}\Delta_i)_i] \cdot \Delta_i\right\}\chi_i, \varphi - \overline{\psi}_i\right]$$

$$- (\phi(\rho_t^{n-1/2} - \partial\widetilde{\rho}^n), \varphi) - \sum_i (\phi_i(\sigma_t^{n-1/2} - \partial\hat{\sigma}^n), \psi)_i$$

$$+ (2(\rho^{n-1/2} - \widetilde{\rho}^{n-1/2})\underline{x}, \underline{\nabla}\varphi) + \sum_i (2(\sigma^{n-1/2} - \hat{\sigma}^{n-1/2})\underline{x}_i, \underline{\nabla}\psi)_i$$

$$- \sum_i \frac{1}{|\Omega_i|}\left[\left\{(\phi_i(\sigma_t^{n-1/2} - \partial\hat{\sigma}^n), 1)_i + [(\phi_i(\sigma_t^{n-1/2} - \partial\hat{\sigma}^n), \Delta_i)_i\right.\right.$$

$$- (\sigma^{n-1/2} - \hat{\sigma}^{n-1/2}, \underline{\nabla}\cdot\kappa_i\underline{\nabla}\Delta_i)_i$$

$$\left.\left. - (2(\sigma^{n-1/2} - \hat{\sigma}^{n-1/2})\underline{x}_i, \underline{\nabla}\Delta_i)_i] \cdot \Delta_i\right\}\chi_i, \varphi - \overline{\psi}_i\right],$$

$$\text{for all } \varphi \in \mathbf{M} \text{ and } \psi \in \mathbf{N}_i^* \text{ (all i)}, n=1,\dots,N.\tag{5.7}$$

(Remember that $\overline{\psi}_i$ is the linear function associated to $\psi|_{\Omega_i}$.)

It will be convenient to write down the boundary and initial conditions. In light of (4.4b), the difference of (2.6) and (4.7b) yields the boundary condition

$$\xi^n = \sigma^n - V^n$$

$$= \frac{1}{|\Omega_i|}[(\zeta^n, \chi_i) + (\zeta^n, \Delta_i\chi_i)\cdot\Delta_i] + \frac{1}{|\Omega_i|}[(\rho^n - \tilde{\rho}^n, \chi_i) + (\rho^n - \tilde{\rho}^n, \Delta_i\chi_i)\cdot\Delta_i],$$

$$\underline{x}\in\partial\Omega_i, \, n=1,...,N. \quad (5.8)$$

The initial conditions are

$$\zeta^0 = 0, \qquad \underline{x}\in\Omega, \qquad\qquad\qquad\qquad\qquad\qquad (5.9)$$

and

$$\xi^0 = 0, \qquad \underline{x}\in\Omega_m, \qquad\qquad\qquad\qquad\qquad\qquad (5.10)$$

by (4.5) and (4.6).

We are now ready to derive our error estimates. First, take $\varphi=\zeta^{n-1/2}$ and $\psi=\xi^{n-1/2}$ in (5.7). In this case,

$$\overline{\psi}_i = \frac{1}{|\Omega_i|}[(\zeta^{n-1/2}, \chi_i) + (\zeta^{n-1/2}, \Delta_i\chi_i)\cdot\Delta_i]$$

$$+ \frac{1}{|\Omega_i|}[(\rho^{n-1/2} - \tilde{\rho}^{n-1/2}, \chi_i) + (\rho^{n-1/2} - \tilde{\rho}^{n-1/2}, \Delta_i\chi_i)\cdot\Delta_i]$$

$$- \frac{\delta_{n,1}}{2|\Omega_i|}[(\rho^0 - \tilde{\rho}^0, \chi_i) + (\rho^0 - \tilde{\rho}^0, \Delta_i\chi_i)\cdot\Delta_i], \qquad (5.11)$$

where $\delta_{n,1}$ is the Kronecker delta symbol. The terms in (5.7) containing the expression $\varphi-\overline{\psi}_i$ do not vanish; however, the effect of $\varphi=\zeta^{n-1/2}$ cancels with the effect of the first term on the right side of (5.11). Hence, (5.7) becomes

$$\frac{1}{2\Delta t}[(\phi\zeta^n, \zeta^n) - (\phi\zeta^{n-1}, \zeta^{n-1})] + \frac{1}{2\Delta t}\sum_i[(\phi_i\xi^n, \xi^n)_i - (\phi_i\xi^{n-1}, \xi^{n-1})_i]$$

$$+ (\underline{\kappa}\underline{\nabla}\zeta^{n-1/2}, \underline{\nabla}\zeta^{n-1/2}) + \sum_i(\underline{\kappa}_i\underline{\nabla}\xi^{n-1/2}, \underline{\nabla}\xi^{n-1/2})_i$$

$$= (2\zeta^{n-1/2}\underline{\chi}, \underline{\nabla}\zeta^{n-1/2}) + \sum_i(2\xi^{n-1/2}\underline{\chi}_i, \underline{\nabla}\xi^{n-1/2})_i$$

$$+ \sum_i\frac{1}{|\Omega_i|}\left[\left\{(\phi_i\partial\xi^n, 1)_i + [(\phi_i\partial\xi^n, \Delta_i)_i\right.\right.$$

$$+ (\underline{\kappa}_i\underline{\nabla}\xi^{n-1/2} - 2\xi^{n-1/2}\underline{\chi}_i, \underline{\nabla}\Delta_i)_i]\cdot\Delta_i\}\chi_i,$$

$$\frac{1}{|\Omega_i|}[(\rho^{n-1/2} - \tilde{\rho}^{n-1/2}, \chi_i) + (\rho^{n-1/2} - \tilde{\rho}^{n-1/2}, \Delta_i\chi_i)\cdot\Delta_i]$$

$$\left.- \frac{\delta_{n,1}}{2|\Omega_i|}[(\rho^0 - \tilde{\rho}^0, \chi_i) + (\rho^0 - \tilde{\rho}^0, \Delta_i\chi_i)\cdot\Delta_i]\right]$$

$$- (\phi(\rho_t^{n-1/2} - \partial\tilde{\rho}^n), \zeta^{n-1/2}) - \sum_i(\phi_i(\sigma_t^{n-1/2} - \partial\hat{\sigma}^n), \xi^{n-1/2})_i$$

$$+ (2(\rho^{n-1/2} - \widetilde{\rho}^{n-1/2})\underline{\chi}, \underline{\nabla}\zeta^{n-1/2}) + \sum_i (2(\sigma^{n-1/2} - \hat{\sigma}^{n-1/2})\underline{\chi}_i, \underline{\nabla}\xi^{n-1/2})_i$$

$$+ \sum_i \frac{1}{|\Omega_i|} \left[\left\{ (\phi_i(\sigma_t^{n-1/2} - \partial\hat{\sigma}^n), 1)_i + [(\phi_i(\sigma_t^{n-1/2} - \partial\hat{\sigma}^n), \Delta_i)_i \right. \right.$$

$$- (\sigma^{n-1/2} - \hat{\sigma}^{n-1/2}, \underline{\nabla}\cdot\mathbf{\kappa}_i\underline{\nabla}\Delta_i)_i$$

$$- (2(\sigma^{n-1/2} - \hat{\sigma}^{n-1/2})\underline{\chi}_i, \underline{\nabla}\Delta_i)_i]\cdot\Delta_i \Big\} \chi_i,$$

$$\frac{1}{|\Omega_i|}[(\rho^{n-1/2} - \widetilde{\rho}^{n-1/2}, \chi_i) + (\rho^{n-1/2} - \widetilde{\rho}^{n-1/2}, \Delta_i\chi_i)\cdot\Delta_i]$$

$$- \frac{\delta_{n,1}}{2|\Omega_i|}[(\rho^0 - \widetilde{\rho}^0, \chi_i) + (\rho^0 - \widetilde{\rho}^0, \Delta_i\chi_i)\cdot\Delta_i]\Big]$$

$$\leq C \Big\{ \|\rho_t^{n-1/2} - \partial\widetilde{\rho}^n\|^2_{H^{-1}(\Omega)} + \|\sigma_t^{n-1/2} - \partial\hat{\sigma}^n\|^2_{H^{-1}(\Omega_m)}$$

$$+ \delta_{n,1}[1 + (\Delta t)^{-1}]\|\rho^0 - \widetilde{\rho}^0\|^2_{L^2(\Omega)}$$

$$+ \|\rho^{n-1/2} - \widetilde{\rho}^{n-1/2}\|^2_{L^2(\Omega)} + \|\sigma^{n-1/2} - \hat{\sigma}^{n-1/2}\|^2_{L^2(\Omega_m)}$$

$$+ (\phi\zeta^{n-1/2}, \zeta^{n-1/2}) + \sum_i (\phi_i\xi^{n-1/2}, \xi^{n-1/2})_i \Big\}$$

$$+ 1/2\{(\mathbf{\kappa}\underline{\nabla}\zeta^{n-1/2}, \underline{\nabla}\zeta^{n-1/2}) + \sum_i (\mathbf{\kappa}_i\underline{\nabla}\xi^{n-1/2}, \underline{\nabla}\xi^{n-1/2})_i\}$$

$$+ \frac{\delta_{n,1}}{8\Delta t}\sum_i (\phi_i\xi^1, \xi^1)_i$$

$$+ \sum_i \frac{1}{|\Omega_i|^2}\left[\{(\phi_i\partial\xi^n, 1)_i + (\phi_i\partial\xi^n, \Delta_i)_i\cdot\Delta_i\}\chi_i, \right.$$

$$(\rho^{n-1/2} - \widetilde{\rho}^{n-1/2}, \chi_i) + (\rho^{n-1/2} - \widetilde{\rho}^{n-1/2}, \Delta_i\chi_i)\cdot\Delta_i\Big]. \quad (5.12)$$

Now, sum on n from 1 to m. After some manipulation, we see that

$$(\phi\zeta^m, \zeta^m) + \sum_i (\phi_i\xi^m, \xi^m)_i$$

$$+ \sum_{n=1}^m (\mathbf{\kappa}\underline{\nabla}\zeta^{n-1/2}, \underline{\nabla}\zeta^{n-1/2})\Delta t + \sum_{n=1}^m \sum_i (\mathbf{\kappa}_i\underline{\nabla}\xi^{n-1/2}, \underline{\nabla}\xi^{n-1/2})_i\Delta t$$

$$\leq C \Big\{ \sum_{n=1}^m \|\rho_t^{n-1/2} - \partial\widetilde{\rho}^n\|^2_{H^{-1}(\Omega)}\Delta t + \sum_{n=1}^m \|\sigma_t^{n-1/2} - \partial\hat{\sigma}^n\|^2_{H^{-1}(\Omega_m)}\Delta t$$

$$+ \|\rho - \widetilde{\rho}\|^2_{L^\infty(J;L^2(\Omega))} + \|\sigma - \hat{\sigma}\|^2_{L^\infty(J;L^2(\Omega_m))}$$

$$+ \sum_{n=1}^m (\phi\zeta^{n-1/2}, \zeta^{n-1/2})\Delta t + \sum_{n=1}^m \sum_i (\phi_i\xi^{n-1/2}, \xi^{n-1/2})_i\Delta t \Big\}$$

$$+ 1/4\sum_i (\phi_i\xi^1, \xi^1)_i$$

$$+ 2\sum_{n=1}^{m} \sum_{i} \frac{1}{|\Omega_i|^2}\left[\{(\phi_i\partial\xi^n, 1)_i + (\phi_i\partial\xi^n, \Delta_i)_i\cdot\Delta_i\}\chi_i,\right.$$

$$\left. (\rho^{n-1/2} - \tilde{\rho}^{n-1/2}, \chi_i) + (\rho^{n-1/2} - \tilde{\rho}^{n-1/2}, \Delta_i\chi_i)\cdot\Delta_i\right]\Delta t.$$

(5.13)

The last term above can be summed by parts. The result is

$$2\sum_{i} \frac{1}{|\Omega_i|^2}\left[\{(\phi_i\xi^m, 1)_i + (\phi_i\xi^m, \Delta_i)_i\cdot\Delta_i\}\chi_i, (\rho^m - \tilde{\rho}^m, \chi_i) + (\rho^m - \tilde{\rho}^m, \Delta_i\chi_i)\cdot\Delta_i\right]$$

$$- 2\sum_{n=1}^{m} \sum_{i} \frac{1}{|\Omega_i|^2}\left[\{(\phi_i\xi^{n-1/2}, 1)_i + (\phi_i\xi^{n-1/2}, \Delta_i)_i\cdot\Delta_i\}\chi_i,\right.$$

$$\left. (\partial\rho^n - \partial\tilde{\rho}^n, \chi_i) + (\partial\rho^n - \partial\tilde{\rho}^n, \Delta_i\chi_i)\cdot\Delta_i\right]\Delta t$$

$$\leq C\left\{\|\rho - \tilde{\rho}\|^2_{L^\infty(J;L^2(\Omega))} + \sum_{n=1}^{m}\|\partial\rho^n - \partial\tilde{\rho}^n\|^2_{H^{-1}(\Omega)}\Delta t\right.$$

$$+ \sum_{n=1}^{m}\sum_{i}(\phi_i\xi^{n-1/2}, \xi^{n-1/2})_i\Delta t\bigg\}$$

$$+ \tfrac{1}{4}\sum_{i}(\phi_i\xi^m, \xi^m)_i,$$

(5.14)

where $\chi_i \in H^1(\Omega)$, for all i, has been used to obtain the optimal bound on $\partial\rho^n - \partial\tilde{\rho}^n$.

Finally, if Δt is not too large, the discrete Gronwall inequality can be applied to (5.13) (5.14) to yield the estimates

$$\max_{0\leq n\leq N}\|\zeta^n\|^2_{L^2(\Omega)} + \max_{0\leq n\leq N}\|\zeta^n\|^2_{L^2(\Omega_m)}$$

$$+ \sum_{n=1}^{N}\|\nabla\zeta^{n-1/2}\|^2_{L^2(\Omega)}\Delta t + \sum_{n=1}^{N}\|\nabla\zeta^{n-1/2}\|^2_{L^2(\Omega_m)}\Delta t$$

$$\leq C\left\{\sum_{n=1}^{N}[\|\rho_t^{n-1/2} - \partial\rho^n\|^2_{H^{-1}(\Omega)} + \|\partial\rho^n - \partial\tilde{\rho}^n\|^2_{H^{-1}(\Omega)}]\Delta t\right.$$

$$+ \sum_{n=1}^{N}\|\sigma_t^{n-1/2} - \partial\tilde{\sigma}^n\|^2_{H^{-1}(\Omega_m)}\Delta t$$

$$+ \|\rho - \tilde{\rho}\|^2_{L^\infty(J;L^2(\Omega))} + \|\sigma - \hat{\sigma}\|^2_{L^\infty(J;L^2(\Omega_m))}\bigg\}$$

$$\leq C\left\{\|\rho - \tilde{\rho}\|^2_{L^\infty(J;L^2(\Omega))} + \|\rho_t - \tilde{\rho}_t\|^2_{L^2(J;H^{-1}(\Omega))}\right.$$

$$+ \|\sigma - \hat{\sigma}\|^2_{L^\infty(J;L^2(\Omega_m))} + \|\sigma_t - \hat{\sigma}_t\|^2_{L^2(J;H^{-1}(\Omega_m))}$$

$$+ \left[\left\|\frac{\partial^3\rho}{\partial t^3}\right\|^2_{L^2(J;H^{-1}(\Omega))} + \left\|\frac{\partial^3\sigma}{\partial t^3}\right\|^2_{L^2(J;H^{-1}(\Omega_m))}\right](\Delta t)^4\bigg\}.$$

It is also valuable to take the test functions $\varphi = \partial \zeta^n$ and $\psi = \partial \xi^n$ in (5.7). Then

$$\overline{\Psi}_i = \frac{1}{|\Omega_i|}[(\partial\zeta^n, \chi_i) + (\partial\zeta^n, \Delta_i\chi_i)\cdot\Delta_i]$$

$$+ \frac{1}{|\Omega_i|}[(\partial\rho^n - \partial\widetilde{\rho}^n, \chi_i) + (\partial\rho^n - \partial\widetilde{\rho}^n, \Delta_i\chi_i)\cdot\Delta_i]$$

$$+ \frac{\delta_{n,1}}{\Delta t|\Omega_i|}[(\rho^0 - \widetilde{\rho}^0, \chi_i) + (\rho^0 - \widetilde{\rho}^0, \Delta_i\chi_i)\cdot\Delta_i]. \tag{5.16}$$

The terms in (5.7) containing $\varphi - \overline{\Psi}_i$ estimate directly , since we now have (5.15). It is known how to treat all the other terms; in particular, the four gravitational terms containing the expressions $\nabla\partial\zeta^n$ and $\nabla\partial\xi^n$ can be treated by a summation by parts argument analogous to the one given above. The final result is

$$\sum_{n=1}^{N}\|\partial\zeta^n\|^2_{L^2(\Omega)}\Delta t + \sum_{n=1}^{N}\|\partial\xi^n\|^2_{L^2(\Omega_m)}\Delta t$$

$$+ \max_{0\leq n\leq N}\|\underline{\nabla}\zeta^n\|^2_{L^2(\Omega)} + \max_{0\leq n\leq N}\|\underline{\nabla}\xi^n\|^2_{L^2(\Omega_m)}$$

$$\leq C\{\sum_{n=1}^{N}[\|\rho_t^{n-1/2} - \partial\rho^n\|^2_{L^2(\Omega)} + \|\partial\rho^n - \partial\widetilde{\rho}^n\|^2_{L^2(\Omega)}]\Delta t$$

$$+ \sum_{n=1}^{N}[\|\sigma_t^{n-1/2} - \partial\sigma^n\|^2_{L^2(\Omega_m)} + \|\partial\sigma^n - \partial\hat{\sigma}^n\|^2_{L^2(\Omega_m)}]\Delta t$$

$$+ \|\rho - \widetilde{\rho}\|^2_{L^\infty(J;L^2(\Omega))} + \|\sigma - \hat{\sigma}\|^2_{L^\infty(J;L^2(\Omega_m))}\}$$

$$\leq C\Big\{\|\rho - \widetilde{\rho}\|^2_{L^\infty(J;L^2(\Omega))} + \|\rho_t - \widetilde{\rho}_t\|^2_{L^2(J;L^2(\Omega))}$$

$$+ \|\sigma - \hat{\sigma}\|^2_{L^\infty(J;L^2(\Omega_m))} + \|\sigma_t - \hat{\sigma}_t\|^2_{L^2(J;L^2(\Omega_m))}$$

$$+ \Big[\Big\|\frac{\partial^3\rho}{\partial t^3}\Big\|^2_{L^2(J;L^2(\Omega))} + \Big\|\frac{\partial^3\sigma}{\partial t^3}\Big\|^2_{L^2(J;L^2(\Omega_m))}\Big](\Delta t)^4\Big\}. \tag{5.17}$$

The following theorem is a combination of the estimates (5.1), (5.2), (5.15), and (5.17).

Theorem:

If the data and solution of the double porosity model are sufficiently smooth, and if Δt is not too large, then the solution of the finite element method approximates the solution of the model as follows:

$$\max_{0 \le n \le N} \| p^n - U^n \|_{L^2(\Omega)} + \max_{0 \le n \le N} \| \sigma^n - V^n \|_{L^2(\Omega_m)}$$

$$\le C \Big\{ [\| p \|_{L^\infty(J;H^r(\Omega))} + \| p_t \|_{L^2(J;\overline{H}^{r-k}(\Omega))}] h^r$$

$$+ \sum_i [\| \sigma \|_{L^\infty(J;H^{s_i}(\Omega_i))} + \| \sigma_t \|_{L^2(J;H^{s_i}(\Omega_i))}] h_i^{s_i}$$

$$+ \Big[\Big\| \frac{\partial^3 p}{\partial t^3} \Big\|_{L^2(J;H^{-1}(\Omega))} + \Big\| \frac{\partial^3 \sigma}{\partial t^3} \Big\|_{L^2(J;H^{-1}(\Omega_m))} \Big] (\Delta t)^2 \Big\}, \tag{5.18}$$

$$\max_{0 \le n \le N} \| p^n - U^n \|_{H^1(\Omega)} + \max_{0 \le n \le N} \| \sigma^n - V^n \|_{H^1(\Omega_m)}$$

$$\le C \Big\{ [\| p \|_{L^\infty(J;H^r(\Omega))} + \| p_t \|_{L^2(J;\overline{H}^{r-1}(\Omega))}] h^{r-1}$$

$$+ \sum_i [\| \sigma \|_{L^\infty(J;H^{s_i}(\Omega_i))} + \| \sigma_t \|_{L^2(J;\overline{H}^{s_i-1}(\Omega_i))}] h_i^{s_i-1}$$

$$+ \Big[\Big\| \frac{\partial^3 p}{\partial t^3} \Big\|_{L^2(J;L^2(\Omega))} + \Big\| \frac{\partial^3 \sigma}{\partial t^3} \Big\|_{L^2(J;L^2(\Omega_m))} \Big] (\Delta t)^2 \Big\}, \tag{5.19}$$

$$\Big\{ \sum_{n=1}^N \| p_t^{n-1/2} - \partial U^n \|^2_{L^2(\Omega)} \Delta t \Big\}^{1/2} + \Big\{ \sum_{n=1}^N \| \sigma_t^{n-1/2} - \partial V^n \|^2_{L^2(\Omega_m)} \Delta t \Big\}^{1/2}$$

$$\le C \Big\{ [\| p \|_{L^\infty(J;H^r(\Omega))} + \| p_t \|_{L^2(J;H^r(\Omega))}] h^r$$

$$+ \sum_i [\| \sigma \|_{L^\infty(J;H^{s_i}(\Omega_i))} + \| \sigma_t \|_{L^2(J;H^{s_i}(\Omega_i))}] h_i^{s_i}$$

$$+ \Big[\Big\| \frac{\partial^3 p}{\partial t^3} \Big\|_{L^2(J;L^2(\Omega))} + \Big\| \frac{\partial^3 \sigma}{\partial t^3} \Big\|_{L^2(J;L^2(\Omega_m))} \Big] (\Delta t)^2 \Big\}, \tag{5.20}$$

where $1 \le r \le R$, $1 \le s_i \le S_i$ (all i), $k=1$ (except $k=0$ if $R=2$), and $\overline{H}^\ell = \overline{H}^{\max(1,\ell)}$.

Note that the error estimates are optimal with respect to the discretization parameters. The regularity required of p and σ is also optimal, except in (5.18), where σ (and p if $R=2$) must be slightly smoother.

References

[1] T. Arbogast, *Analysis of the simulation of single phase flow through a naturally fractured reservoir*, to appear.

[2] T. Arbogast, proposed Ph. D. thesis, University of Chicago, Chicago, Illinois, 1987.

[3] G. I. Barenblatt, Iu. P. Zheltov, and I. N. Kochina, *Basic concepts in the theory of seepage of homogeneous liquids in fissured rocks [strata]*, Prikl. Mat. Mekh., v. 24, 1960, pp. 852-864. J. Appl. Math. Mech., v. 24, 1960, pp. 1286-1303.

[4] J. Douglas, Jr., P. J. Paes Leme, T. Arbogast, and T. Schmitt, *Simulation of flow in naturally fractured reservoirs*, paper SPE 16019 presented at the ninth Society of Petroleum Engineers Symposium on Reservoir Simulation in San Antonio, Texas, Feb. 1-4, 1987.

[5] H. Kazemi, *Pressure transient analysis of naturally fractured reservoirs with uniform fracture distribution*, Soc. Pet. Eng. J., Dec. 1969, pp. 451-462.

[6] S. J. Pirson, *Performance of fractured oil reservoirs*, Bull. Amer. Assoc. Petrol. Geologists, v. 37, 1953, pp. 232-244.

[7] F. Sonier, P. Souillard, and F. T. Blaskovich, *Numerical simulation of naturally fractured reservoirs*, paper SPE 15627 presented at the 61st Annual Technical Conference and Exhibition of the Society of Petroleum Engineers, New Orleans, Louisiana, Oct. 5-8, 1986.

[8] A. de Swaan O., *Analytic solutions for determining naturally fractured reservoir properties by well testing*, Soc. Pet. Eng. J., June 1976, pp. 117-122.

[9] J. E. Warren and P. J. Root, *The behavior of naturally fractured reservoirs*, Soc. Pet. Eng. J., Sept. 1963, pp. 245-255.

[10] M. F. Wheeler, *A priori L_2 error estimates for Galerkin approximations to parabolic partial differential equations*, SIAM J. Numer. Anal., v. 10, 1973, pp. 723-759.

TWO-PHASE IMMISCIBLE FLOW IN NATURALLY FRACTURED RESERVOIRS

Todd Arbogast

Department of Mathematics, University of Chicago, Chicago, IL 60637;
Institute for Mathematics and its Applications, University of Minnesota, Minneapolis, MN 55455

Jim Douglas, Jr.

Department of Mathematics, University of Chicago, Chicago, IL 60637;
Institute for Mathematics and its Applications, University of Minnesota, Minneapolis, MN 55455

Juan E. Santos

Yacimientos Petroliferos Fiscales and Universidad de Buenos Aires, Buenos Aires, Argentina;
Institute for Mathematics and its Applications, University of Minnesota, Minneapolis, MN 55455

Abstract

A model is defined to simulate a waterflood in a multidimensional, naturally fractured petroleum reservoir. The imbibition process is correctly modelled as a boundary condition on each matrix block. The model is presented in terms of a saturation and the global pressure of Chavent. The numerical method is based on the use of a mixed finite element method for the pressure and standard Galerkin methods for the saturations in the fractures and the blocks. Optimal order asymptotic convergence of the approximate solution to that of the differential system is established under the assumption of nondegeneracy of the relative permeability functions.

1. Introduction

A new approach to modelling fluid flow in naturally fractured petroleum reservoirs has recently been introduced [1], [2], [11] with the object being to improve the treatment of the interaction between the matrix blocks and the fractures. The flow of a single-phase, compressible liquid of constant compressibility has been considered in a relatively complete fashion; i. e., a model has been defined, shown to be well posed, and discretized in a manner that is both practically feasible and asymptotically convergent to the true solution at an optimal rate [1], [2]. A simple linear waterflood has also been modelled and discretized by means of finite differences [11], though analysis of the discretization has not yet been given.

Here we shall consider the waterflood in a more general setting. We shall derive a model for a two-phase, immiscible, incompressible waterflood in a domain in $\mathbb{R}^d$, $d \leq 3$; however, we shall ignore the effects of gravity. The model can be considered to be applicable to a horizontal linear or planar reservoir or to one that is thin in the vertical direction. This is a serious assumption, since gravity affects the imbibition of water into the martrix blocks in an important way. The boundary conditions relating the behavior of the fluids in the fractures to that in the blocks change significantly when gravity is taken into account. Such a model is under development.

As in the case of the linear waterflood model, we shall assume that the blocks are small; as a consequence, we shall neglect the effect of the viscous forces in the fractures in our treatment of the blocks. Thus, we concentrate on modelling the dominant effect of the imbibition process on the blocks. The viscous forces can be treated as they were in [2], but gravity should not be ignored in that case, as its effect exceeds that of the potential drop across a block.

The model to be presented herein will be formulated in terms of a saturation and the global pressure introduced by Chavent [6]. One advantage of this choice is that the differential equations appear in a form quite similar to those for miscible displacement, which has received somewhat greater attention in the mathematical literature in recent years; our numerical model is a variant of techniques first applied to the miscible problem. The second advantage is that, as a consequence of the neglect of the viscous forces in the fractures in the treatment of the blocks, the global pressure will be seen to be a constant over each block at any fixed time, so that the differential system needed to describe the flood in a block will reduce to a single equation for the saturation in the block.

Douglas [7] has described and analyzed finite difference procedures for approximating waterflood problems on standard, unfractured reservoirs; he also used the global pressure in place of a phase pressure. We shall approximate the solution of our problem by adapting a known finite element procedure for miscible displacement problems [9], [10]. We shall combine a mixed finite element method for approximating the pressure and a total flow rate in the fractures with standard Galerkin methods for the saturation both in the fractures and in the individual blocks. This procedure will be analyzed under the same assumptions as were made in [7], where their reasonableness was discussed. In particular, we shall assume that the relative permeabilities remain positive, so that the differential equations stay nondegenerate. Also, we shall take the external flow to be smoothly distributed, instead of being concentrated at wells. These assumptions imply coercivity and regularity for the problem.

An outline of the paper is as follows. In the next section we derive the differential model, and the equations will be put in weak form in the following section. The finite element procedure will be introduced in Section 4 and analyzed in Section 5, which represents about one half of the manuscipt. Finally, in Section 6 a list of otherwise undefined notation is given. We have tried to maintain a consistency in notation here with that used in [7] and [14]. When a quantity exists for both the fracture and matrix block systems, we have chosen to denote the quantity in the fractures by a capital letter and in the blocks by the same symbol in lower case.

2. The Two-Phase, Immiscible Model

Let $\Omega \subset \mathbb{R}^d$, $d \leq 3$, be the fractured reservoir and let $\Omega_i \subset \Omega$ be the ith matrix block. Let $\Omega_m = \cup_i \Omega_i$. The usual equations describing two-phase, immiscible, incompressible displacement in Ω_i when the effect of gravity is omitted are given by

$$\varphi s_{o,t} - \nabla \cdot (kk_{ro}(s_o)\mu_o^{-1}\nabla p_o) = 0, \qquad (x,t) \epsilon \Omega_i \times J, \qquad (2.1)$$

$$\varphi s_{w,t} - \nabla \cdot (kk_{rw}(s_w)\mu_w^{-1}\nabla p_w) = 0, \qquad (x,t) \epsilon \Omega_i \times J, \qquad (2.2)$$

where $J = (0,T]$ and the subscript t denotes differentiation with respect to time. Let $s = s_o = 1-s_w$. The pressures in the two phases are related by the capillary pressure

$$p_c = p_c(s) = p_o - p_w, \qquad (2.3)$$

and it is the case that $p_c'(s) > 0$. Let

$$\lambda(s) = k_{ro}\mu_o^{-1} + k_{rw}\mu_w^{-1}, \qquad \lambda_o(s) = k_{ro}\mu_o^{-1}\lambda^{-1}, \qquad \lambda_w(s) = k_{rw}\mu_w^{-1}\lambda^{-1}.$$

Chavent's global pressure variable [6] is given by

$$p = \tfrac{1}{2}(p_o+p_w) + \tfrac{1}{2}\int_0^{p_c} (\lambda_o-\lambda_w)(p_c^{-1}(\xi))d\xi. \qquad (2.4)$$

As in [7], adding (2.1) and (2.2) gives the pressure equation

$$-\nabla \cdot (k\lambda\nabla p) = 0, \qquad (2.5)$$

and subtracting (2.2) from (2.1) gives the saturation equation

$$\varphi s_t + \lambda_o' u \cdot \nabla s - \nabla \cdot (k\lambda\lambda_o\lambda_w p_c' \nabla s) = 0, \qquad (2.6)$$

where

$$u = -k\lambda\nabla p. \qquad (2.7)$$

The initial and boundary conditions for a matrix block are chosen as follows. Let $\{\chi_i\}$ be a partition of unity on Ω such that

$$\Sigma_i \chi_i = 1, \qquad \int_\Omega \chi_i dx = |\Omega_i|, \qquad \chi_i \geq 0,$$

and the support of χ_i is near Ω_i. Let

$$\hat{\psi}_i = |\Omega_i|^{-1} \int_\Omega \psi \chi_i dx.$$

Assume that the pressure variation in the fractures across a block can be ignored. Then the boundary conditions on the block, which reflect continuity of the water and oil pressures, will be taken to be

$$p_o(x,t) = \hat{P}_{oi}(t) \quad \text{and} \quad p_w(x,t) = \hat{P}_{wi}(t), \qquad (x,t)\epsilon\partial\Omega_i \times J. \tag{2.8}$$

With the analogous fracture quantities $S = S_o = 1 - S_w$ and $P_c(S) = P_o - P_w$, these two relations then imply that

$$p_c(s(x,t)) = p_o(x,t) - p_w(x,t) = \hat{P}_{oi}(t) - \hat{P}_{wi}(t) = \hat{P}_c(S(\cdot,t))_i,$$

so that

$$s(x,t) = p_c^{-1}(\hat{P}_c(S(\cdot,t))_i), \qquad (x,t)\epsilon\partial\Omega_i \times J. \tag{2.9}$$

The physical assumption (2.8) ignores an effect that is of magnitude $O(\text{diam}(\Omega_i))$ and since

$$\hat{P}_c(S(\cdot,t))_i - \hat{P}_c(\hat{S}_i(t)) = O(\text{diam}(\Omega_i))^2,$$

we shall take

$$s(x,t) = p_c^{-1}(P_c(\hat{S}_i(t))), \qquad (x,t)\epsilon\partial\Omega_i \times J, \tag{2.10}$$

in place of (2.9). Below it can be seen that the imposition of continuity of the capillary pressure and the global pressure, rather than the phase pressures, leads to (2.10).

A consequence of the assumption above to ignore the pressure drop in the fractures across a block is that the initial conditions on a block, which reflect initial equilibrium with the surrounding fractures, are given by

$$p_w(x,0) = \hat{P}_{wi}(0) \quad \text{and} \quad p_o(x,0) = \hat{P}_{oi}(0), \qquad x\epsilon\Omega_i, \tag{2.11}$$

so that

$$p_c(s(x,0)) = \hat{P}_c(S(\cdot,0))_i, \qquad x\epsilon\Omega_i. \tag{2.12}$$

Again, instead of (2.12), we shall choose as the block initial saturation the value

$$s(x,0) = p_c^{-1}(\hat{P}_c(\hat{S}_i(0)), \qquad x\epsilon\Omega_i . \tag{2.13}$$

Note that from (2.4) and (2.11) it follows for $(x,t)\epsilon\partial\Omega_i\times J$ that

$$p(x,t) = \tfrac{1}{2}(\hat{P}_{oi}(t) + \hat{P}_{wi}(t)) + \tfrac{1}{2}\int_0^{P_c(\hat{S}_i(t))} (\lambda_o - \lambda_w)(p_c^{-1}(\xi))d\xi \tag{2.14}$$

depends only on the time t. Thus, for each time $t\epsilon J$, p is constant on Ω_i by (2.5), and it follows from (2.7) that $u\equiv0$. Let $\sigma = p_c^{-1}\circ P_c$. Then the differential system for the flow in the matrix block Ω_i becomes the single equation

$$\varphi s_t - \nabla\cdot(k\lambda\lambda_o\lambda_w p_c'\nabla s) = 0, \qquad (x,t)\epsilon\Omega_i\times J, \tag{2.15}$$

with the boundary condition

$$s(x,t) = \sigma(\hat{S}_i(t)), \qquad (x,t)\epsilon\partial\Omega_i\times J, \tag{2.16}$$

and the initial condition

$$s(x,0) = \sigma(\hat{S}_i(0)), \qquad x\epsilon\Omega_i . \tag{2.17}$$

Let

$$v_0 = -kk_{ro}\mu_0^{-1}\nabla p_0$$

be the Darcy velocity of the oil phase. Then the block Ω_i transmits through its surface $\partial\Omega_i$ a space-averaged flow rate of oil given by

$$Q_{oi}(t) = -|\Omega_i|^{-1}\int_{\partial\Omega_i} v_0\cdot\nu \, da = -|\Omega_i|^{-1}\int_{\Omega_i}\nabla\cdot v_0 \, dx = -|\Omega_i|^{-1}\int_{\Omega_i}\varphi s_t dx. \tag{2.18}$$

Thus, we define a total matrix source term by

$$Q_{om}(x,t) = \Sigma_i \, Q_{oi}(t)\chi_i(x) , \qquad (x,t)\epsilon\Omega\times J. \tag{2.19}$$

Now consider the differential system for the fractures. Assume that each matrix block interacts only with the fracture system; thus, the blocks do not interact with each other and they do not interact with external sources and sinks (i.e., wells). Then Darcy's law, conservation of mass, the assumed absence of gravitational terms, and the assumption above imply that

$$\Phi S_{ot} - \nabla\cdot(KK_{ro}(S_0)\nabla P_0) = Q_{oe} + Q_{om} , \qquad (x,t)\epsilon\Omega\times J, \tag{2.20}$$

$$\Phi S_{wt} - \nabla\cdot(KK_{rw}(S_w)\nabla P_w) = Q_{we} + Q_{wm}. \qquad (x,t)\epsilon\Omega\times J, \tag{2.21}$$

where the assumed incompressibility of the fluids (and of the matrix rock) implies that the matrix water sink term Q_{wm} satisfies the relation

$$Q_{wm} + Q_{om} = 0. \tag{2.22}$$

Let

$$\Lambda(S) = K_{ro}\mu_0^{-1} + K_{rw}\mu_w^{-1}, \qquad \Lambda_0(S) = K_{ro}\mu_0^{-1}\Lambda^{-1}, \qquad \Lambda_w(S) = K_{rw}\mu_w^{-1}\Lambda^{-1}.$$

The global pressure for the fracture system is defined analogously by

$$P = \tfrac{1}{2}(P_0+P_w) + \tfrac{1}{2}\int_0^{P_c} (\Lambda_0-\Lambda_w)(P_c^{-1}(\xi))\,d\xi. \tag{2.23}$$

Again, adding and subtracting (2.20) and (2.21) leads to the system

$$\nabla\cdot U \qquad\qquad\qquad = Q_e\,, \qquad (x,t)\epsilon\Omega\times J, \tag{2.24}$$

$$U + K\Lambda\nabla P \qquad\qquad = 0\,, \qquad (x,t)\epsilon\Omega\times J, \tag{2.25}$$

$$\Phi S_t + \Lambda_0' U\cdot\nabla S - \nabla\cdot(K\Lambda\Lambda_0\Lambda_w P_c'\nabla S) = Q_{om} - \Lambda_0 Q_e^*\,, \qquad (x,t)\epsilon\Omega\times J, \tag{2.26}$$

where $Q_e = Q_{oe} + Q_{we}$ and Q_e^* is its positive part. To obtain the right-hand side of (2.26) above, we made the assumptions [7] that

$$Q_{we}=Q_e \text{ and } Q_{oe}=0 \text{ if } Q_e\geq 0; \qquad Q_{we}=\Lambda_w Q_e \text{ and } Q_{oe}=\Lambda_0 Q_e \text{ if } Q_e<0. \tag{2.27}$$

The boundary conditions for the fracture system will be chosen so as to impose no flow across $\partial\Omega$. If

$$V_0 = -K\Lambda\Lambda_0\nabla P_0 \qquad\text{and}\qquad V_w = -K\Lambda\Lambda_w\nabla P_w$$

are the Darcy velocities of the oil and water phases, respectively, we ask that

$$V_0\cdot\nu = V_w\cdot\nu = 0, \qquad (x,t)\epsilon\partial\Omega\times J, \tag{2.28}$$

which in turn requires that

$$-K\Lambda\Lambda_0\Lambda_w P_c'\nabla S\cdot\nu = 0, \qquad (x,t)\epsilon\partial\Omega\times J, \tag{2.29}$$

and

$$U\cdot\nu \qquad\qquad = 0, \qquad (x,t)\epsilon\partial\Omega\times J. \tag{2.30}$$

Finally, we have the initial condition

$$S(x,0) = S^0(x), \qquad x\epsilon\Omega. \tag{2.31}$$

Equations (2.15)-(2.19), (2.24)-(2.26), and (2.29)-(2.31) completely define the behavior of a waterflood in the simplified naturally fractured reservoir that we admit in this study; the pressure P is determined up to an additive constant.

3. A Weak Form of the Problem

Let $H(\text{div};\Omega) = \{v \in L^2(\Omega)^d : \nabla \cdot v \in L^2(\Omega)\}$ and set

$$V = H(\text{div};\Omega) \cap \{v \cdot \nu = 0 \text{ on } \partial\Omega\}, \quad W = L^2(\Omega)/\{w \equiv \text{constant on } \Omega\}.$$

Let $(\cdot,\cdot)$ denote the $L^2(\Omega)$ or $L^2(\Omega)^d$ inner product and $(\cdot,\cdot)_i$ the corresponding inner products over Ω_i.

Assume boundedness of the coefficients in the differential system and of the components of its solution, both in the fractures and in the blocks. Then, a weak form of the system can be set up as follows. Solving for the global pressure in the fractures is equivalent to finding a map $\{U,P\}:J \to V \times W$ such that

$$([K\Lambda(S)]^{-1}U,v) - (\nabla \cdot v, P) = 0, \qquad v \in V, \tag{3.1}$$

$$(\nabla \cdot U, w) = (Q_e, w), \qquad w \in W. \tag{3.2}$$

To put the saturation equations in weak form first let

$$\overline{\psi}_i = |\Omega_i|^{-1} \int_{\Omega_i} \psi dx, \quad D(x,S) = K\Lambda\Lambda_o\Lambda_w P_c', \quad d(x,s) = k\lambda\lambda_o\lambda_w P_c'.$$

The equations (2.15) and (2.26) can be tested against $H^1(\Omega)$ and $H_0^1(\Omega_i)$, respectively, and our weak form of the saturation equations consists of finding the two maps $s:J \to U_i\{H_0^1(\Omega_i) + \sigma(\hat{S}_i(t))\}$ and $S:J \to H^1(\Omega)$ such that

$$(\varphi s_t, \zeta)_i + (d(s)\nabla s, \nabla \zeta)_i = 0, \qquad \zeta \in H_0^1(\Omega_i), \tag{3.3}$$

$$(\Phi S_t, \theta) + \Sigma_i \varphi_i \overline{s}_{ti} \hat{\theta}_i |\Omega_i| + (\Lambda_o'(S)U \cdot \nabla S, \theta) + (D(S)\nabla S, \nabla\theta) = -(\Lambda_o(S)Q_e^*, \theta), \qquad \theta \in H^1(\Omega), \tag{3.4}$$

where, for simplicity, the porosity of each block has been assumed to be a constant.

Equations (3.1)-(3.4) together with the initial conditions (2.17) and (2.31) and the boundary conditions (2.16) define the weak form of the model. As mentioned in the Introduction we shall make some additional assumptions about the coefficients and the source and sink terms in the convergence analysis. The diffusion coefficients $D(S)$ and $d(s)$ vanish at $S = 1-S_{res,w}$ and $S = S_{res,o}$ and at $s = 1-s_{res,w}$ and $s = s_{res,o}$, respectively; however, they will be assumed to be bounded below and above by positive constants:

$$0 < D_1 \leq D(x,S) \leq D_2 < \infty, \quad 0 < d_1 \leq d(x,s) \leq d_2 < \infty. \tag{3.5}$$

Also, the smoothness needed of the various coefficients, the source terms, and the solution itself will be implicit in the argument.

4. A Finite Element Approximation Procedure

For $0 < h_f < 1$ and $0 < h_m < 1$, let $\mathcal{T}_{h_f}(\Omega)$ and $\mathcal{T}_{i,h_m}(\Omega_i)$ be quasiregular partitions of Ω and $\{\Omega_i\}$ into simplices or rectangles of diameters bounded by h_f and h_m, respectively. Let $\mathcal{M} = \mathcal{M}_{h_f} \subset H^1(\Omega)$ and $\eta_i^0 = \eta_{i,h_m}^0 \subset H_0^1(\Omega_i)$ be standard C^0 finite element spaces associated with $\mathcal{T}_{h_f}(\Omega)$ and $\mathcal{T}_{i,h_m}(\Omega_i)$, respectively, such that

$$\inf\{\|u-\theta\|_{W^{j,p}(\Omega)} : \theta \in \mathcal{M}\} \le C\|u\|_{W^{q,p}(\Omega)} h_f^{q-j}, \qquad 1 \le q \le q^* + 1, \quad j = 0,1, \quad p = 2,4, \tag{4.1}$$

$$\inf\{\|v-\zeta\|_{j,\Omega_i} : \zeta \in \eta_i^0\} \le C\|v\|_{r,\Omega_i} h_m^{r-j}, \qquad 1 \le r \le r^* + 1, \quad j = 0,1, \tag{4.2}$$

for any $u \in W^{q,p}(\Omega)$ and $v \in H^r(\Omega_i) \cap H_0^1(\Omega_i)$. Let $\tilde{V}_{h_f} \times \tilde{W}_{h_f}$ denote either the Raviart-Thomas-Nedelec [13], [15], the Brezzi-Douglas-Fortin-Marini [4], the Brezzi-Douglas-Marini [5] (if $d=2$), or the Brezzi-Douglas-Durán-Fortin [3] (if $d=3$) space associated with $\mathcal{T}_{h_f}(\Omega)$ of index such that V is approximated by $\tilde{V}_{h_f}$ to order $q^* + 1$. Set $V_f = V_{h_f} = \{v \in \tilde{V}_{h_f} : v \cdot \nu = 0$ on $\partial\Omega\} \subset V$ and $W_f = W_{h_f} \subset W$. It is known that

$$\inf\{\|u-v\|_0 : v \in V_f\} \le C\|u\|_q h_f^q, \qquad 0 \le q \le q^* + 1, \tag{4.3}$$

$$\inf\{\|\nabla \cdot (u-v)\|_0 : v \in V_f\} \le C\|\nabla \cdot u\|_q h_f^q, \qquad 0 \le q \le q^{**}, \tag{4.4}$$

$$\inf\{\|p-w\|_0 : w \in W_f\} \le C\|p\|_q h_f^q, \qquad 0 \le q \le q^{**}, \tag{4.5}$$

for all $\{u,p\} \in V \times W$, where $q^{**} = q^* + 1$ for the first two spaces ([4], [13], [15]) and $q^{**} = q^*$ for the other two spaces ([3], [5]). Of course, one could take $V_f \times W_f$ over a quasiregular partition $\mathcal{T}_h(\Omega)$ analogous to but different from $\mathcal{T}_{h_f}(\Omega)$. In that case, the analysis of the next section will show that it would be expedient to choose the approximation order $R^* + 1$ in such a way that h^{R^*} and $h_f^{q^*}$ are of the same order.

Let L be a positive integer and $\Delta t = T/L$, and set

$$\psi^n = \psi(n\Delta t), \qquad \partial\psi^n = (\psi^n - \psi^{n-1})/\Delta t$$

for appropriate n.

Following the ideas in [9], [10] and [1], [11], we define a finite element procedure as the solution of the following systems:

i) $\{U_h^n, P_h^n\} \in V_f \times W_f$, $n=0,1,\ldots,L$:

$$([K\wedge(S_h^n)]^{-1} U_h^n, v) - (\nabla \cdot v, P_h^n) = 0, \qquad v \in V_f, \qquad (4.6)$$

$$(\nabla \cdot U_h^n, w) = (Q_e^n, w), \qquad w \in W_f \qquad (4.7)$$

ii) $S_h^n \in \mathfrak{M}$, $n=1,\ldots,L$:

$$(\Phi \partial S_h^n, \theta) + \Sigma_i \varphi_i \overline{\partial S}_{hi}^n \, \hat{\theta}_i |\Omega_i| + (\Lambda_o'(S_h^{n-1}) U_h^{n-1} \cdot \nabla S_h^n, \theta) + (D(S_h^{n-1}) \nabla S_h^n, \nabla \theta)$$
$$= -(\Lambda_o(S_h^{n-1}) Q_e^{*,n}, \theta), \qquad \theta \in \mathfrak{M} \qquad (4.8)$$

iii) $s_{1h}^n \in \eta_i^o + \sigma(\hat{S}_{h,i}^{n-1})$, $n=1,\ldots,L$:

$$(\varphi\{s_{1h}^n - s_h^{n-1}\}/\Delta t, \zeta)_i + (d(s_h^{n-1}) \nabla s_{1h}^n, \nabla \zeta)_i = 0, \qquad \zeta \in \eta_i^o \qquad (4.9)$$

iv) $s_{2h}^n \in \eta_i^o + \Delta t$, $n=1,\ldots,L$:

$$(\varphi s_{2h}^n/\Delta t, \zeta)_i + (d(s_h^{n-1}) \nabla s_{2h}^n, \nabla \zeta)_i = 0, \qquad \zeta \in \eta_i^o, \qquad (4.10)$$

where, with the notation $\gamma_i^{n-1} = \sigma'(\hat{S}_{h\,i}^{n-1})$,

v) $s_h^n \in \eta_i^o + [\sigma(\hat{S}_{h\,i}^{n-1}) + \gamma_i^{n-1}(\hat{S}_{hi}^n - \hat{S}_{h\,i}^{n-1})]$, $n=1,\ldots,L$:

$$s_h^n = s_{1h}^n + \gamma_i^{n-1} \cdot \partial \hat{S}_{hi}^n \cdot s_{2h}^n. \qquad (4.11)$$

Note that s_h^n satisfies the equation

$$(\varphi \partial s_h^n, \zeta)_i + (d(s_h^{n-1}) \nabla s_h^n, \nabla \zeta)_i = 0, \qquad \zeta \in \eta_i^o. \qquad (4.12)$$

To start the procedure, approximate S^0 by $S_h^0 \in \mathfrak{M}$ in any fashion such that

$$\|S^0 - S_h^0\|_0 + h_f \|S^0 - S_h^0\|_1 \le C \|S^0\|_q h_f^q, \qquad 1 \le q \le q^* + 1, \qquad (4.13)$$

and take

$$s_h^0 = \sigma(\hat{S}_i^0). \qquad (4.14)$$

After startup, for $n=1,\ldots,L$, the equations are solved in the following order. First, using S_h^{n-1} and (4.6)-(4.7), evaluate $\{U_h^{n-1}, P_h^{n-1}\}$. Next, from S_h^{n-1}, s_h^{n-1}, and (4.9)-(4.10), obtain s_{1h}^n and s_{2h}^n. Now, using U_h^{n-1}, S_h^{n-1}, s_h^{n-1}, s_{1h}^n, s_{2h}^n, (4.11), and (4.8), compute S_h^n. Finally, s_h^n itself is evaluated from s_{1h}^n, s_{2h}^n, S_h^{n-1}, S_h^n, and (4.11).

5. Analysis of the Error

Let $\eta = U-U_h$ and $\pi = P-P_h$. It is known [9] how to estimate the size of these errors in terms of the error $S-S_h$. The estimates (5.4) of [9] can be improved by an application of the ideas in [12] (in particular, the duality lemmas). Such improvements have appeared there and in [3], [4], and [5]. Combining these results with the estimates (6.2) of [9] leads to the refined estimates

$$\|\eta^n\|_0 \leq C[\|U^n\|_{q}h_f^q + \|\tilde{U}^n\|_{L^\infty(\Omega)}\|S^n-S_h^n\|_0], \qquad 1\leq q\leq q^*+1, \qquad (5.1)$$

$$\|\nabla\cdot\eta^n\|_0 \leq C\|\nabla\cdot U^n\|_{q}h_f^q, \qquad 1\leq q\leq q^{**}, \qquad (5.2)$$

$$\|\pi^n\|_0 \leq C[\|P^n\|_{\max(2,q)}h_f^q + \|\tilde{U}^n\|_{L^\infty(\Omega)}\|S^n-S_h^n\|_0], \qquad 1\leq q\leq q^{**}, \qquad (5.3)$$

where $\tilde{U}$ is a projection of U that is bounded in $L^\infty(\Omega)$ if u is, and C depends on $\|S^n\|_{W^{1,\infty}(\Omega)}$ (in (5.3)). Next, let us introduce the elliptic projections $\tilde{S}_h:J\to\mathbb{M}$ and $\tilde{s}_h:J\to\eta_i^0 + \sigma(\hat{S}_i(t))$ defined by

$$(D(S)\nabla\{S-\tilde{S}_h\},\nabla\theta) + (\Lambda_0'(S)U\cdot\nabla\{S-\tilde{S}_h\},\theta) + (\lambda\{S-\tilde{S}_h\},\theta) = 0, \quad \theta\in\mathbb{M}, \qquad (5.4)$$

where for coercivity

$$\lambda \geq D_1^{-1}\|\Lambda_0'\|^2_{L^\infty(\Omega)}\|U\|^2_{L^\infty(J;L^\infty(\Omega))},$$

and

$$(d(s)\nabla\{s-\tilde{s}_h\},\nabla\zeta)_i = 0, \qquad \zeta\in\eta_i^0, \qquad (5.5)$$

$$\tilde{s}_h = s = \sigma(\hat{S}_i(t)) \text{ on } \partial\Omega_i. \qquad (5.6)$$

It follows from the usual analyses of Galerkin methods for elliptic problems that

$$\|S-\tilde{S}_h\|_0 + h_f\|S-\tilde{S}_h\|_1 \leq C\|S\|_q h_f^q, \qquad 1\leq q\leq q^*+1, \qquad (5.7)$$

$$\|S_t-\tilde{S}_{ht}\|_0 \leq C[\|S\|_q + \|S_t\|_q]h_f^q, \qquad 1\leq q\leq q^*+1, \qquad (5.8)$$

$$\|S-\tilde{S}_h\|_{L^4(\Omega)} \leq C\|S\|_{W^{q,4}(\Omega)}h_f^q, \qquad 1\leq q\leq q^*+1, \qquad (5.9)$$

$$\|s-\tilde{s}_h\|_{0,\Omega_i} + h_m\|s-\tilde{s}_h\|_{1,\Omega_i} \leq C\|s\|_{r,\Omega_i}h_m^r, \qquad 1\leq r\leq r^*+1, \qquad (5.10)$$

$$\|s_t-\tilde{s}_{ht}\|_{0,\Omega_i} \leq C[\|s\|_{r,\Omega_i} + \|s_t\|_{r,\Omega_i}]h_m^r, \qquad 1\leq r\leq r^*+1. \qquad (5.11)$$

Thus, the convergence analysis reduces to obtaining optimal order bounds for the errors $\zeta_h = \tilde{S}_h-S_h$ and $\zeta_{mh} = \tilde{s}_h-s_h$.

For simplicity, choose the initial condition

$$S_h^0 = \tilde{S}_h^0,$$

the elliptic projection of the initial condition S^0, so that $\xi_h^0 = 0$. Then, combine (3.4), (4.8), and (5.4) to obtain the relation

$$(\Phi\partial\xi_h^n,\theta) + \Sigma_i\varphi_i\partial\overline{\xi}_{mhi}^n\,\hat{\theta}_i|\Omega_i| + (\Lambda_0'(S_h^{n-1})U_h^{n-1}\cdot\nabla\xi_h^n,\theta) + (D(S_h^{n-1})\nabla\xi_h^n,\nabla\theta)$$
$$= -(\Phi\{S_t^n-\partial\tilde{S}_h^n\},\theta) + (Q_e^{*,n}\{\Lambda_0(S_h^{n-1})-\Lambda_0(S^n)\},\theta) - ([D(S^n)-D(S_h^{n-1})]\nabla\tilde{S}_h^n,\nabla\theta)$$
$$+ (\lambda\{S^n-\tilde{S}_h^n\},\theta) - ([\Lambda_0'(S^n)-\Lambda_0'(S_h^{n-1})]U^n\cdot\nabla\tilde{S}_h^n,\theta)$$
$$- (\Lambda_0'(S_h^{n-1})\{U^n-U_h^{n-1}\}\cdot\nabla\tilde{S}_h^n,\theta) - \Sigma_i\varphi_i(\overline{S}_i^n-\partial\overline{\tilde{S}}_{hi}^n)\,\hat{\theta}_i|\Omega_i|, \qquad \theta\in\mathfrak{M}, \ 1\le n\le L. \quad (5.12)$$

Next, note that (2.17), (4.14), and (5.5)-(5.6) imply that $\xi_{mh}^0 = 0$. Also, combining (3.3), (4.12), and (5.5) leads to the equation

$$(\varphi\partial\xi_{mh}^n,\zeta)_i + (d(s_h^{n-1})\nabla\xi_{mh}^n,\nabla\zeta)_i$$
$$= -(\varphi\{s_t^n-\partial\tilde{s}_h^n\},\zeta)_i - ([d(s^n)-d(s_h^{n-1})]\nabla\tilde{s}_h^n,\nabla\zeta)_i, \qquad \zeta\in\mathfrak{M}_i^0, \ 1\le n\le L. \quad (5.13)$$

Also, it follows from (2.16), (4.11), and (5.6) that

$$\xi_{mh}^n = \sigma(\hat{S}_i^n) - \sigma(\hat{S}_{hi}^{n-1}) - \aleph_i^{n-1}(\hat{S}_{hi}^n-\hat{S}_{hi}^{n-1}), \qquad (x,t)\in\partial\Omega_i\times J. \quad (5.14)$$

In the analysis that will come it is convenient to rewrite the expression above. Note that

$$\sigma(\hat{S}_i^n) = \sigma(\hat{S}_{hi}^{n-1}) + \aleph_i^{n-1}(\hat{S}_i^n-\hat{S}_{hi}^{n-1}) + \delta_i^n,$$

where

$$\delta_i^n = \int_{\hat{S}_{hi}^{n-1}}^{\hat{S}_i^n} (\hat{S}_i^n - \theta)\,\sigma''(\theta)d\theta, \quad (5.15)$$

so that (5.14) becomes

$$\xi_{mh}^n = \aleph_i^{n-1}(\hat{S}_i^n-\hat{S}_{hi}^n) + \delta_i^n$$
$$= \aleph_i^{n-1}[\hat{S}_i^n - \hat{\tilde{S}}_{hi}^n] + \aleph_i^{n-1}\hat{\xi}_{hi}^n + \delta_i^n, \qquad (x,t)\in\partial\Omega_i\times J. \quad (5.16)$$

Choose $\theta = \xi_h^n$ in the error equation (5.12):

$$(\Phi\partial\xi_h^n,\xi_h^n) + \Sigma_i\varphi_i\partial\overline{\xi}_{mhi}^n\,\hat{\xi}_{hi}^n|\Omega_i| + (D(S_h^{n-1})\nabla\xi_h^n,\nabla\xi_h^n)$$
$$= -(\Phi\{S_t^n-\partial\tilde{S}_h^n\},\xi_h^n) + (Q_e^{*,n}\{\Lambda_0(S_h^{n-1})-\Lambda_0(S^n)\},\xi_h^n) - ([D(S^n)-D(S_h^{n-1})]\nabla\tilde{S}_h^n,\nabla\xi_h^n)$$
$$+ (\lambda\{S^n-\tilde{S}_h^n\},\xi_h^n) - ([\Lambda_0'(S^n)-\Lambda_0'(S_h^{n-1})]U^n\cdot\nabla\tilde{S}_h^n,\xi_h^n) - (\Lambda_0'(S_h^{n-1})\{U^n-U_h^{n-1}\}\cdot\nabla\tilde{S}_h^n,\xi_h^n)$$
$$- \Sigma_i\varphi_i(\overline{S}_i^n-\partial\overline{\tilde{S}}_{hi}^n)\hat{\xi}_{hi}^n|\Omega_i| - (\Lambda_0'(S_h^{n-1})U_h^{n-1}\cdot\nabla\xi_h^n,\xi_h^n)$$
$$\equiv T_1^n + \cdots + T_8^n, \qquad\qquad 1\le n\le L. \quad (5.17)$$

First, the left-hand side of (5.17) is bounded below by

$$(2\Delta t)^{-1}[(\Phi\xi_h^n,\xi_h^n) - (\Phi\xi_h^{n-1},\xi_h^{n-1})] + D_1\|\nabla\xi_h^n\|_0^2 + \Sigma_i\varphi_i\partial\xi_{hi}^n\,\xi_{hi}^n|\Omega_i|.$$

Then, notice that all of the terms on the right-hand side of (5.17) with the exception of the last two can be bounded using the ideas in [9], [10], so that here we just briefly summarize the bounds for the first six terms. Now, (5.8) shows that

$$|T_1^n| \le C[\|S_t^n-\partial S^n\|_0 + \|\partial(S-\tilde{S}_h)^n\|_0]\|\xi_h^n\|_0$$

$$\le C\{\Delta t\|S_{tt}\|^2_{L^2(J_n;L^2(\Omega))}$$

$$+ (\Delta t)^{-1}[\|S\|^2_{L^2(J_n;H^q(\Omega))} + \|S_t\|^2_{L^2(J_n;H^q(\Omega))}]h_f^{2q} + \|\xi_h^n\|_0^2\}, \quad 1\le q\le q^*+1,$$

where $J_n = ((n-1)\Delta t, n\Delta t]$. With S suitably smooth, known L^∞ estimates for Galerkin methods for elliptic problems imply that $\|\nabla\tilde{S}_h^n\|_{L^\infty(\Omega)} \le C$. Hence, since $S^n - S_h^{n-1} = (S^n - S^{n-1}) + (S^{n-1} - \tilde{S}_h^{n-1}) + \xi_h^{n-1}$, it follows from (5.7) that

$$|T_2^n| + |T_3^n| + |T_4^n| + |T_5^n|$$

$$\le C[\|S_h^{n-1} - S^n\|_0\|\xi_h^n\|_0 + \|S^n-S_h^{n-1}\|_0\|\nabla\tilde{S}_h^n\|_{L^\infty(\Omega)}\|\nabla\xi_h^n\|_0$$

$$+ \|S^n - \tilde{S}_h^n\|_0\|\xi_h^n\|_0 + \|S^n-S_h^{n-1}\|_0\|U^n\|_{L^\infty(\Omega)}\|\nabla\tilde{S}_h^n\|_{L^\infty(\Omega)}\|\xi_h^n\|_0]$$

$$\le C[\Delta t\|S_t\|^2_{L^2(J_n;L^2(\Omega))} + \|S\|^2_{L^\infty(J;H^q(\Omega))}h_f^{2q} + \|\xi_h^{n-1}\|_0^2 + \|\xi_h^n\|_0^2] + \epsilon\|\nabla\xi_h^n\|_0^2,$$

$$1\le q\le q^*+1,$$

for ϵ as small as we please. To estimate T_6^n, write $U^n - U_h^{n-1}$ as $(U^n - U^{n-1}) + \eta^{n-1}$ and apply (5.1) and (5.7):

$$|T_6^n| \le C(\|U^n-U^{n-1}\|_0 + \|\eta^{n-1}\|_0)\|\nabla\tilde{S}_h^n\|_{L^\infty(\Omega)}\|\xi_h^n\|_0$$

$$\le C\{\Delta t\|U_t\|^2_{L^2(J_n;L^2(\Omega))} + [\|U\|^2_{L^\infty(J;H^q(\Omega))} + \|S\|^2_{L^\infty(J;H^q(\Omega))}]h_f^{2q}$$

$$+ \|\xi_h^{n-1}\|_0^2 + \|\xi_h^n\|_0^2\}, \quad 1\le q\le q^*+1.$$

We take up now the estimation of the new terms T_7^n and T_8^n. Using (5.11), we see that

$$|T_7^n| \le C\Sigma_i\{(\bar{s}_{ti}^n - \partial\bar{\tilde{s}}_{hi}^n)^2 + (\hat{\xi}_{hi}^n)^2\}|\Omega_i|$$

$$\le C[\|s_t^n - \partial\tilde{s}_h^n\|^2_{0,\Omega_m} + \|\xi_h^n\|_0^2]$$

$$\le C\{\Delta t\|s_{tt}\|^2_{L^2(J_n;L^2(\Omega_m))}$$

$$+ (\Delta t)^{-1}[\|s\|^2_{L^2(J_n;H^r(\Omega_m))} + \|s_t\|^2_{L^2(J_n;H^r(\Omega_m))}]h_m^{2r} + \|\xi_h^n\|_0^2\}, \quad 1\le r\le r^*+1.$$

Let $\Pi:V\to V_{h_f}$ denote the vector projection operator defined in one of [3], [4], [5],

[13], or [15], as indicated by the choice of the mixed finite element space made in setting up our numerical technique. We shall employ some of the approximation and boundedness properties of Π. In particular,

$$\|U - \Pi U\|_0 \leq C\|U\|_q h_T^q, \quad 1 \leq q \leq q^* + 1,$$

and ΠU is bounded in $L^\infty(\Omega)$. Then,

$$|T_8^n| \leq |(\Lambda_0'(S_h^{n-1})\Pi U^{n-1} \cdot \nabla \xi_h^n, \xi_h^n)| + |(\Lambda_0'(S_h^{n-1})(\Pi U^{n-1} - U_h^{n-1}) \cdot \nabla \xi_h^n, \xi_h^n)|$$
$$\leq C[\|\nabla \xi_h^n\|_0 \|\xi_h^n\|_0 + \|\Pi U^{n-1} - U_h^{n-1}\|_{L^4(\Omega)} \|\nabla \xi_h^n\|_0 \|\xi_h^n\|_{L^4(\Omega)}].$$

Quasiregularity of the partition of Ω in $\mathbb{R}^2$ or $\mathbb{R}^3$ implies the inverse property

$$\|\Pi U^{n-1} - U_h^{n-1}\|_{L^4(\Omega)} \leq C h_T^{-3/4} \|\Pi U^{n-1} - U_h^{n-1}\|_0.$$

So, with (5.1) and (5.7),

$$\|\Pi U^{n-1} - U_h^{n-1}\|_{L^4(\Omega)} \leq C h_T^{-3/4} [\|U^{n-1} - \Pi U^{n-1}\|_0 + \|\eta^{n-1}\|_0]$$
$$\leq C h_T^{-3/4} \{(\|U^{n-1}\|_q + \|S^{n-1}\|_q) h_T^q + \|\xi_h^{n-1}\|_0\}, \quad 1 \leq q \leq q^* + 1.$$

Also, Hölder's inequality and the Sobolev embedding theorem imply that

$$\|\xi_h^n\|_{L^4(\Omega)} \leq \|\xi_h^n\|_0^{1/4} (\|\xi_h^n\|_{L^6(\Omega)})^{3/4} \leq C\|\xi_h^n\|_0^{1/4} \|\xi_h^n\|_1^{3/4}.$$

Thus,

$$|T_8^n| \leq C\{\|\xi_h^n\|_0 \|\nabla \xi_h^n\|_0 + h_T^{-3/4} [(\|U^{n-1}\|_q + \|S^{n-1}\|_q) h_T^q + \|\xi_h^{n-1}\|_0] \|\xi_h^n\|_0^{1/4} \|\xi_h^n\|_1^{7/4}\}$$
$$\leq C\{1 + \{h_T^{-3/4} [(\|U\|_{L^\infty(J;H^q(\Omega))} + \|S\|_{L^\infty(J;H^q(\Omega))}) h_T^q + \|\xi_h^{n-1}\|_0]\}^8\} \|\xi_h^n\|_0^2$$
$$+ \epsilon \|\nabla \xi_h^n\|_0^2, \quad 1 \leq q \leq q^* + 1.$$

For convenience in notation set

$$T_i^2 = \|S_t\|^2_{L^2(J;L^2(\Omega))} + \|S_{tt}\|^2_{L^2(J;L^2(\Omega))} + \|S_{tt}\|^2_{L^2(J;L^2(\Omega_m))} + \|U_t\|^2_{L^2(J;L^2(\Omega))}, \quad (5.18)$$

$$F_i^2(q) = \|S\|^2_{L^\infty(J;H^q(\Omega))} + \|S_t\|^2_{L^2(J;H^q(\Omega))} + \|U\|^2_{L^\infty(J;H^q(\Omega))}, \quad (5.19)$$

$$M_i^2(r) = \|s\|^2_{L^2(J;H^r(\Omega_m))} + \|s_t\|^2_{L^2(J;H^r(\Omega_m))}, \quad (5.20)$$

Multiply (5.17) by Δt and sum from $n=1$ to $n=\beta$. Apply the estimates obtained above to see that

$$\tfrac{1}{2}(\Phi\xi_h^\beta,\xi_h^\beta) + (D_1-2\epsilon)\Sigma_{n=1}^\beta\|\nabla\xi_h^n\|_0^2\Delta t + \Sigma_{n=1}^\beta\Sigma_i\,\varphi_i\partial\overline\xi_{mhi}^n\,\hat\xi_{hi}^n|\Omega_i|\Delta t$$

$$\leq C\{\,\mathbf{T}_i^2(\Delta t)^2 + \mathbf{F}_i^2(q)h_T^{2q} + \mathbf{M}_i^2(r)h_m^{2r}$$

$$+ \Sigma_{n=1}^\beta\{1+[h_T^{-3/4}(\mathbf{F}_1(q)h_T^q+\|\xi_h^{n-1}\|_0)]^8\}\|\xi_h^n\|_0^2\Delta t\,\},\quad 1\leq q\leq q^*+1,\ 1\leq r\leq r^*+1. \quad (5.21)$$

Next, choose the test function $\zeta = \{\xi_{mh}^n - [\aleph_i^{n-1}(\hat\xi_{hi}^n + (\hat S_i^n-\hat{\tilde S}_{hi}^n)) + \delta_i^n]\}[\aleph_i^{n-1}]^{-1}\,\epsilon\,\eta_u^0$ in (5.13), multiply by Δt, and sum i and n to obtain the equation

$$\Sigma_{n=1}^\beta\Sigma_i\{([\aleph_i^{n-1}]^{-1}\varphi\partial\xi_{mh}^n,\xi_{mh}^n)_i - \varphi_i\partial\overline\xi_{mhi}^n\hat\xi_{hi}^n|\Omega_i| + ([\aleph_i^{n-1}]^{-1}d(s_h^{n-1})\nabla\xi_{mh}^n,\nabla\xi_{mh}^n)_i\}\Delta t$$

$$= \Sigma_{n=1}^\beta\Sigma_i\{\varphi_i\partial\overline\xi_{mhi}^n(\hat S_i^n-\hat{\tilde S}_{hi}^n)|\Omega_i| - (\varphi[\aleph_i^{n-1}]^{-1}\{s_t^n-\partial\tilde s_{hi}^n\},\xi_{mh}^n)_i + \varphi_i(\overline s_{ti}^n-\partial\overline{\tilde s}_{hi}^n)\hat\xi_{hi}^n|\Omega_i|$$

$$+ \varphi_i(\overline s_{ti}^n-\partial\overline{\tilde s}_{hi}^n)(\hat S_i^n-\hat{\tilde S}_{hi}^n)|\Omega_i| - ([\aleph_i^{n-1}]^{-1}\{d(s^n)-d(s_h^{n-1})\}\nabla\tilde s_h^n,\nabla\xi_{mh}^n)_i$$

$$+ [\aleph_i^{n-1}]^{-1}\varphi_i\partial\overline\xi_{mhi}^n\delta_i^n|\Omega_i| + [\aleph_i^{n-1}]^{-1}\varphi_i(\overline s_{ti}^n-\partial\overline{\tilde s}_{hi}^n)\delta_i^n|\Omega_i|\}\Delta t$$

$$\equiv t_1^\beta + \cdots + t_7^\beta \qquad\qquad (5.22)$$

Our implicit assumptions on the smoothness of the coefficients must include the boundedness of σ', $[\sigma']^{-1}$, and σ'' in order to deal with $\aleph_i^{n-1}$ and δ_i^n. These assumptions are consistent with the assumption that the relative permeabilities are bounded below by a positive constant. The left-hand side of (5.22) can then be bounded below by

$$C_1[\,\|\xi_{mh}^\beta\|_{0,\Omega_m}^2 + \Sigma_{n=1}^\beta\|\nabla\xi_{mh}^n\|_{0,\Omega_m}^2\Delta t\,] - \Sigma_{n=1}^\beta\Sigma_i\,\varphi_i\partial\overline\xi_{mhi}^n\,\hat\xi_{hi}^n\,\Delta t.$$

Set

$$\mathbf{T}^2 = \mathbf{T}_i^2 + \|s_t\|^2_{L^2(J;L^2(\Omega_m))} + \|s_t\|^4_{L^4(J;L^4(\Omega))}, \qquad\qquad (5.23)$$

$$\mathbf{F}^2(q) = \mathbf{F}_i^2(q) + \|s\|^4_{L^\infty(J;W^{q,4}(\Omega))}, \qquad\qquad (5.24)$$

$$\mathbf{M}^2(r) = \mathbf{M}_i^2(r) + \|s\|^2_{L^\infty(J;H^r(\Omega_m))}. \qquad\qquad (5.25)$$

We shall now bound the terms t_j^β, $1\leq j\leq 7$.

First, summation by parts on t_1^β gives

$$|t_1^\beta| = |\,\Sigma_i\{\varphi_i\overline\xi_{mhi}^\beta(\hat S_i^\beta-\hat{\tilde S}_{hi}^\beta) - \Sigma_{n=1}^\beta\varphi_i\overline\xi_{mhi}^{n-1}(\partial\hat S_i^n-\partial\hat{\tilde S}_{hi}^n)\Delta t\,\}|\Omega_i|\,|$$

$$\leq \epsilon\|\xi_{mh}^\beta\|_{0,\Omega_m}^2 + C[\mathbf{F}_i^2(q)h_T^{2q} + \Sigma_{n=1}^\beta\|\xi_{mh}^n\|_{0,\Omega_m}^2\Delta t\,], \qquad 1\leq q\leq q^*+1,$$

by (5.7) and (5.8). The next three terms are bounded easily with (5.7) and (5.11) by

$$|t_2^\beta| + |t_3^\beta| + |t_4^\beta|$$

$$\leq C\Sigma_{n=1}^\beta \Sigma_i \{\|s_i^n - \partial \tilde{s}_h^n\|_{0,\Omega_i} \|\xi_{mh}^n\|_{0,\Omega_i}$$

$$+ |\bar{s}_i^n - \partial \overline{\tilde{s}}_{h,i}^n||\xi_{h,i}^n||\Omega_i| + |\bar{s}_i^n - \partial \overline{\tilde{s}}_{h,i}^n||\hat{S}_i^n - \tilde{S}_{h,i}^n||\Omega_i|\} \Delta t$$

$$\leq C\Sigma_{n=1}^\beta \{\|s_i^n - \partial \tilde{s}_h^n\|_{0,\Omega_m}^2 + \|S^n - \tilde{S}_h^n\|_0^2 + \|\xi_{mh}^n\|_{0,\Omega_m}^2 + \|\xi_h^n\|_0^2\} \Delta t$$

$$\leq C[T_i^2(\Delta t)^2 + M_i^2(r)h_m^{2r} + F_i^2(q)h_m^{2q}$$

$$+ \Sigma_{n=1}^\beta \|\xi_{mh}^n\|_{0,\Omega_m}^2 \Delta t + \Sigma_{n=1}^\beta \|\xi_h^n\|_0^2 \Delta t], \quad 1 \leq r \leq r^*+1, \quad 1 \leq q \leq q^*+1.$$

In order to estimate the term t_5^β, we need the known L^∞ estimates for Galerkin methods for elliptic problems that imply the bound $\|\nabla \tilde{s}_h^n\|_{L^\infty(\Omega_m)} \leq C$. So, (5.10) gives the bound

$$|t_5^\beta| \leq C\Sigma_{n=1}^\beta \Sigma_i \|s^n - s_h^{n-1}\|_{0,\Omega_i} \|\nabla \tilde{s}_h^n\|_{L^\infty(\Omega_i)} \|\nabla \xi_{mh}^n\|_{0,\Omega_i} \Delta t$$

$$\leq \epsilon \Sigma_{n=1}^\beta \|\nabla \xi_{mh}^n\|_{0,\Omega_m}^2 \Delta t + C[T^2(\Delta t)^2 + M^2(r)h_m^{2r} + \Sigma_{n=1}^\beta \|\xi_{mh}^n\|_{0,\Omega_m}^2 \Delta t], \quad 1 \leq r \leq r^*+1.$$

The sixth term on the right-hand side of (5.22) must be treated in a different way since we do not have a direct estimate of either $\partial \xi_{mh}^n$ or $\partial \xi_h^n$. We will use the naive estimate

$$\|\partial \xi_{mh}^n\|_{0,\Omega_m} \leq [\|\xi_{mh}^n\|_{0,\Omega_m} + \|\xi_{mh}^{n-1}\|_{0,\Omega_m}](\Delta t)^{-1}.$$

This loss of a power of Δt will be offset by δ_i^n which, as can be seen from the expression (5.15), is bounded by the square of the error:

$$\delta_i^n \leq C|\hat{S}_i^n - S_{h,i}^{n-1}|^2$$

$$\leq C[|\hat{S}_i^n - S_i^{n-1}|^2 + |S_i^{n-1} - \tilde{S}_{h,i}^{n-1}|^2 + |\xi_{h,i}^{n-1}|^2]$$

$$\leq C\{|\Omega_i|^{-1/2}[\|(S^n - s^{n-1})\chi_i^{1/4}\|_{L^4(\Omega)}^2 + \|(S^{n-1} - \tilde{S}_{h,i}^{n-1})\chi_i^{1/4}\|_{L^4(\Omega)}^2]$$

$$+ |\Omega_i|^{-1}\|\xi_h^{n-1}\chi_i^{1/2}\|_0^2\}.$$

Unfortunately, the induction argument to be given below will just fail unless we are more careful. Note that, as indicated in [8], there exists $z \in \eta_i^0 + 1$ such that

$$\|z\|_{0,\Omega_i}^2 \leq C|\partial\Omega_i|h_m \quad \text{and} \quad \|z\|_{1,\Omega_i}^2 \leq C|\partial\Omega_i|h_m^{-1}. \tag{5.26}$$

Choose $\zeta = [\partial_i^{n-1}]^{-1}(1-z)\delta_i^n \in \eta_i^0$ in (5.13), multiply by Δt, and add over i and n to obtain the relation

$$t_6^\beta + t_7^\beta = \sum_{n=1}^\beta \sum_i [\partial_t^{n-1}]^{-1} [\varphi_i \partial \xi_{mhi}^n + \varphi_i(\overline{s}_{ti}^n - \partial \widetilde{\overline{s}}_{hi}^n)] \delta_i^n |\Omega_i| \Delta t$$

$$= \sum_{n=1}^\beta \sum_i [\partial_t^{n-1}]^{-1} \{ (\varphi \partial \xi_{mh}^n, z)_i + (d(s_h^{n-1})\nabla \xi_{mh}^n, \nabla z)_i$$

$$+ (\varphi \{s_t^n - \partial \widetilde{s}_h^n\}, z)_i + (\{d(s^n) - d(s_h^{n-1})\}\nabla \widetilde{s}_h^n, \nabla z)_i \} \delta_i^n \Delta t.$$

Hence, with (5.7) and (5.9)–(5.11),

$$|t_6^\beta| + |t_7^\beta|$$

$$\leq C \sum_{n=1}^\beta \sum_i \{ \|\partial \xi_{mh}^n\|_{0,\Omega_i} \|z\|_{0,\Omega_i} + \|\nabla \xi_{mh}^n\|_{0,\Omega_i} \|\nabla z\|_{0,\Omega_i} + \|s_t^n - \partial \widetilde{s}_h^n\|_{0,\Omega_i} \|z\|_{0,\Omega_i}$$

$$+ \|s^n - s_h^{n-1}\|_{0,\Omega_i} \|\nabla s_h^n\|_{L^\infty(\Omega_i)} \|\nabla z\|_{0,\Omega_i} \} |\delta_i^n| \Delta t$$

$$\leq \epsilon \sum_{n=1}^\beta [\|\xi_{mh}^n\|_{0,\Omega_m}^2 + \|\xi_{mh}^{n-1}\|_{0,\Omega_m}^2 + \|\nabla \xi_{mh}^n\|_{0,\Omega_m}^2 + \|s_t^n - \partial \widetilde{s}_h^n\|_{0,\Omega_m}^2$$

$$+ \|s^n - s_h^{n-1}\|_{0,\Omega_m}^2] \Delta t$$

$$+ C \sum_{n=1}^\beta \sum_i |\partial \Omega_i| \{ h_m [(\Delta t)^{-2} + 1] + h_m^{-1} \} (\delta_i^n)^2 \Delta t$$

$$\leq C [T^2(\Delta t)^2 + M^2(r) h_m^{2r} + \sum_{n=1}^\beta \|\xi_{mh}^n\|_{0,\Omega_m}^2 \Delta t] + \epsilon \sum_{n=1}^\beta \|\nabla \xi_{mh}^n\|_{0,\Omega_m}^2 \Delta t$$

$$+ C \sum_{n=1}^\beta \sum_i |\partial \Omega_i| h_m ((\Delta t)^{-2} + h_m^{-2}) \{ |\Omega_i|^{-1} [\|(S^n - S^{n-1})\chi_i^{1/4}\|_{L^4(\Omega)}^4$$

$$+ \|(S^{n-1} - \widetilde{S}_i^{n-1})\chi_i^{1/4}\|_{L^4(\Omega)}^4] + |\Omega_i|^{-2} \|\xi_h^{n-1}\|_0^2 \|\xi_h^{n-1}\chi_i^{1/2}\|_0^2 \}$$

$$\leq C [T^2(\Delta t)^2 + M^2(r) h_m^{2r} + \sum_{n=1}^\beta \|\xi_{mh}^n\|_{0,\Omega_m}^2 \Delta t] + \epsilon \sum_{n=1}^\beta \|\nabla \xi_{mh}^n\|_{0,\Omega_m}^2 \Delta t$$

$$+ C((\Delta t)^{-2} + h_m^{-2})\{ C^{**}[T^2(\Delta t)^4 + F^2(q) h_\gamma^{4q}] + C^* h_m \sum_{n=1}^\beta \|\xi_h^{n-1}\|_0^4 \Delta t \},$$

$$1 \leq r \leq r^* + 1, \quad 1 \leq q \leq q^* + 1,$$

where

$$C^* = \max_i |\partial \Omega_i| |\Omega_i|^{-2}, \qquad C^{**} = h_m \max_i |\partial \Omega_i| |\Omega_i|^{-1}.$$

The constant C^* is fixed but large; it will not appear in the final error estimates. The constant C^{**} is of moderate size.

Set

$$E^2(q,r) = T^2(\Delta t)^2 [1 + (\Delta t)^2 h_m^{-2}] + F^2(q) h_\gamma^{2q} [1 + h_\gamma^{2q}((\Delta t)^{-2} + h_m^{-2})] + M^2(r) h_m^{2r}.$$

Then the bounds for t_j^β, $1 \leq j \leq 7$, applied to (5.22) imply that the following inequality holds:

$$(C_1 - 2\epsilon)[\|\xi_{mh}^\beta\|_{0,\Omega_m}^2 + \sum_{n=1}^\beta \|\nabla \xi_{mh}^n\|_{0,\Omega_m}^2 \Delta t] - \sum_{n=1}^\beta \sum_i \varphi_i \partial \xi_{mhi}^n \xi_{hi}^n |\Omega_i| \Delta t$$

$$\leq C_2 [E^2(q,r) + \sum_{n=1}^\beta (\|\xi_{mh}^n\|_{0,\Omega_m}^2 + \|\xi_h^n\|_0^2) \Delta t$$

$$+ h_m C^* ((\Delta t)^{-2} + h_m^{-2}) \sum_{n=1}^\beta \|\xi_h^{n-1}\|_0^4 \Delta t], \quad 1 \leq q \leq q^* + 1, \quad 1 \leq r \leq r^* + 1. \quad (5.27)$$

Thus, adding (5.21) and (5.27) we see that

$$
\|\xi_h^\beta\|_0^2 + \|\xi_{mh}^\beta\|_{0,\,\Omega_m}^2 + \Sigma_{n=1}^\beta \left(\|\nabla\xi_h^n\|_0^2 + \|\nabla\xi_{mh}^n\|_{0,\,\Omega_m}^2 \right) \Delta t
$$

$$
\leq C_3 \Big\{ E^2(q,r) + \Sigma_{n=1}^\beta \left(\|\xi_h^n\|_0^2 + \|\xi_{mh}^n\|_{0,\,\Omega_m}^2 \right) \Delta t
$$

$$
+ \Sigma_{n=1}^\beta \left[h_r^{-3/4} (F_1(q)h_r^q + \|\xi_h^{n-1}\|_0) \right]^8 \|\xi_h^n\|_0^2 \Delta t
$$

$$
+ h_m C^* ((\Delta t)^{-2} + h_m^{-2}) \Sigma_{n=1}^\beta \|\xi_h^{n-1}\|_0^4 \Delta t \Big\}
$$

$$
\leq C_4 \Big\{ E^2(q,r) + \Sigma_{n=1}^\beta \|\xi_{mh}^n\|_{0,\,\Omega_m}^2 \Delta t
$$

$$
+ \tfrac{1}{2}[1 + (h_r^{-3/2}(F_1^2(q)h_r^{2q} + \max_{n\leq\beta-1} \|\xi_h^n\|_0^2))]^4
$$

$$
+ h_m C^* ((\Delta t)^{-2} + h_m^{-2}) \max_{n\leq\beta-1} \|\xi_h^n\|_0^2] \Sigma_{n=1}^\beta \|\xi_h^n\|_0^2 \Delta t \Big\},
$$

$$
1 \leq q \leq q^*+1, \ 1 \leq r \leq r^*+1. \quad (5.28)
$$

Let us make the induction hypothesis that

$$
\max_{n\leq\beta-1} \left(\|\xi_h^n\|_0^2 + \|\xi_{mh}^n\|_{0,\,\Omega_m}^2 \right) + \Sigma_{n=1}^{\beta-1} \left(\|\nabla\xi_h^n\|_0^2 + \|\nabla\xi_{mh}^n\|_{0,\,\Omega_m}^2 \right) \Delta t \leq C_5 E^2(q,r),
$$

$$
1 \leq q \leq q^*+1, \ 1 \leq r \leq r^*+1, \quad (5.29)
$$

where C_5 is fixed and defined below. Note that since $\xi_h^0 = \xi_{m\,h}^0 = 0$, (5.29) holds trivially for $\beta=1$. Using (5.29), (5.28) becomes

$$
\|\xi_h^\beta\|_0^2 + \|\xi_{mh}^\beta\|_{0,\,\Omega_m}^2 + \Sigma_{n=1}^\beta \left(\|\nabla\xi_h^n\|_0^2 + \|\nabla\xi_{mh}^n\|_{0,\,\Omega_m}^2 \right) \Delta t
$$

$$
\leq C_4 \Big\{ E^2(q,r) + \Sigma_{n=1}^\beta \|\xi_{mh}^n\|_{0,\,\Omega_m}^2 \Delta t
$$

$$
+ \tfrac{1}{2}[1 + (h_r^{-3/2}(F_1^2(q)h_r^{2q} + C_5 E^2(q,r)))]^4
$$

$$
+ h_m C^* ((\Delta t)^{-2} + h_m^{-2}) C_5 E^2(q,r)] \Sigma_{n=1}^\beta \|\xi_h^n\|_0^2 \Delta t \Big\},
$$

$$
1 \leq q \leq q^*+1, \ 1 \leq r \leq r^*+1.
$$

Now let us assume that

$$
\{ h_r^{-3/2} + h_m[(\Delta t)^{-2} + h_m^{-2}] \} E^2(q,r) \longrightarrow 0
$$

as Δt, h_r, and $h_m \longrightarrow 0$. This will be satisfied if Δt, h_r^q, and h_m^r are all of the same order as they tend to zero. Choose Δt, h_r, and h_m so small that

$$
\{ h_r^{-3/2}[F_1^2(q)h_r^{2q} + C_5 E^2(q,r)]\}^4 + h_m C^* ((\Delta t)^{-2} + h_m^{-2}) C_5 E^2(q,r) \leq 1.
$$

Then it follows that

$$\|\xi_h^\beta\|_0^2 + \|\xi_{mh}^\beta\|_{0,\,\Omega_m}^2 + \sum_{n=1}^\beta \left(\|\nabla\xi_h^n\|_0^2 + \|\nabla\xi_{mh}^n\|_{0,\,\Omega_m}^2 \right) \Delta t$$

$$\leq C_4 \left\{ E^2(q,r) + \sum_{n=1}^\beta \left(\|\xi_h^n\|_0^2 + \|\xi_{mh}^n\|_{0,\Omega_m}^2 \right) \Delta t \right\}.$$

Gronwall's inequality applied to this expression implies that

$$\|\xi_h^\beta\|_0^2 + \|\xi_{mh}^\beta\|_{0,\,\Omega_m}^2 + \sum_{n=1}^\beta \left(\|\nabla\xi_h^n\|_0^2 + \|\nabla\xi_{mh}^n\|_{0,\,\Omega_m}^2 \right) \Delta t \leq C_6 E^2(q,r),$$

with

$$C_6 = C_4(1 - C_4\Delta t)^{-T/\Delta t} \leq 2C_4 e^{TC_4} \equiv C_5,$$

for Δt not too large, so that the induction argument is complete and (5.29) holds for any $(\beta-1)$ between 1 and L.

The following theorem is a combination of (5.1)-(5.3), (5.7), (5.10), and (5.29).

<u>Theorem.</u> Assuming the nondegeneracy condition (3.5) and sufficient smoothness of the various coefficients, the source terms, and the solution itself, if

$$\Delta t, \; h_f, \text{ and } h_m \longrightarrow 0$$

in such a way that Δt, h_f^q, and h_m^r are all of the same order for some q and r such that $1 \leq q \leq q^* + 1$ and $1 \leq r \leq r^* + 1$, then the error generated by the procedure (4.6)-(4.14) satisfies the estimates

$$\max_{0 \leq n \leq L} \left[\|S^n - S_h^n\|_0 + \|s^n - s_h^n\|_{0,\,\Omega_m} + \|U^n - U_h^n\|_0 \right] \leq C \left[T\Delta t + F(q)h_f^q + M(r)h_m^r \right],$$

$$\left[\sum_{n=0}^L \|\nabla(S^n - S_h^n)\|_0^2 \Delta t \right]^{1/2} \leq C \left[T\Delta t + F(q)h_f^{q-1} + M(r)h_m^r \right],$$

and

$$\left[\sum_{n=0}^L \|\nabla(s^n - s_h^n)\|_{0,\,\Omega_m}^2 \Delta t \right]^{1/2} \leq C \left[T\Delta t + F(q)h_f^q + M(r)h_m^{r-1} \right],$$

where T, $F(q)$, and $M(r)$ are defined in (5.18)-(5.20) and (5.23)-(5.25). Moreover, if $1 \leq q \leq q^{**}$, then

$$\max_{0 \leq n \leq L} \|\nabla\cdot(U^n - U_h^n)\|_0 \leq C \|\nabla\cdot U\|_{L^\infty(J;H^q(\Omega))} h_f^q$$

and

$$\max_{0 \leq n \leq L} \|P^n - P_h^n\|_0 \leq C \left[T\Delta t + (\|P\|_{L^\infty(J;L^2(\Omega))} + F(q))h_f^q + M(r)h_m^r \right].$$

By setting $\varphi \equiv 0$, a finite element procedure for an unfractured reservoir Ω is defined in Section 4. This procedure satisfies the above theorem with the matrix quantities deleted from it. Of course, the proof could be substantially simplified in this case.

6. Some Notation

Symbol		Meaning
(fracture)	(matrix)	
K	k	permeability
$K_{r\theta}$	$k_{r\theta}$	relative permeability of the θ phase
P	p	Chavent's global pressure
P_c	p_c	capillary pressure (oil minus water pressure)
P_θ	p_θ	pressure of the θ phase
Q_e	—	total external volumetric source
$Q_{\theta e}$	—	external θ source
S	s	oil saturation
$S_{res,\theta}$	$s_{res,\theta}$	residual θ saturation
S_θ	s_θ	saturation of the θ phase
U	$u\ (=0)$	total flow rate ("global velocity")
V_θ	v_θ	Darcy velocity of the θ phase
	θ	oil (o) or water (w)
Φ	φ	porosity ($\varphi = \varphi_i$ on Ω_i is assumed to be a constant)
	μ_θ	viscosity of the θ phase

Acknowledgments

The work of Juan E. Santos was supported by the Consejo Nacional de Investigaciones Científicas y Técnicas de la República Argentina and that of Todd Arbogast and Jim Douglas, Jr., in part by the National Science Foundation.

References

[1] T. Arbogast, *Analysis of the simulation of single phase flow through a naturally fractured reservoir*, to appear.

[2] T. Arbogast, *The double porosity model for single phase flow in naturally fractured reservoirs*, these proceedings.

[3]   F. Brezzi, J. Douglas, Jr., R. Durán, and M. Fortin, *Mixed finite elements for second order elliptic problems in three variables*, to appear in Numer. Math.

[4]   F. Brezzi, J. Douglas, Jr., M. Fortin, and L. D. Marini, *Efficient rectangular mixed finite elements in two and three space variables*, to appear.

[5]   F. Brezzi, J. Douglas, Jr., and L. D. Marini, *Two families of mixed finite elements for second order elliptic problems*, Numer. Math., 47 (1985), 217-235.

[6]   G. Chavent, *A new formulation of diphasic incompressible flows in porous media*, Applications of Methods of Functional Analysis to Problems in Mechanics, Lecture Notes in Mathematics 503, Springer-Verlag, New York, 1976.

[7]   J. Douglas, Jr., *Finite difference methods for two-phase incompressible flow in porous media*, SIAM J. Numer. Anal., 20 (1983), 681-696.

[8]   J. Douglas, Jr., T. Dupont, and M. F. Wheeler, *A Galerkin procedure for approximating the flux on the boundary for elliptic and parabolic boundary value problems*, RAIRO, 8 (1974), 47-59.

[9]   J. Douglas, Jr., R. E. Ewing, and M. F. Wheeler, *The approximation of the pressure by a mixed method in the simulation of miscible displacement*, RAIRO Analyse numérique, 17 (1983), 17-33.

[10] J. Douglas, Jr., R. E. Ewing, and M. F. Wheeler, *A time-discretization procedure for a mixed finite element approximation of miscible displacement in porous media*, RAIRO Analyse numérique, 17 (1983), 249-265.

[11] J. Douglas, Jr., P. J. Paes Leme, T. Arbogast, and T. Schmitt, *Simulation of flow in naturally fractured reservoirs*, paper SPE 16019 presented at the Ninth Society of Petroleum Engineers Symposium on Reservoir Simulation in San Antonio, Texas, Feb. 1-4, 1987.

[12] J. Douglas, Jr., and J. E. Roberts, *Global estimates for mixed methods for second order elliptic equations*, Math Comp., 44 (1985), 39-52.

[13] J. C. Nedelec, *Mixed finite elements in R^3*, Numer. Math., 35 (1980), 315-341.

[14] D. W. Peaceman, Fundamentals of Numerical Reservoir Simulation, Elsevier, New York, 1977.

[15] P. A. Raviart and J. M. Thomas, *A mixed finite element method for 2nd order elliptic problems*, Mathematical Aspects of the Finite Element Method, Lecture Notes in Mathematics 606, Springer-Verlag, New York, 1977.

REACTION-INFILTRATION INSTABILITIES[1]

J. Chadam[2] and P. Ortoleva

Department of Mathematics & Statistics	Departments of Chemistry & Geology
McMaster University	Indiana University
Hamilton, Ontario, Canada, L8S 4K1	Bloomington, Indiana, U.S.A. 47405

When reactive waters flow through a porous medium they can dissolve the minerals and cause changes in the porosity. This, through Darcy's law, can alter the flow, giving rise to a feedback mechanism which can cause instabilities in the shape of the porosity level surfaces. This mechanism most certainly is important in many geochemical situations (e.g. the diagenesis and evolution of mineral, oil and gas reservoirs, the dynamics of nuclear and chemical waste repositories, in situ coal gasification, etc.). Our own interest in the subject arose from trying to understand the occurence of so-called roll-front redox mineral deposits[1,2,3] but the ideas may be useful in enhanced oil recovery if the injected chemicals or microbes interact substantially with the rock or petroleum. No doubt the coupling of this reaction-infiltration instability to the more widely studied water-steam-oil flow instabilities should lead to a very rich area for further research.

Typically, the essential geochemical processes of relevance to each of the above situations can be modelled mathematically as a system of coupled, highly non-linear reaction-transport equations[1,2]. In general, however, even the simplified versions of these equations arising from overly simplified physical models are too complicated to be studied abstractly or analytically[4]. Our approach here (Section 1) is to restrict attention to a physically important class of problems for which the effective reaction zone (where the serious complications appear) is much smaller and less interesting that the scale of the phenomenon being studied. The resulting set of reaction-transport equations can then be studied using matched asymptotics[4] to obtain a more amenable moving free boundary problem for the reaction interface[2,4] (Section 2). This will allow us to give[3,4,5] (Section 3) a mathematical treatment of the evolution of the shape of the reaction interface in terms of bifurcation and stability theory. We shall also present (Section 4) the results of some preliminary numerical studies. A more complete code, REACTRAN, is being developed which extends to higher spatial dimensions our present one-dimensional code which handles complicated geochemistry with many aqueous species and minerals even when the reaction zones are very narrow.

---
[1] Parts of this work were supported by the DOE (USA) and NSERC (Canada).

[2] Lecturer and preparer of this report. Others who contributed to various parts of the work described here are: G. Auchmuty, J. Hettmer, D. Hoff, E. Merino, C. Moore, E. Ripley and A. Sen.

1. A Simple Reaction-Infiltration Model

Consider an aquifer consisting of an insoluble porous matrix (e.g. quartz sandstone) with some soluble mineral (e.g. calcite) partially filling the pores. If water is forced through this porous medium, the soluble component will be dissolved out upstream and the water will become saturated sufficiently far downstream. Between these extremes there is a dissolution zone across which the

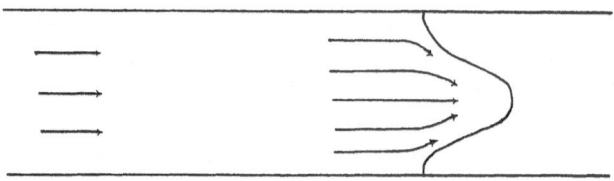

Fig. 1. Focusing of flow to tip of porosity level curve.

soluble mineral content – and hence the porosity – changes from its original downstream value to the final, altered value upstream. The question of interest is whether the shape of this dissolution zone is stable. Notice that if a bump (in the porosity level curves) in the reaction zone exists at some time, the flow of the undersaturated waters tends to be focused to the tip of the bump via Darcy's law since inside the bump (on the upstream side) the permeability is greater than in the neighbouring regions (see Fig. 1). Thus dissolution is enhanced at the tip causing it to advance more rapidly. This is the porosity change/flow destabilization mechanism. On the other hand, diffusion from the sides of the tip raises the concentration of the solute in the water which is focusing at the tip and hence will decelerate this advancement. The competition between these two processes can lead to decay of the bump, restabilization to a morphologically more complicated dissolution zone or possibly to complete destabilization.

2. Two Mathematical Models

In this section we write down without details (c.f. references 3,4 for derivations) the models we shall subsequently treat analytically and numerically.

2.1. Coupled ODE/PDE Model

The rate of increase of the porosity ϕ (equivalently the rate of dissolution of the soluble mineral) is proportional to the reaction rate:

$$\varepsilon\frac{\partial\phi}{\partial t} = -\,(\phi_f - \phi)^{2/3}(\gamma - 1)(= -\,R(\phi,\gamma)) \tag{2.1}$$

Here ϕ_f is the final porosity after complete dissolution, γ is the scaled concentration of solute in water (with equilibrium concentration being 1) and $\varepsilon = c_{eq}/\rho \ll 1$ is the ratio of the original equilibrium concentration to the density of the soluble mineral. The 2/3-power indicates that we are considering surface reactions. The solute concentration per rock volume, $\phi\gamma$, satisfies a mass conservation equation:

$$\varepsilon\frac{\partial(\phi\gamma)}{\partial t} = \nabla\cdot[\phi D(\phi)\nabla\gamma + \phi\lambda(\phi)\gamma\nabla p] + \frac{1}{\varepsilon} R(\phi,\gamma) \qquad (2.2)$$

where $D(\phi)$, $\lambda(\phi)$ are the porosity dependent, scaled diffusion coefficient and permeability respectively, and p is the pressure. Darcy's law has been used in the convective term of (2.2). It is also used in combination with the continuity equation to give:

$$\nabla\cdot[\phi\lambda(\phi)\nabla p] = \frac{\partial\phi}{\partial t} \ (= \frac{1}{\varepsilon} R(\phi,\gamma)). \qquad (2.3)$$

In addition we impose the asymptotic conditions:

$$\gamma \rightarrow 0, \ \phi \rightarrow \phi_f \ \text{and} \ \frac{\partial p}{\partial x} \rightarrow \frac{\kappa_f p_f'}{D_f} = -\frac{v_t}{D_f} \ \text{as} \ x \rightarrow -\infty \qquad (2.4)$$

and

$$\gamma \rightarrow 1, \ \phi \rightarrow \phi_0, \frac{\partial p}{\partial x} \rightarrow ? \ \text{as} \ x \rightarrow +\infty. \qquad (2.5)$$

These indicate that far upstream the water is fresh ($\gamma = 0$) and the mineral has been completely dissolved out ($\phi = \phi_f$). Also the pressure gradient (equivalently the velocity through Darcy's law $v_f = -\kappa_f p_f'$) is specified as in (2.4) with the effects of the scaling appearing explicitly. Far downstream the water is saturated ($\gamma = 1$), the porosity is still at its original, unaltered value ($\phi = \phi_0$) and the pressure gradient (equivalently the velocity) is to be determined. Equations (2.1-5), along with given initial data and zero flux boundary conditions on the transverse boundaries, form a complete problem for the unknowns γ, ϕ, p. Unfortunately, nothing can be calculated analytically from these equations except the velocity of a travelling planar dissolution zone. On the other hand, they form the basis of our numerical simulations which will be discussed later.

2.2. Moving Free Boundary Model

In order to obtain an analytically tractable problem, we take the large solid density limit $\varepsilon = c_{eq}/\rho \rightarrow 0$. The dissolution zone, typically of width $\varepsilon^{1/2}$, collapses to a dissolution interface located at $x = R(y,t)$, with R unknown. Then, off this interface there is no reaction and the only consistent way to satisfy equations (2.1-3) to all orders of ε is as follows. Upstream of the dissolution interface where from scaling $\lambda(\phi_f)$ and $D(\phi_f) = 1$, one has

$$
\left.
\begin{aligned}
\nabla\gamma + \nabla\gamma\cdot\nabla p &= 0 \\
\phi &= \phi_f \\
\Delta p &= 0
\end{aligned}
\right\} \quad \text{in } x < R(y,t),\ 0 < y < L.
$$

$$\tag{2.6}$$
$$\tag{2.7}$$
$$\tag{2.8}$$

while downstream one obtains

$$
\left.
\begin{aligned}
\gamma &\equiv 1 \\
\phi &= \phi_0 \\
\Delta p &= 0
\end{aligned}
\right\} \quad \text{in } x > R(y,t),\ 0 < y < L
$$

$$\tag{2.9}$$
$$\tag{2.10}$$
$$\tag{2.11}$$

where we have taken $\phi_0(x,y,0) = \phi_0$, constant, to show that the morphological instabilities will even occur in this spatially homogeneous situation. Besides the asymptotic conditions (2.4,5) one also obtains, via matched asymptotics, boundary conditions on the unknown moving dissolution interface. Specifically, one has

$$
\left.
\begin{aligned}
\gamma &= 1 \\
p- &= p+ \\
\frac{\partial p-}{\partial y} - \frac{\partial p-}{\partial y}\frac{\partial R}{\partial y} &= \Gamma\left(\frac{\partial p+}{\partial x} - \frac{\partial p+}{\partial y}\cdot\frac{\partial R}{\partial y}\right) \\
\frac{\partial\gamma}{\partial x} - \frac{\partial\gamma}{\partial y}\cdot\frac{\partial R}{\partial y} &= (1 - \phi_0/\phi_f)R_t
\end{aligned}
\right\}
\begin{aligned}
&\text{on} \\
&x = R(y,t) \\
&0 < y < L
\end{aligned}
$$

$$\tag{2.12}$$
$$\tag{2.13}$$
$$\tag{2.14}$$
$$\tag{2.15}$$

where $0 < \Gamma = \phi_0\kappa_0/\phi_f\kappa_f < 1$ is a measure of the porosity change. Equation (2.15) relates the rate of advancement of the moving dissolution interface to the flux of the concentration and is called a Stefan condition. A final scaling $x' = \frac{\pi}{L}x$, $y' = \frac{\pi}{L}y$, $t' = (\frac{\pi}{L})^2(1 - \phi_0/\phi_f)^{-1}t$ with $R' = \frac{\pi}{L}R$ (and dropping the primes) makes the transverse dimension $0 < y < \pi$, and results in the two changes

$$
\gamma \to 0,\ \phi \to \phi_f,\ \frac{\partial p}{\partial x} \to -\frac{v_f L}{D_f\pi} \quad \text{as } x \to -\infty
$$

$$\tag{2.4'}$$

and

$$
\frac{\partial\gamma}{\partial x} - \frac{\partial\gamma}{\partial y}\cdot\frac{\partial R}{\partial y} = R_t \quad \text{on } x = R(y,t),\ 0 < y < \pi
$$

$$\tag{2.15'}$$

Problem (2.4',5,...14,15') with initial conditions and zero flux transverse conditions on $y = 0,\pi$ is the version we shall examine analytically in the next section. Notice that only two essential parameters remain in the problem, the dynamical parameter $v_f = v_f L/D_f$ and the measure of the porosity change $\Gamma = \phi_0\kappa_0/\phi_f\kappa_f$.

3. Shape Instabilities

In this section we describe our analytical results[4,5] in the context of the large solid density problem (2.4',5,...14,15'). This free boundary problem is different from but not unrelated to those which arise by similar limiting procedures in combustion[6], solidification[7], electrochemical forming and machining[8], etc., and is tractable by similar techniques. Here the planar, constant velocity solution can be obtained explicitly and completely, including the concentration and pressure profiles which were not available for the more general coupled ODE/PDE model. The linearized stability of this solution is then described, giving a precise value of the parameter v_f (in terms of Γ) for which the planar solution looses stability to another, more structured, solution. In the language of bifurcation theory, we determine the critical parameter value for which the spectrum of the linearized problem changes sign from negative to positive, thus determining the location of a possible bifurcation point. Finally we sketch the local bifurcation analysis to show that the linear instabilities are restabilized by the nonlinearities to a morphologically more complicated solution. More specifically we shall obtain a Landau equation for the amplitude of the linearly unstable mode thus indicating a standard pitchfork bifurcation diagram as in Fig. 2.

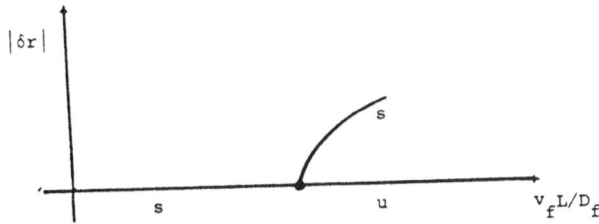

Fig. 2. Pitchfork stability diagram indicating loss of stability of planar ($\delta r=0$) to nonplanar ($\delta r \neq 0$) solutions in terms of the bifurcation parameter $v_f L/D_f$.

3.1. Planar Solution

Denoting the planar state quantities with a super bar, one can easily check[4,5] that the following constant velocity solution satisfies problem (2.4',5,...14,15'):

$$\bar{R}(t) = \bar{V}t \tag{3.1}$$

$$\bar{\gamma}(x,t) = \begin{cases} e^{-\bar{v}_f(x-\bar{V}t)} & x < \bar{V}t \tag{3.2a} \\ 1 & x > \bar{V}t \end{cases} \tag{3.2b}$$

$$\bar{p}(x,t) = \begin{cases} -\bar{v}_f(x-\bar{V}t) & x < \bar{V}t \tag{3.3a} \\ -\bar{v}_0(x-\bar{V}t) & x > \bar{V}t. \end{cases} \tag{3.3b}$$

where $\bar{\nu}_f = \nu_f/\pi = v_f L/D_f \pi$, $\bar{\nu}_0 = \phi_f \bar{\nu}_f/\phi_0$ (from (2.5) and (2.14)) and the velocity of the planar interface $\bar{V} = \bar{\nu}_f$ from (2.15').

3.2. Linear Shape Instability

In order to examine the stability of the above planar solutions (3.1,3) with respect to bumps we consider perturbations of the type (i.e. a generic term in the Fourier decomposition — only cosine terms appear because of the zero flux transverse boundary conditions)

$$R(y,t) = \bar{V}t + \delta\, r_{1m}(t)\, \cos my \qquad (3.4a)$$

$$\gamma(x,y,t) = \bar{\gamma}(x,t) + \delta\, \gamma_{1m}(x)\, r_{1m}(t)\, \cos my \qquad (3.4b)$$

$$p(x,y,t) = \bar{p}(x,t) + \delta\, p_{1m}(x)\, r_{1m}(t)\, \cos my. \qquad (3.4c)$$

Considering δ to be small, the linearized version of equations (2.4',5,...14,15') can be derived[4] (i.e. retain terms in first power of δ). These can be solved explicitly for γ_{1m} and p_{1m} and the Stefan condition gives[4,5] the following condition on the amplitude $r_{1m}(t)$ of the cos my bump:

$$r'_{1m}(t) = \frac{\bar{\nu}_f}{1 + \Gamma}\, [\bar{\nu}_f - (\bar{\nu}_f^2 + 4m^2)^{1/2} + (1 - \Gamma)|m|]\, r_{1m}(t) \qquad (3.5a)$$

This differential equation indicates that the amplitude of the bump grows or decays depending on the sign of the coefficient. The connection with the equivalent, more conventional viewpoint follows by expressing $r_{1m}(t) = r e^{\sigma(m)t}$ in terms of the spectrum $\sigma(m)$ of the linearized problem and obtaining from (3.5a) the dispersion relation

$$\sigma(m) = \frac{\bar{\nu}_f}{1 + \Gamma}\, [\bar{\nu}_f - (\bar{\nu}_f^2 + 4m^2)^{1/2} + (1 - \Gamma)|m|]. \qquad (3.5b)$$

The m-dependence of σ is shown in Fig. 3 revealing clearly that the planar solution (3.1,3) is linearly unstable to long wavelength perturbations (because $\Gamma < 1$) and stable to short wavelength perturbations.

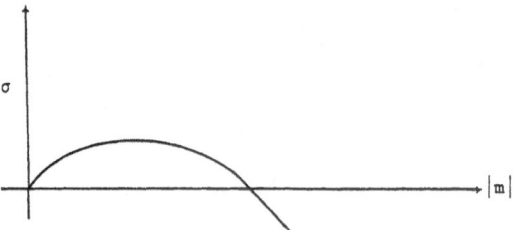

Fig. 3.   Graph of dispersion relation (3.5b).

The critical wave number ($|m_0|$ at which $\sigma(m_0) = 0$) is given by

$$|m_0| = \frac{2(1 - \Gamma)}{(3 - \Gamma)(1 + \Gamma)} \bar{\nu}_f \tag{3.6}$$

Since our channel width has been normalized to π, the first mode which can be carried is $|m_0| = 1$ giving, from (3.6), the critical parameter value (of $\nu_f = \nu_f L/D_f$)

$$\nu_c = \nu_c(\Gamma) = \frac{(3 - \Gamma)(1 + \Gamma)\pi}{2(1 - \Gamma)}. \tag{3.7}$$

From this we see that the instability does indeed arise analytically and that, as is physically realistic, larger flow speeds, larger transverse dimensions, larger porosity/permeability changes promote the instability while larger diffusion coefficients inhibit the instability (i.e. diffusion is stabilizing, as mentioned earlier). The limit of $\Gamma \rightarrow 1$ (i.e. no porosity change) suggests it is very difficult to produce instabilities. This has been verified by a separate analysis[3]. Thus this instability can occur only if "significant" amounts of the soluble mineral are dissolved.

3.3. Non-linear Restabilization

We begin by scaling the independent variables. Because the instability occurs at finite wavelength none is required for the spatial variables while, as is common for a (anticipated) pitchfork bifurcation,

$$t_2 = \varepsilon^2 t. \tag{3.8}$$

Additionally, we write

$$\nu_f = \nu_c + \varepsilon \nu_1 + \varepsilon^2 \nu_2 + \text{---}. \tag{3.9}$$

We find[5] at $O(\varepsilon^2)$ that $\nu_1 = 0$ (as usual for pitchfork bifurcations) so that the physical significance of the small parameter ε is

$$\varepsilon \simeq (\nu_f - \nu_c)^{1/2} \tag{3.10}$$

where we have taken, without loss of generality, $\nu_2 = 1$. Thus

$$\nu_f = \nu_c + \varepsilon^2 + \varepsilon^3 \nu_3 + \text{---}. \tag{3.11}$$

The stability calculation then proceeds by expanding all of the dependent variables in terms of ε (suppressing the sub-2 in the new t_2 variable):

$$R(y,t) = \bar{V}(\varepsilon)t$$
$$+ \varepsilon(r_{10}(t) + r_{11}(t)\cos y + r_{12}(t)\cos 2y + \text{---})$$
$$+ \varepsilon^2(r_{20}(t) + r_{21}\cos y + r_{22}(t)\cos 2y + \text{---})$$
$$+ \varepsilon^3(r_{30}(t) + r_{31}\cos y + r_{23}(t)\cos 2y + \text{---})$$
$$+ 0(\varepsilon^4), \tag{3.12a}$$

$$\gamma(x,y,t) = \bar{\gamma}(x,t;\varepsilon)$$
$$+ \varepsilon(\gamma_{10}(x,t) + \gamma_{11}(x,t)\cos y + \gamma_{12}(x,t)\cos 2y + \text{---})$$
$$+ \varepsilon^2(\gamma_{20}(x,t) + \gamma_{21}(x,t)\cos y + \gamma_{22}(x,t)\cos 2y + \text{---})$$
$$+ \varepsilon^3(\gamma_{30}(x,t) + \gamma_{31}(x,t)\cos y + \gamma_{32}(x,t)\cos 2y + \text{---})$$
$$+ 0(\varepsilon^4), \tag{3.12b}$$

and similarly for $p(x,y,t)$. Following the prescription outlined in standard weakly nonlinear stability analysis (except that the equations are solved directly rather than obtaining conditions from the orthogonality of inhomogeneous terms and solutions of the homogeneous problem) one obtains a Landau differential equation for the amplitude of the unstable mode, $r_{11}(t)$:

$$r'_{11}(t) = w\, r_{11}(t) - \Lambda\, r_{11}(t)^3 \tag{3.13}$$

where

$$w = w(\Gamma) = \frac{v_c}{1+\Gamma}\, \frac{[v_c^2 + 4)^{1/2} - v_c]}{(v_c^2 + 4)^{1/2}} \geqslant 0 \tag{3.14}$$

and the Landau constant, $\Lambda = \Lambda(\Gamma)$, which is algebraically very complicated, is given in Fig. 4. The positivity of Λ indicates that in the vicinity of the

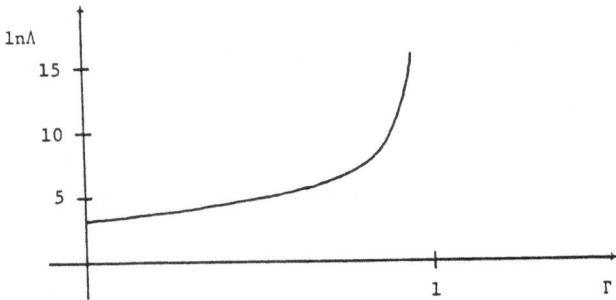

Fig. 4. Graph of the logarithm of the Landau constant versus $\Gamma = \phi_0\kappa_0/\phi_f\kappa_f$, a measure of the porosity change.

critical point the linearized instabilities (from $w \geqslant 0$) are restabilized by the nonlinearities at the next highest order, and (from (3.13)) that the bifurcation diagram is the symmetric pitchfork. The asymptotic amplitude of the bump can be obtained from (3.13) to be $(w/\Lambda)^{1/2}$.

4. Numerical Simulations

The actual shape of the stabilized bump, especially far from the critical point (to which the analysis of the last section does not apply) must be obtained from numerical simulations. Because interface tracking is a difficult problem we return to the coupled ODE/PDE model with $\varepsilon = c_{eq}/\rho$ small (= 0.05) but not zero. Using parameter values suggested by the analytical results of the previous section and standard numerical methods[4] for solving equations (2.1-5), we investigated the three cases $\nu_f \ll \nu_c$, $\nu_f \simeq \nu_c$ and $\nu_f \gg \nu_c$. Figs. 5a), b), c) depict these cases

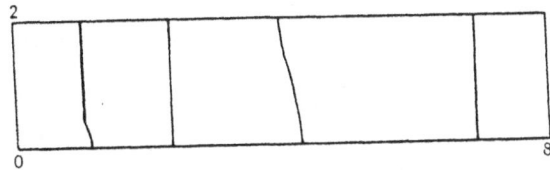

Fig. 5a). $\nu_f \ll \nu_c$. Evolution of porosity level curve for times 0.0, 1.8, 3.6, 5.4.

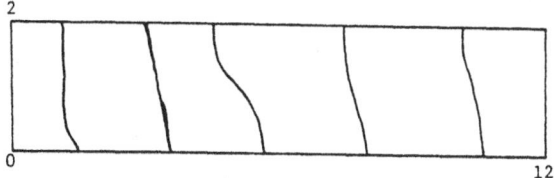

Fig. 5b). $\nu_f \simeq \nu_c$. Evolution of porosity level curve for times 0.0, 1.5, 3.0, 3.75.

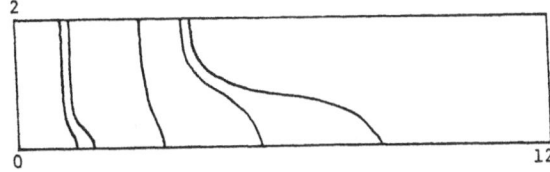

Fig. 5c). $\nu_f \gg \nu_c$. Evolution of porosity level curves for times 0.0, .24, .48, .72, .96.

respectively, indicating stability of the planar front, restabilization to a new shape and a highly unstable dissolution zone respectively.

REFERENCES

1. J. Chadam, P. Ortoleva, E. Merino and C. Moore, "Geochemical Self-Organization I: Feedback Mechanisms and Modeling Approach", accepted for publication, Amer. J. Sci. (1987).

2. J. Chadam, P. Ortoleva, E. Merino and A. Sen, "Self-Organization in Water-Rock Interaction Systems, II: The Reactive-Infiltrate Instability", accepted for publication, Amer. J. Sci. (1987).

3. J. Chadam, G. Auchmuty, E. Merino, P. Ortoleva and E. Ripley, "The structure and stability of moving redox fronts", SIAM J. Appl. Math. 46 (1986), 588-604.

4. J. Chadam, P. Ortoleva and A. Sen, "Reactive infiltration Instabilities", accepted for publication, IMA J. of Appl. Math (1986).

5. J. Chadam, P. Ortoleva and A. Sen, "Weakly nonlinear stability of reaction-percolation interfaces", accepted for publication, SIAM J. of App. Math. (1986).

6. B. J. Matkowsky and G. I. Shivashinsky, "Propogation of a Pulsating Reaction Front in Solid Fuel Combustion", SIAM J. Appl. Math. 35 (1978), 465-478.

7. J. Chadam and P. Ortoleva, "The stabilizing effect of surface tension on the development of the free boundary in a planar one-dimensional Cauchy-Stefan problem", IMA J. of Appl. Math. 30 (1983).

8. J. A. McGeough, Principles of Electrochemical Machining, Chapman and Hall, London (1974).

CHARACTERISTIC PETROV-GALERKIN SUBDOMAIN METHODS FOR CONVECTION-DIFFUSION PROBLEMS

Helge K.Dahle and Magne S.Espedal
Department of Mathematics
University of Bergen, Norway

Richard E.Ewing
Departments of Mathematics, Petroleum Engineering and Chemical Engineering
University of Wyoming
Laramie, Wyoming 82071, U.S.A.

1. Introduction

In a recent paper[1] , characteristic Petrov-Galerkin subdomain methods are developed for the solution of a two-dimensional, two-phase immiscible flow problem.

A preconditioner, based on substructuring is constructed to solve the nonlinear parabolic equation for the saturation of water, S:

$$\phi \frac{\partial}{\partial t} S + \frac{d}{dS} \bar{f}(S, \mathbf{u}) \cdot \nabla S + \nabla \cdot (\mathbf{b}(S, \mathbf{u})S) - \varepsilon \nabla \cdot (D(S)\nabla S) = q_2(\mathbf{x}, t) \tag{1}$$

$$0 < \varepsilon << 1.$$

The pressure equations:

$$\nabla \cdot \mathbf{u} = q_1 \tag{2}$$

$$\mathbf{u} = -a(\mathbf{x}, S)(\nabla p - \rho(S)\nabla z)$$

are solved by mixed finite element methods[2-7], which give the volumetric flow $\mathbf{u}$, to optimal order of approximation.

A new operator splitting technique[8-10], is developed to solve Equation 1.

The fractional flow term is divided into two parts; $\mathbf{f}(S, \mathbf{u}) = \bar{\mathbf{f}}(S, \mathbf{u}) + \mathbf{b}(S, \mathbf{u})S$, such that $\bar{\mathbf{f}}$ gives the unique physical velocity for established shocks and $\bar{\mathbf{f}} = \mathbf{f}$ outside a shock. The nonlinear part $\mathbf{b}(S, \mathbf{u})$ balances the diffusion at the shock once the shock has been established. We treat convection by timestepping along the characteristics of the associated pure convection problem, defined by $\bar{f}$, and use Petrov-Galerkin finite element methods to determine the effects of the nonlinearities in the b term and the diffusion.

We define a coarse grid, adequate for the slow variation of u and the characteristic solution. The splitting of f(S,u) allows us to use long timesteps in the characteristic solution procedure, such that a shock is moved on the order of one coarse gridblock in one timestep.

A substructuring of this grid is then given such that we get a proper resolution of the saturation front. A local a *priori* error estimate determines an area of coarse gridblocks, where substructuring is necessary. Recently, a new class of preconditioners, based on substructuring, has been developed for symmetric elliptic boundary-value problems[11-13]. We have used these techniques to construct a preconditioner for our problem.

It is important to notice that the solution procedure allows very long timesteps without reduction in accuracy. Further, because most of the computations can be performed independently on each of the coarse gridblocks, the algorithms are well suited for parallel computer architectures.

In this paper we will illustrate the features of the method with one-dimensional calculation. Here, the saturation equation (1) decouples from the pressure equation (2), which simplifies the problem. However, the saturation equation still contains the problems with shock solutions[14,15] and the one-dimensional model shows the capability of the method to resolve these sharp fronts, using long timesteps without introducing serious discretization errors, where the physics of the problem makes this possible.

In the following, we will give an outline of the theory. We will define the operator splitting which gives a well behaved hyperbolic problem and an unsymmetrical elliptic problem. In order to get a solution to optimal order in the last problem, optimal testfunctions have to be defined. Also some numerical examples are presented.

2. Operator Splitting

As noted above, at time-level $t = t_m$, we divide the fractional flow function into two parts[1]:

$$f(S) = \bar{f}^m(S) + b^m(S)S ; \qquad f(0) = 0 ; \qquad f(1) = 1, \qquad (3)$$

where

$$\bar{f}^m(S) = \begin{cases} \frac{1}{S_a^m}f(S_a^m)S , & 0 \le S \le S_a^m , \\ f(S) , & S_a^m \le S \le 1 . \end{cases} \qquad (4)$$

S_a^m is the top saturation of a shock at $t = t_m$, $0 < S_a^m \le S_b < 1$, which means that a shock may develop for $S \le S_b$. Further, $b^m(S_a^m) = 0$; $b^m(S) < 0$ for $0 \le S < S_a^m$.

Now, we solve Equation 1, using the operator splitting [1]:

$$\phi\frac{\partial\bar{S}}{\partial t} + \frac{d}{dS}\bar{f}^m(\bar{S})\frac{\partial}{\partial x}\bar{S} = \phi\frac{d}{d\tau}\bar{S} = 0 \quad, \tag{5}$$

$$\phi\frac{\partial S}{\partial\tau} + \frac{\partial}{\partial x}\left(b^m(S)S\right) - \varepsilon\frac{\partial}{\partial x}\left(D(S)\frac{\partial}{\partial x}S\right) = q_2(x,t) \quad, \tag{6}$$

$$t_m \leq t \leq t_{m+1} \quad ; \quad x \geq 0 \, ,$$

$$S(0,x) = S_0(x) \quad ; \quad S(t,0) = 1 \quad ; \quad \frac{\partial}{\partial x}S(x,t)\Big|_{x=1} = 0 \, . \tag{7}$$

In Equation 6., $\frac{\partial}{\partial\tau}$ is calculated along the approximate characteristic defined by Equation 5.

We may note that Equation 5. together with the initial and boundary conditions 7., produce the same unique physical solution as $\phi\frac{\partial S}{\partial t} + \frac{\partial}{\partial x}f(S) = 0$ with an entropy[14,15] condition imposed.

Integrating Equation 5. backwards along the characteristic from a fixed gridpoint $x = x_i$, we get[1,4,8,10] :

$$\hat{x} = x - \frac{1}{\phi}\frac{d}{dS}\bar{f}(S^m(x))\,\Delta t \, , \tag{8}$$

$$\hat{S}^m(\hat{x}) = S(\hat{x}^m) \, , \tag{9}$$

$$\phi\frac{\partial}{\partial\tau}S = \phi\frac{S^{m+1}(x) - \hat{S}^m(\hat{x})}{\Delta t} + O(\varepsilon) \, . \tag{10}$$

Furthermore, we approximate the coefficients[1]:

$$\begin{aligned}
b(S^{m+1}) &= b(\hat{S}^m) + O(\varepsilon\Delta t) = b(x,t^m) + O(\varepsilon\Delta t) \, , \\
D(S^{m+1}) &= D(\hat{S}^m) + O(\varepsilon\Delta t) = D(x,t^m) + O(\varepsilon\Delta t) \, .
\end{aligned} \tag{11}$$

We may note that from the formulation of equations for immiscible flow, we have:

$$D(0) = D(1) = 0.$$

3. Optimal Testfunctions.

Using Equations 10. and 11., we solve Equation 6. by a Petrov-Galerkin method:

$$B(S_h^{m+1}, \psi_i) \equiv (S_h^{m+1}, \psi_i) - \left(\frac{\Delta t}{\phi} b(x, t^m) S_h^{m+1}, \frac{\partial}{\partial x} \psi_i\right) + \left(\frac{\varepsilon \Delta t}{\phi} D(x, t^m) \frac{\partial}{\partial x} S_h^{m+1}, \frac{\partial}{\partial x} \psi_i\right) =$$

$$+(\hat{S}_h^m, \psi_i) + \left(\frac{\Delta t}{\phi} q_2(x, t^m), \psi_i\right) + O(\varepsilon \Delta t) , \qquad i = 1, 2, ..., N , \qquad (12)$$

$$S_h^{m+1} \in M_h \subset H^1(0, 1) \quad ; \qquad \psi_i \in N_h \subset H^1(0, 1) ,$$

where M_h and N_h are the trial and test spaces spanned by $\{\theta_i\}$ and $\{\psi_i\}$, $i = 1, 2, 3...N$, respectively.

$B(\cdot, \cdot)$ from Equation 11., is an unsymmetrical bilinear form with spatially-dependent coefficients.

We will define the testfunctions ψ_i by a symmetrization procedure, developed by Barrett and Morton[16], which yields:

$$B(S^{m+1}, \psi_i) = B^*(S^{m+1}, \theta_i) \equiv (a S_x^{m+1}, \theta_{ix}) \qquad (13)$$

where $0 < a < K$.

The testfunctions defined by Equation 13. have nonlocal support. However, a procedure for localization is given by Demkowitcz and Oden[17-18]. We may note that, since the new bilinear form, $B^*(\cdot, \cdot)$, is symmetric and positive definite, we obtain optimal approximation properties when ψ_i is determined through Equation 13. Also, we obtain superconvergence[17] at the nodes.

Now, we define a coarse grid on $[0, 1]$ with mesh spacing $d > 0$ by the partition: $x = Id$, $I = 0, 1, 2, ...N$. On each coarse gridblock $\Gamma_I = [X_I, X_{I+1}]$, we define a refined grid by: $x = jh$, $j = 0, 1, 2, ...N - I$, where $d > h > 0$. Further, we choose θ_i as the hat-function, such that:

θ_I: trialfunctions with support on the coarse grid elements with node at $x = Id$,

θ_i: trialfunctions with support on the refined grid elements, with node at $x = ih$.

ψ_I and ψ_i are defined in a similar fashion.

The local optimal testfunctions[17], ψ_i, are defined by:

$$B(\omega, \psi_i) = 0 , \qquad \text{with} \qquad \omega(v_i) = 0 \qquad (14)$$

at all nodes v_i.

In our case, approximate analytical forms are given by:

$$\psi_i = \begin{cases} \left[e^{\gamma^+} - e^{\gamma^-}\right]^{-1} \left[e^{\gamma^+(x-x_{i-1})/h} - e^{\gamma^-(x-x_{i-1})/h}\right] , & x_{i-1} \leq x \leq x_i \\ \left[e^{-\gamma^+} - e^{-\gamma^-}\right]^{-1} \left[e^{\gamma^+(x-x_{i+1})/h} - e^{\gamma^-(x-x_{i+1})/h}\right] , & x_i \leq x \leq x_{i+1} \end{cases} \tag{15}$$

where

$$\gamma^\pm = -\frac{\bar{b}_i h}{2\varepsilon \bar{D}_i} \pm h \left[\frac{\bar{b}_i^2}{4} + \frac{\phi}{\varepsilon \Delta t \bar{D}_i}\right]^{1/2} , \tag{16}$$

and $\bar{b}_i$ and $\bar{D}_i$ denote averages.

Further, we may show that[1] ψ_I may be approximated by:

$$\psi_I = \delta(x - x_I) ,$$

which means that we may use the solution of Equation 5. at the coarse grid nodes.

The expression 15. is hard to use in computations. We will approximate ψ_i by the following form[16] :

$$\bar{\psi}_i = \theta_i + \sigma_i \begin{cases} \alpha_i - \delta_i , & x_{i-1} \leq x \leq x_i \\ \alpha_i + \delta_i , & x_i \leq x \leq x_{i+1} \end{cases} \tag{17}$$

$$\sigma_i = \begin{cases} \frac{1}{h^2}(x - x_{i-1})(x - x_i) , & x_{i-1} \leq x \leq x_i \\ -\frac{1}{h^2}(x - x_i)(x - x_{i+1}) , & x_i \leq x \leq x_{i+1} \end{cases}$$

where α_i and δ_i are determined by comparing $B(S_h^m, \psi_i)$ and $B(S_h^m, \bar{\psi}_i)$.

In order to resolve the fairly sharp shock-solutions of Equation 1., the refined grid spacing h, has to be small such that $h \ll d$. Further d may be chosen as $\frac{\Delta t}{\phi}|b(\frac{S_a^m}{2})| \approx d$, which gives:

$$\frac{\Delta t}{\phi}\left|b(\frac{S_a^m}{2})\right| \gg h. \tag{18}$$

Equation 18. implies that the (S_h^m, ψ_i) term in $B(\cdot, \cdot)$ plays a minor role in the calculation of the front shape. This term may be important only at the top of the saturation front where $\frac{\Delta t}{\phi}|b(S)| \lesssim h$. Here, however, θ_i approximate the dominating part of the testfunctions.

Using Equation 18. together with $(\varepsilon \frac{\Delta t}{\phi} \bar{D}_i)^{1/2} \ll d$, we get the following results:

$$\bar{\psi}_i = \theta_i + \alpha_i \sigma_i + O\left(h, \sqrt{\varepsilon \Delta t \bar{D}_i / \phi}\right), \tag{19}$$

$$B(S_h^m, \psi_i) = B(S_h^m, \theta_i + \alpha_i \sigma_i) + O\left(h, \sqrt{\varepsilon \Delta t \bar{D}_i / \phi}\right), \tag{20}$$

where

$$\alpha_i = 3\left[\frac{2\varepsilon \bar{D}_i}{\bar{b}_i h} - \coth\left(\frac{\bar{b}_i h}{2\varepsilon \bar{D}_i}\right)\right]. \tag{21}$$

The part retained in $\bar{\psi}_i$ given by Equation 19. expresses the balance between the nonlinearities of the fractional flow function and the diffusion in the shock region[19], which determines the broadening of the front. From an upstream weighting point of view, the $\frac{\Delta t}{\phi} b$ term adds enough diffusion to give the correct width of the front and a well behaved numerical calculation.

Barrett and Morton[16] give an expression for the error introduced by aproximate symmetrization for an advection-diffusion equation with constant coefficients. Equivalent expressions can be worked out for Equation 20.

So far, no rigorous error estimate for the total error originating from the approximate symmetrization and the operator splitting given by Equations 5. and 6., has been given. However, work is under way on this problem.

4. Numerical Examples

In the numerical examples given by Figures 1-5, $f(S)$ and $D(S)$ are defined as

$$f(S) = S^3(S^3 + (1 - S)^3)^{-1}; \qquad D(S) = 4S(1 - S).$$

Further, ψ_i is approximated by $\bar{\psi}_i$, given by Equations 19. and 21.

The calculation presented, shows the effects of the variation of the small parameter ε and different refinements of the coarse grid, (Figures 1-5).

The first step in the procedure[1] is to calculate the saturation at the coarse grid nodes. Normally the shock is located within one of the coarse gridblocks. This means that Equation 12. only has to be solved locally on this grid block[1], using a refined grid

When a front is established, the timestep is chosen such that the front moves close to one coarse grid size d. The validity of using long timesteps is clearly demonstrated since the shape of the front is very stable. This means that the error introduced by the time-discretization, becomes very small.

It turns out that the procedure is fairly robust. When the refined gridsize is too big to resolve the front, we of course introduce an error in the form of a numerical brodening, but the solution is otherwise well behaved, (Figure 3).

In all the runs, the massbalance error, mainly introduced via time-stepping backward along the characteristic, is less than one per cent.

We define:

N: Number of coarse gridblocks on the interval $[0, 1]$.

$N_i - 1$: Number of internal nodes in a coarse gridblock.

F: $\Delta t = F \cdot (\,$ timestep needed to move the shock one coarse gridblock $\,)$.

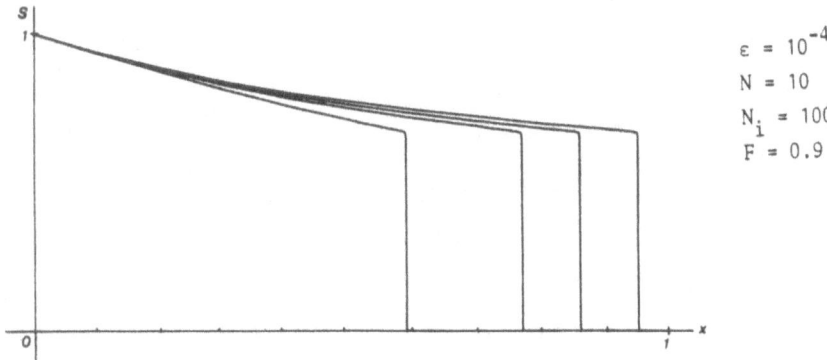

Fig. 1: We get as expected almost a Buckley - Leverett solution. The front is just barely resolved in this case, but this causes little extra diffusion.

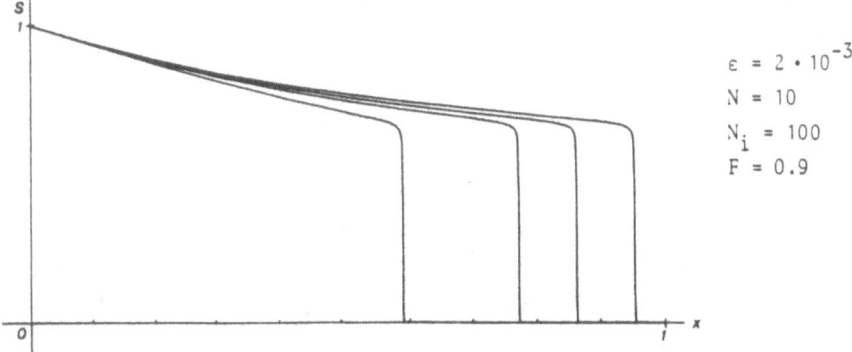

Fig. 2: More structure is starting to show up in the solution. The shock should be rather well resolved in this case.

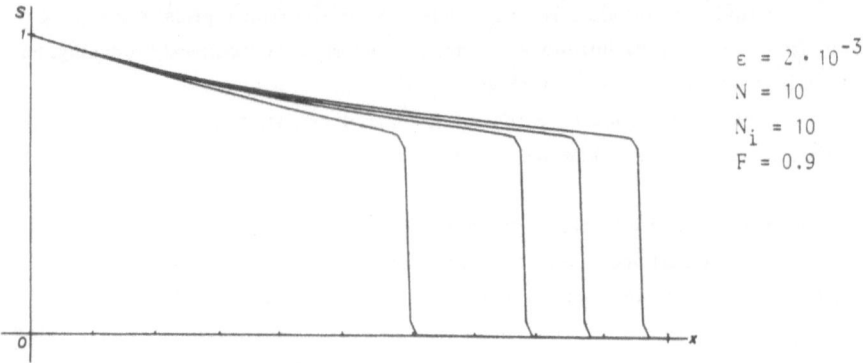

Fig. 3: Except for numerical broadening, the solution is wellbehaved.

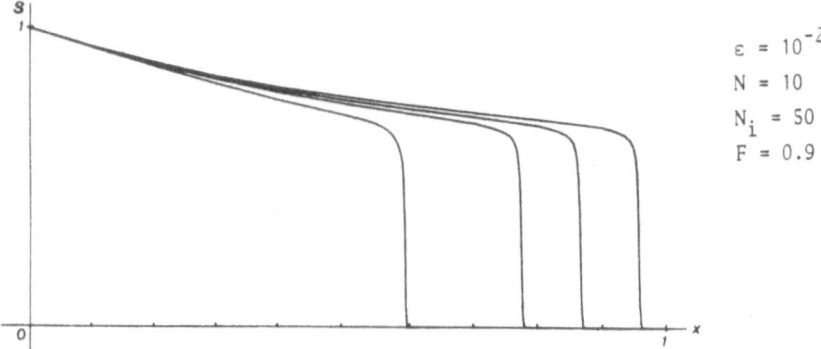

Fig. 4: The solution is well resolved for this choice of parameters.

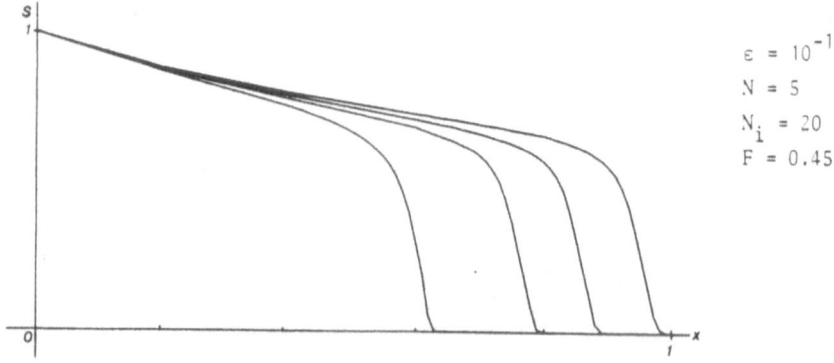

Fig. 5: Even with a fairly big ε , very little distortion of the front appear. The front is moved 0,45 d in one timestep here.

5. Conclusions

The success of the computations illustrates the potential of the operator splitting techniques presented in this paper. A very coarse spatial grid and a long timestep are quite adequate to describe the transport part of the operator. Without using these concepts of characteristics, much smaller timesteps would have been needed due to the large temporal truncation errors around moving fronts induced by standard timestepping methods. The operator splitting allows fully implicit timestepping which is important for accurate approximation of the nonlinear diffusion and balancing terms. The resolution of very narrow fronts has been achieved with greatly reduced computational expence through the use of substructuring and preconditioning techniques. The techniques extend naturally to higher space dimensions.

The use of Petrov-Galerkin methods has effectively stabilized the fine-grid solution with transport terms. If the fine grid is not sufficiently fine to resolve the front, the Petrov-Galerkin method spreads the front over a minimal number of grid blocks instead of causing oscillations like standard Galerkin methods do on coarse grids. The local approximation of the optimal test functions appears to be very effective in stabilizing the computations.

A large part of the computation is performed separately on uniform local refinements of coarse grid cells. Therefore the techniques are well-suited for parallel and vector computer architectures. This will be very important in extending these methods to multidimensional applications.

6. Acknowledgement

This research was supported in part by the Norwegian Research Council for Science and Humanities (NAVF) and Statoil under the basic research program VISTA. (Dahle, Espedal).

Further the research was supported in part by U.S. Army Research Office Contract No. DAAG29-84-K-0002, by U.S. Air Force Office of Scientific Research Contract No. AFOSR-85-0117, and by The National Science Foundation Grant No. DMS-8504360. (Ewing).

We also greatfully acknowledge support from the Institute For Mathematics and its Applications, University of Minnesota.

References

1. M.S.Espedal and R.E.Ewing. Characteristic Petrov-Galerkin subdomain methods for two-phase immiscible flow, Comp. Math. in Appl. Mech. and Eng. (to appear).

2. R.E.Ewing. Mathematics of Reservoir Simulation, Research Frontiers in Applied Mathematics, Vol.1, SIAM, Philadelphia, Pennsylvania, 1984.

3. T.F.Russell. Finite elements with characteristics for two-component incompressible miscible displacement. Sixth SPE Symp. on Reservoir Simulation (SPE 10500), New Orleans, 1982.

4. R.E.Ewing, T.F.Russell, and M.F.Wheeler. Convergence analysis of an approximation of miscible displacement in porous media by mixed finite elements and a modified method of characteristics. Computer Meth. in Appl. Mech. and Eng., 47 (1984), pp. 73-92.

5. G.Chavent, G.Cohen, and J.Jaffre. Discontinuous upwinding and mixed finite elements for two-phase flows in reservoir simulation. Comp. Meth. in Appl. Mech. and Eng., 47 (1984), pp.93-118.

6. J.Douglas, Jr., R.E.Ewing, and M.F.Wheeler. The approximation of the pressure by a mixed method in the simulation of miscible displacement. RAIRO Analyse Numerique, 17 (1983), pp. 17-33.

7. P.A.Raviart and J.M.Thomas. A mixed finite element method for second order elliptic problems. Mathematical Aspects of Finite Element Method, Lecture Notes in Mathematics 606, Springer, 1977.

8. J.Douglas, Jr. and T.F.Russell. Numerical methods for convection–dominated diffusion problems based on combining the method of characteristics with finite element or finite difference procedure. SIAM J.Numer. Anal., 19 (1982), pp. 871-885.

9. R.E.Ewing and T.F.Russell. Efficient time stepping methods for miscible displacement problems in porous media. SIAM J.Numer. Anal., 19 (1982), pp.1-67.

10. T.F.Russell. Galerkin time stepping along characteristics for Burgers' equation. Scientific Computing, R.Stepleman, et al. (editors), IMACS/North Holland Publishing Company.

11. J.H.Bramble, J.E.Pasciak, and A.H.Schatz. The construction of preconditioners for elliptic problems by substructuring, J.Math. Comp. (to appear)

12. P.E.Bjorstad and O.B.Widlund. Iterative methods for the solution of elliptic problems on regions partitioned into substuctures. Preprint, Courant Institute of Mathematical Sciences, NYU, September, 1984.

13. J.H.Bramble, R.E.Ewing, J.E.Pasciak, and A.H.Schatz. A preconditioning technique for the efficient solution of problems with local grid refinement. Comp. Meth. in Appl. Mech. and Eng. (to appear)

14. O.A.Olienik. Uniqueness and stability of the generalized solution of the Cauchy problem for a quasilinear equation. Uspehi Mat. Nauk., 14 (1959), PP. 165-170 (Eng. transl. Amer. Math. Soc. Transl., Ser. 2, 33 (1963), pp. 285-290.

15. W.B.Lindquist. The scalar Riemann problem in two spatial dimensions: Piecewise smoothness of solutions and its breakdown. New York University, DOE Research and Development Report, DOE/ER/03077-227, September 1984.

16. J.W.Barret and K.W.Morton. Approximate symmetrization and Petrov-Galerkin methods for diffusion-convection problems. Comp. Meth. in Appl. Mech. and Eng., 45 (1984), pp. 97-122.

17. L.Demkowitcz and J.T.Oden. An adaptive characteristic Petrov-Galerkin method for convection-dominated linear and nonlinear parabolic problems in one space variable. Preprint, The University of Texas at Austin, Spring 1985, J.Comp. Phys. (to appear).

18. L.Demkowitcz and J.T.Oden. An adaptive characteristic Petrov-Galerkin finite element method for convection–dominated linear and nonlinear parabolic problems in two space variables. Comp. Meth. in Appl. Mech. and Eng., 55 (1986), pp. 63-87.

19. J.Kevorkian and J.D.Cole. Perturbation methods in applied mathematics. Applied Mathematical Sciences, 34, Springer–Verlag, 1980.

On The Simulation of Heterogeneous Petroleum Reservoirs

Prabir Daripa [1]

James Glimm [1, 2, 3]

Brent Lindquist [1]

Mohsen Maesumi

Department of Mathematics
Courant Institute of Mathematical Sciences
New York University
New York, NY 10012

Oliver McBryan [1, 3, 4]

Center for Theory and Simulation
Cornell University
Ithaca, NY 14853

1. Introduction

The study of petroleum reservoirs is characterized by strongly nonlinear equations, complex physical and chemical processes, strong spatial variation or discontinuities in key reservoir parameters, uncertain or statistical geological data and unstable fluid regimes. Numerical simulation is one of the accepted methods for the study of petroleum reservoirs and improvements in numerical methods is one route which may allow progress in such studies. No single method or set of numerical ideas is sufficient at the present time. In fact computational simulation is used for many distinct length scales, and to suppress or represent accurately a wide range of details in the reservoir and fluid description. The appropriate numerical method then depends on the level of description required and the purpose of the computation. Similarly, we believe that a variety of improved methods might each be useful, possibly in distinct contexts or facets of the reservoir simulation problem. In the same vein,

1. Supported in part by the Applied Mathematical Sciences subprogram of the Office of Energy Research. U. S. Department of Energy, under contract DE-AC02-76ER03077.
2. Supported in part by the Army Research Office, grant DAAG29-85-0188.
3. Supported in part by the National Science Foundation, grant DMS-83-12229.
4. Permanent address: Courant Institute of Mathematical Sciences. New York University. New York. NY 10012

we believe that for any proposed idea or method, reservoir parameters and computational problems can be found for which it is ill-suited.

The authors and coworkers have proposed [2, 10, 11, 13, 12, 8] the front tracking method as useful in applications to petroleum reservoir simulation. A variety of tests of a numerical analysis nature were performed for the method, verifying convergence under mesh refinement and absence of mesh orientation effects [13]. The ability to handle complex interface bifurcation [8], fingering instabilities [9, 13] and polymer injection [3, 4] (as an example of tertiary oil recovery) indicates a level of robustness in this method. The main purpose of this paper is to report on two features which will allow further series of tests by enabling a more realistic description of reservoir heterogeneities.

The reservoir equations and the numerical method. The equations governing the flow of two incompressible phases (oil and water) in a porous medium can be approximated by

$$(\phi s)_t + \nabla \cdot v\, f(s) = 0 , \tag{1.1}$$

$$v = -\, K\lambda \cdot \nabla P , \tag{1.2}$$

$$\nabla \cdot v = \nabla \cdot K\lambda \cdot \nabla P = 0 , \tag{1.3}$$

where $s = s(x,t)$ is the fractional volume of the water phase, $\phi(x)$ is the porosity of the medium, $v(x,t)$ is the total fluid (oil plus water) velocity, $P(x,t)$ is the pressure, $K(x)$ is the porous medium (rock) permeability tensor, $\lambda = \lambda(s)$ is the total fluid relative transmissibility function, and $f(s)$ is the so-called fractional flow function relating the water phase velocity to the total fluid velocity. Equations (1.1) and (1.3) represent conservation of the fluids, (1.2) is Darcy's law. We shall consider flow in two spatial dimensions. In order to concentrate on specific physical questions, we have omitted the effects of gravity, capillary pressure (surface tension), variable medium depth, and flow sources from (1.1) through (1.3). See [24, 25] for further details.

There are two flow regimes associated with the form of $f(s)$. The case of a fractional flow function linear in s describes miscible flow; a non-linear function describes immiscible flow. For miscible flow the fluid discontinuities (shocks) are actually contact discontinuities and the shock propagates at the fluid particle velocity. This is not the case for immiscible flow and in that case the fluid particles pass through the shock front. As a consequence, though the hyperbolic equation (1.1) has a simpler wave structure for miscible flow, it is inherently a much more unstable flow regime then immiscible flow. In the linearized (small perturbation) regime, the stability of a jump discontinuity for (1.1) - (1.3) is shown to be determined by the frontal mobility ratio

$$m \equiv \frac{\lambda(s_b)}{\lambda(s_a)} \tag{1.4}$$

where s_a (s_b) is the state on the ahead (behind) side of the traveling shock. The case $m = 1$ is the limit of linear stability, with $m > 1$ corresponding to the unstable regime. For miscible flow m has the potential of becoming infinite, for immiscible flow it is bounded by a constant value.

The analyses in this paper are sparked by the application of the front tracking method to reservoir flow problems. The front tracking method is a hybrid which combines special adaptive methods for the enhanced resolution of discontinuities with conventional differencing schemes for the solution in the region between discontinuities. The method as currently implemented is sequential; the pressure (elliptic) equation (1.3) and the conservation (hyperbolic) equation (1.1) are solved separately every time step, each solution involving data from the previous solution of the other. We note that the difference in characteristic time scales between (1.1) and (1.3) provides sufficient justification for this splitting of the system (1.1), (1.3). Employing a sequential method has the distinct advantage of allowing different solution techniques to be used for each equation. The pressure equation is solved by the method of finite elements, which is well suited to the mathematical character of elliptic equations. Similarly a combination of finite differences and analytical Riemann problem solutions is used for (1.1), which is appropriate in view of its hyperbolic nature.

The presence of phase or other types of discontinuities in the physical problem gives rise to discontinuous coefficients in the elliptic pressure equation. A special adaptive grid, which is modified at each time step, is used to resolve these features and compute the flow field accurately [23]. Each such adaptive grid, which we shall refer to as the 'elliptic grid', has the index structure of a regular rectangular N_e x M_e grid and hence the numerical solution can be accelerated by fast solution techniques.

In the hyperbolic equation, the discontinuities have the mathematical structure of shock waves, and are propagated by jump relations which relate the shock speed to the magnitude of certain discontinuities across the shock. This part of the computation can be viewed as a hybrid of finite differences and of the method of characteristics or of moving point methods [5]. For a discussion of front tracking in greater depth, see [2,13,14]. The solution of (1.1) in the regions between fronts uses a finite difference scheme with respect to a regular rectangular N_h x M_h grid which covers the entire computational region, and which shall be referred to as the hyperbolic grid.

In the next section, we describe the use of front tracking to represent geological discontinuities such as layers or faults. This material is a preliminary report on a portion of the Ph.D. thesis of one of us (M.M.). Next we report on the effect of porosity variation on reservoir fingering. The main conclusion is that porosity is less significant than permeability as a cause of interface instabilities. A final section contains some comments on M. Shearer's no-go theorem.

2. Geological Layers

2.1. Introduction

In this section we study the effects of a discontinuous permeability tensor, K, brought about by the presence of distinct geological layers. We concentrate on the problem of a single discontinuity (a zero width transition zone) separating two homogeneous rock layers of different, constant, permeabilities. We assume the porosity of both layers is the same, and constant, and scale it out by redefining the time variable, t. We are interested in obtaining an analytical understanding of the propagation of an oil-water phase bank (the front) through this geological discontinuity, and in the subsequent development of a numerical algorithm to embody the analytical results and allow calculations through media of greater variation in layering. The point of interaction between a moving front and the stationary rock discontinuity will be referred to as the "node". In the version reported on here, the rock discontinuity shall be assumed to be either horizontal or vertical.

Such an interaction falls into a much broader class of interacting discontinuities of hyperbolic systems. In such interactions one is interested in the general problems of bifurcation, deflection and evolution of the intersection point. Glimm and Sharp [6] studied this phase bank - layer discontinuity interaction as an example of elementary waves and classified the possible exact solutions for the deflection of a front by such a rock discontinuity. Assuming finite, leading order data, they found two solutions. The first consisted of a one parameter family of solutions in which the flow is parallel to the front and the node remains stationary in space. The second is a solution in which flow is normal to the front, and hence the node propagates in space. In this case, the angle of incidence for the front on the rock discontinuity is restricted to a fixed angle given in terms of the ratio of permeabilities for the two geological layers. Both of these solutions are too restrictive to give an indication of how an interaction between the front and the rock discontinuity develops, though they give possible insight into steady state solutions. A third solution was also obtained which allowed the shock to cross the layer tangentially.

This problem has been analyzed with greater generality [21]. The theoretical solution of this problem for general angles and boundary conditions is complicated by the fact that the elliptic equation has a singularity at the node (the velocity is (usually) either zero or infinity). Nevertheless an approximate deflection law can be obtained by introducing an averaging length scale on the scale of the ignored physical phenomena, e.g. capillarity. The resulting equations will be discussed on another occasion [21]. In the next section we describe a simple algorithm that will give an approximate method for deflection, evolution, and lateral bifurcation of a front as it passes a layer. We make two simplifying assumptions.

1) The fractional flow function is the same in both layers (i.e. the same in both rock types).

2) The flow is miscible.

The node propagation routine

In the front tracking method the propagation of a front (in the hyperbolic step of the sequential scheme) is generally accomplished by splitting the hyperbolic operator along perpendicular directions in a local coordinate frame. The front is advanced by solving a Riemann problem in the direction normal to the front and the state variable (in this case the saturation) is first updated according to the normal movement of the front. This step is followed by solving the hyperbolic equation, in a direction locally tangential to the front. This is done by a (one dimensional) finite difference scheme, thus updating the state variable at the points which define the front, and accounting for flow tangential to the front. Near a point of interaction of fronts (a node) these steps cannot be taken in the usual way since the problem is inherently two dimensional and splitting into local normal and tangential directions is no longer well defined.

An algorithm for the propagation of the front in the vicinity of a layer has been developed and is given below. Consider a point on the front that is to be propagated past the rock discontinuity (i.e. from the upstream side (layer) to the downstream side (layer) of the discontinuity). The algorithm divides the propagation into two parts. First the point is propagated *using the component of the velocity normal to the front* until it reaches the discontinuity. Then an angle of deflection, velocity and the new normal direction for the propagation into the downstream layer is computed *using only the information in the upstream layer*. The point is then propagated for the remaining time of the timestep using the new propagation direction and velocity. The deflection of the front direction at the discontinuity is analogous to Fermat's principle. We re-iterate the obvious, namely that the point in question does not represent a moving fluid particle but a point on a shock surface along which there is tangential slip.

2.2. Cross flow

As an application of the algorithm discussed above, we study a case of interest which occurs when the flow is mainly parallel to the layer. Consider a miscible flow initialized as in Fig. 2.1a . Assuming that the left layer is of higher permeability than the right, and assuming no cross flow between the layers, at a later time the solution is shown in Fig. 2.1b . Cross flow between the layers can be taken into account qualitatively by noting that the pressure distribution in the no-cross flow solution is piecewise linear in each layer. For the unstable flow regime ($m > 1$) these pressure distributions are shown in Fig. 2.1c . Thus for $y < c$ the pressure distributions would favor cross flow into the high permeability layer and for $y > c$ the cross flow would be into the low permeability layer. (See [29].) Under the approximation that in the cross flow case the pressure distributions can be accurately represented by Fig. 2.1c, it can be shown that

$$\frac{c - a}{b - a} = \frac{1}{1 + m \left(\frac{L}{a} - 1 \right)} , \tag{2.1}$$

where L is the length of the computational region in the y direction.

Fig. 2.2 shows the numerical solution of this problem using the algorithm developed above. For our choice of parameters the ratio (2.1) is about 0.1 . The direction of flow agrees with above approximation; over the major portion of vertical part of the front the cross flow is toward the slow region. Therefore one expects the cross-over point C to be much closer to the slower part of the front as shown in Fig. 2.2 . One might expect that the portion of the front between A and C should move into the fast region, as shown in Fig. 2.1d, but this was not observed numerically. Even when the front was initialized as in Fig. 2.1d, the front did not persist in this configuration.

The finger formation is an indication of a singularity at the node. Initially the singularity becomes stronger as the finger becomes sharper at B, thereby accelerating the runaway behavior. At A, the reverse happens; as the angle of the front with the discontinuity moves away from the normal, the velocity singularity becomes weaker.

3. Porosity Variation

3.1. Introduction

One of the main objectives in oil recovery is to suppress the fingering and channeling instabilities which are initiated by small and large scale disturbances through the nonuniformities in the medium. The nonuniformities that we have in mind are the variations in the permeability and the porosity of the medium. In [3] the fingering problem associated with heterogeneity in the permeability field was studied. See also [20] for a discussion including the effects of capillary pressure. A partial remedy to this effect through the use of polymer flooding was analyzed in [4]. In this section we address the fingering problem associated with variable porosity.

To gain an insight into the effect of porosity variation, consider (1.1) - (1.3). The effect of porosity can be qualitatively understood as follows. If the porosity were constant, it can be removed from (1.1) (leaving (1.2) and (1.3) unchanged), by redefining the time variable, $t \to \bar{t} \equiv t / \phi$. As a consequence, the speed of any wave that would appear in the solution to the hyperbolic equation (1.1) would be modified by a constant factor inversely proportional to the porosity. In the case of variable porosity the above argument can be applied locally, to leading order approximation, thus implying a spatial variation of wave speeds. As in the case of nonuniform permeability, this variation can act to produce fingering.

If we introduce a new velocity field, $\bar{v} = \dfrac{v}{\phi}(x)$, (1.1) - (1.3) can be rewritten as

$$s_t + \nabla \cdot (\bar{v} \, f(s)) = f(s) \, \nabla \cdot \bar{v} \,, \tag{3.1}$$

$$\bar{v} = - \frac{K}{\phi} \lambda \cdot \nabla P \,, \tag{3.2}$$

$$\nabla \cdot K\lambda \cdot \nabla P = 0 \,. \tag{3.3}$$

From (3.2) we see that the effective rock permeability for the velocity field $\bar{v}$ is K / ϕ. It is well known that regions of local maxima in the permeability field serve as nuclei for finger growth in porous media flow (see, for example, [3]). Thus (3.2) implies that regions of low porosity will have an effect similar to that of high permeability. However, since the porosity does not enter into the elliptic equation (3.3), its effect will be milder than variations in the permeability field itself. In addition we note that regions of high reservoir porosity commonly have high permeability. Thus porosity variations will normally provide a partial offset to the fingering tendency of permeability variations. .

To explore further the effects of permeability, we consider the equations (1.1)-(1.3) in one spatial dimension. This is particularly appropriate when one thinks of the movement of discontinuities of the hyperbolic solution as a two step procedure (as in the algorithm used in the front tracking method), namely the propagation of the discontinuity in a locally normal direction followed by a step in which the tangential slip of the fluid along the interface is accounted for. In one space dimension, the seepage velocity v is constant as seen from elliptic equation (1.3) and setting it to unity, without any loss of generality, reduces (1.1) to

$$(\phi s)_t + f(s)_x = 0 . \tag{3.4}$$

Expressing (3.4) in terms of the conserved quantity ϕs,

$$(\phi s)_t + \frac{f_s}{\phi}(\phi s)_x = -\frac{f_s}{\phi} s \, \phi_x , \tag{3.5}$$

reveals that, along the characteristic lines

$$\frac{dx}{dt} = \frac{f_s}{\phi(x)} , \tag{3.6}$$

ϕs is not constant but changes by

$$\frac{d(\phi s)}{dt} = -\frac{f_s}{\phi} s \, \phi_x . \tag{3.7}$$

Indeed, the effect of the source term is to force the saturation, s, to be constant along characteristics, as can be seen by expressing (3.4) in the non-conservative form

$$s_t + \frac{f_s}{\phi(x)} s_x = 0 . \tag{3.8}$$

Clearly, the variable porosity affects the curvature of the characteristics in space time (equation (3.6)) and the "time to shock formation" when starting with smooth data. It is also seen from (3.6) that the characteristics will not be smooth at a point of discontinuity in ϕ.

In [22], source terms such as found in (3.5) are seen to lead to additional standing waves in the solution of the one-dimensional Riemann problem associated with a hyperbolic equation. Such waves do not arise in the present case due to the special form of the source terms and, in particular, the constancy of s along characteristics.

3.2. Code modification for inclusion of permeability effects

As the variable porosity enters into the system only through the hyperbolic equation (3.1), the adaptation of the front tracking method to the case of variable porosity requires incorporating its effect in the solution of the hyperbolic equation only. Before we discuss the incorporation of the porosity, it will be helpful to describe briefly the basic ideas behind solving the hyperbolic equation in our front tracking method. At any fixed time, the (bounded) spatial domain of the computation consists of a number of regions in which the solution is smooth. These regions are separated by discontinuities across which the saturation is discontinuous. The numerical algorithm to advance the solution of the hyperbolic equation (3.1) from time t to t + dt is done by a spatial splitting of the hyperbolic operator, solving separately for the propagation of the discontinuities (the "front") which includes the solution for the saturation immediately on each side of the discontinuities, and for the solution in the smooth "interior" regions.

The position and shape of each discontinuity is resolved by a finite number of points. Each point of a discontinuity is advanced in a direction (locally) normal to the discontinuity by solving the Riemann problem associated with the normal component of (3.1). The Riemann problem solution gives both the saturation information immediately ahead and behind the propagated point, and its speed of propagation. The solution for these saturations is dictated solely by the flux function, f(s). Since this flux function does not depend on the porosity, the states across the discontinuity are unaffected by the variation in porosity. However, as shown above, the speed of propagation is proportional to the inverse of the local porosity. Inclusion of porosity effects in this step thus consists of incorporating the porosity in the speed of each front point.

The tangential flow of fluid along each discontinuity is incorporated through the solution of the tangential component of the equation (3.1). This is accomplished by employing standard one dimensional finite difference methods (i.e. upwind) using the stencil determined by the representative points. This finite difference solution is obtained separately for each side of the discontinuity, and these solutions are easily modified to take variable porosity into account.

The solution of the two dimensional hyperbolic equation in the smooth regions uses spatial x-y operator splitting and standard one-dimensional finite difference schemes which, as just mentioned, are easily modified to take into account the variation of porosity.

3.3. Test problems

We compare the effects of a variation in the porosity and in the rock permeability fields on fingering in an areal, quarter five-spot flood. Let $\chi(x, y)$ be a gaussian random variable of mean one, and standard variation σ. Let ϕ and K be constant values for rock porosity and permeability respectively. Then $\phi \chi(x, y)$ is a gaussian random field for the porosity, with mean ϕ and variation $\sigma \phi$. A similar statement holds for $K \chi(x, y)$. Using a particular choice for the random variable that allows a specification of a given length scale for the variation (see [20]), we consider two calculations for the unstable flow regime of immiscible displacement; the first with a variation in the porosity field given by $\phi \, \chi(x, y)$ with K fixed, the second with ϕ constant and the variation in the permeability field given by $K \, \chi(x, y)$. Fig. 3.1 shows the growth of fingers due to the variation in porosity. The local maxima and minima in the porosity field are shown respectively by $+$ and $-$ signs. The fingering can be seen to be initiated in regions of local minima of the porosity. Fig. 3.2 shows the effect of variable permeability with fixed porosity. The permeability variation causes much stronger fingering than the porosity variation.

4. Shearer's Theorem

A mathematical analysis [16,17,26,27] of the equations for three phase (oil, gas, water) flow has revealed a very complex pattern of nonlinear waves and wave interactions. The no-go theorem of M. Shearer states that under very general hypotheses, the "worst case" complications occur generically. This requires a careful examination of both the hypotheses and the conclusions. In this section we give a preliminary analysis of the conclusions.

An elliptic region must occur generically in the three phase flow equations, according to Shearer's theorem. Since the Cauchy problem is ill-posed for elliptic equations and since the Cauchy problem must be solved, an elliptic region could be regarded as a deficiency in the equations. Here we adopt an opposite point of view, and argue that one can learn to live with the elliptic regime [7].

A mathematical analysis of the Riemann problem for related equations reveals no pathology or nonphysical behavior in the solution [15]. A numerical solution of three phase flow equations reaches the same conclusion [1]. The elliptic region is a bounded, interior subset of the state space. The elliptic instability appears to manifest itself by causing the solution to exit from this region, whereupon it enters the (stable) hyperbolic region. Thus the equations could be viewed as predominantly hyperbolic with a non-infinitesimal, nonlinear stabilization of their infinitesimally unstable elliptic region.

In two and three space dimensions, we expect this hyperbolic stabilization to behave in the manner of a phase transition. The fluid will prefer to flow in hyperbolic portions of state space. If an elliptic concentration of phases were somehow initialized, we expect the solution would segregate itself into spatially coherent blobs of mixtures, with each blob located within the hyperbolic portion of phase space. For a first order phase transition described by a single order parameter, the coherent blobs (pure phases) in state space at the edge of the mixed phase region are joined pairwise by tie lines. The tie lines then uniquely sweep out the mixed phase region, and any point in the mixed phase region decomposes into the pure phases at the end of the tie line it lies upon.

In the present case, one would look for tie lines by finding a unique solution of the Riemann problem. However the reasoning is circular, as the Riemann problem is probably not unique in exactly this region. In this case the nonuniqueness is resolved by appeal to more fundamental equations, including capillary pressure diffusion equations, and finally tie lines (or some more complex solution behavior) will be determined as a consequence. It would be most useful to derive an associated order parameter.

Since the elliptic region has little effect on the solution, it is tempting to "eliminate" it by deforming the equations within a fixed topological equivalence class. What are the resulting hyperbolized model equations? Evidently if there are tie lines, they should each be shrunk to a point, leaving a line of umbilic points where the two hyperbolic wave speeds coincide. In this case the hyperbolic behavior in a neighborhood of the elliptic region will coincide with that studied for a polymer flood oil reservoir [19,28,18]. This latter observation has been made previously by B. Keyfitz. However for parameters typical of real reservoirs, the elliptic region is very small. Thus the above line of degenerate hyperbolic points can, in a further approximation, be shrink to a point. Doing this leads an isolated point of degeneracy, as in the models studied by Marchesin and coauthors.

5. Conclusions

The front tracking code developed by the authors and coworkers has been subjected to a series of tests in the petroleum reservoir application. These tests are continuing and are becoming increasingly representative of realistic engineering practice. Fundamental progress in numerical algorithms (e.g. the front untangling and bifurcation algorithms [8]) and in mathematical theory (e.g. the solution of Riemann problems in one and two space dimensions) was necessary for the success of these tests.

References

1. J. B. Bell, J. A. Trangenstein, and G. R. Shubin, "Conservation Laws of Mixed Type Describing Three-Phase Flow in Porous Media," *SIAM J. Appl. Math.*, vol. 46, pp. 1000-1017, 1986.

2. I-L. Chern, J. Glimm, O. McBryan, B. Plohr, and S. Yaniv, "Front Tracking for Gas Dynamics," *J. Comp. Phys.*, vol. 62, pp. 83-110, 1986.

3. P. Daripa, J. Glimm, J. Grove, W. B. Lindquist, and O. McBryan, "Reservoir Simulation by the Method of Front Tracking," *Proc. of the IFE/SSI seminar on Reservoir Description and Simulation with Emphasis on EOR*, Oslo, Sept. 1986.

4. P. Daripa, J. Glimm, W. B. Lindquist, and O. McBryan, *Polymer Floods: A Case Study of Nonlinear Wave Analysis and of Instability in Tertiary Oil Recovery*, DOE Research and Development Report DOE/ER/03077-275, Nov. 1986.

5. A. O. Gardner, D. W. Peaceman, and A. L. Pozzi, *Numerical calculation of multidimensional miscible displacements by the method of characteristics*, SPE Reprint series no. 8 Miscible Processes.

6. James Glimm and D. H. Sharp, "Elementary Waves for Hyperbolic Equations in Higher Space Dimensions: An Example from Petroleum Reservoir Modeling," Proceedings of the Special Session on Nonstrictly Hyperbolic Conservation Laws, AMS Contemporary Mathematics Series, Jan., 1987.

7. J. Glimm, "Fueling the Twenty-First Century," *SIAM News*, vol. 20, no. 1, January 1987.

8. J. Glimm, J. Grove, W. B. Lindquist, O. A. McBryan, and G. Tryggvason, "The Bifurcation of Tracked Scalar Waves," *SIAM J. Sci. Stat. Comp.*, To appear.

9. J. Glimm, E. Isaacson, W. B. Lindquist, O. McBryan, and S. Yaniv, "Statistical Fluid Dynamics II: The Influence of Geometry on Surface Instabilities," in *Frontiers in Applied Mathematics*, vol. 1, SIAM, Philadelphia, 1983. Ed. R. Ewing

10. J. Glimm, E. Isaacson, D. Marchesin, and O. McBryan, "A Shock Tracking Method for Hyperbolic Systems," ARO Report 80-3, Feb. 1980.

11. J. Glimm, E. Isaacson, D. Marchesin, and O. McBryan, "Front Tracking for Hyperbolic Systems," *Adv. in Appl. Math.*, vol. 2, pp. 91-119, 1981.

12. J. Glimm, C. Klingenberg, O. McBryan, B. Plohr, D. Sharp, and S. Yaniv, "Front Tracking and Two Dimensional Riemann Problems," *Adv. in Appl. Math.*, vol. 6, pp. 259-290, 1985.

13. J. Glimm, W. B. Lindquist, O. McBryan, and L. Padmanabhan, "A Front Tracking Reservoir Simulator I: 5-Spot Validation Studies and the Water Coning Problem," in *Frontiers in Applied Mathematics*, Glimm, vol. 1, SIAM, Philadelphia, 1983.

14. J. Glimm, W. B. Lindquist, O. McBryan, and G. Tryggvason, "Sharp and Diffuse Fronts in Oil Reservoirs: Front Tracking and Capillarity," *SIAM, Proc. Math. and Comp. Methods in Seismic Exploration and Reservoir Modeling*, Houston, Jan, 1985.

15. H. Holden, "The Riemann Solution for a Prototype 2x2 System of Conservation Laws for Stone's Model in Oil Reservoir Simulation," *Comm. Pure Appl. Math.*, To appear.

16. E. Isaacson, D. Marchesin, B. Plohr, and J. B. Temple, *The Classification of Solutions of Quadratic Riemann Problems (I)*, PUC/RJ Report Mat 12/85 and MRC Technical Summary Report #2891, 1985.

17. E. Isaacson, D. Marchesin, B. Plohr, and J. B. Temple, *The Classification of Solutions of Quadratic Riemann Problems (II)*, PUC/RJ Report 1985 and MRC Technical Summary Report #2892, 1985.

18. E. Isaacson, "Global Solution of a Riemann Problem for a Nonstrictly Hyperbolic System of Conservation Laws Arising in Enhanced Oil Recovery," *J. Comp. Phys.*, To appear.

19. B. Keyfitz and H. Kranser, "A system of non-strictly hyperbolic conservation laws arising in elasticity theory," *Arch. Ratl. Mech. Anal.*, vol. 72, 1980.

20. M. J. King, W. B. Lindquist, and L. Reyna, "Stability of Two Dimensional Immiscible Flow to Viscous Fingering," *DOE Research and Development Report DOE/ER/03077-244*, March, 1985.

21. M. Maesumi. PhD thesis - in preparation

22. D. Marchesin and P. J. Paes-Leme, *A Riemann Problem in Gas Dynamics with Bifurcation*, PUC/RJ Report Mat 02/84 , 1984.

23. O. McBryan, "Elliptic and Hyperbolic Interface Refinement in Two Phase Flow," in *Boundary and Interior Layers - Computational and Asymptotic Methods*, ed. J. Miller, Boole Press, Dublin, 1980.

24. D. W. Peaceman, *Fundamentals of Numerical Reservoir Simulation*, Elsevier, Amsterdam, 1977.

25. G. A. Pope, "The Application of Fractional Flow Theory to Enhanced Oil Recovery," *Soc. Pet. Eng. J.*, pp. 191-205, June 1980 .

26. D. Schaeffer and M. Shearer, "The Classification of 2x2 Systems of Non-Strictly Hyperbolic Conservation Laws with Application to Oil Recovery," *Comm. Pure Appl. Math.*, 1986, To appear.

27. M. Shearer, D. Schaeffer, D. Marchesin, and P. J. Paes-Leme, "Solution of the Riemann Problem for a Prototype 2x2 System of Non-Strictly Hyperbolic Conservation Laws," Submitted to *Arch. Rat. Mech. Anal.*, 1985.

28. J. B. Temple, "Global Existence of the Cauchy Problem for a Class of 2x2 Nonstrictly Hyperbolic Conservation Laws," *Adv. Appl. Math.*, vol. 3, pp. 335-375, 1982.

29. V. J. Zapata, *A Theoretical Analysis of Viscous Crossflow*, Univ. of Texas at Austin Report No. UT 81-3, August, 1981.

101

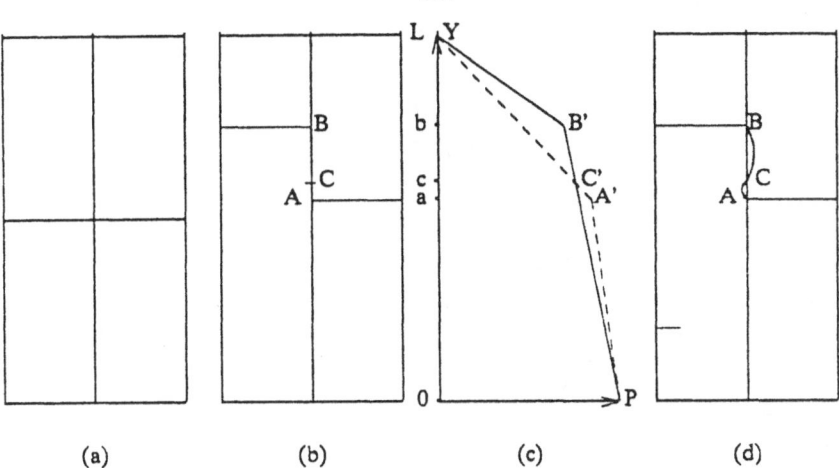

Fig. 2.1 (a) The initial setup for a miscible flow run through two rock layers separated by a sharp boundary. The initial oil-water bank is horizontal, the higher permeability layer is on the left. (b) The (one dimensional) solution assuming no flow between the layers. (c) The pressure solutions for (b). The solid line is for the left layer, the dashed line for the right. (d) The direction of flow expected from (b) and (c) if cross flow is included.

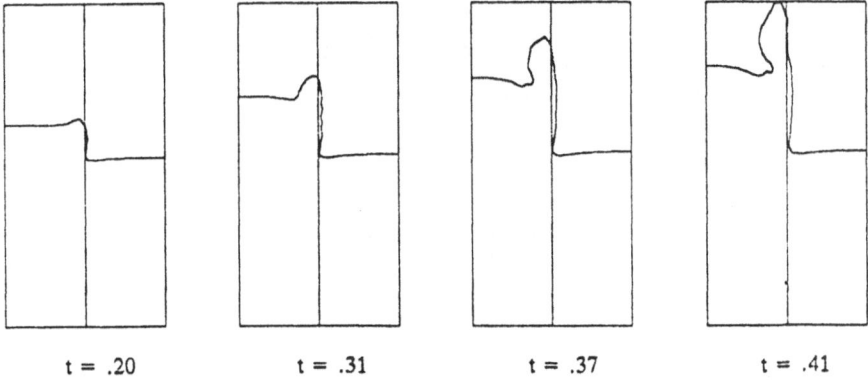

t = .20   t = .31   t = .37   t = .41

Fig. 2.2 The solution for the problem discussed in Fig. 2.1 computed using the Front Tracking Method. The initialization is same as in Fig 2.1(a).

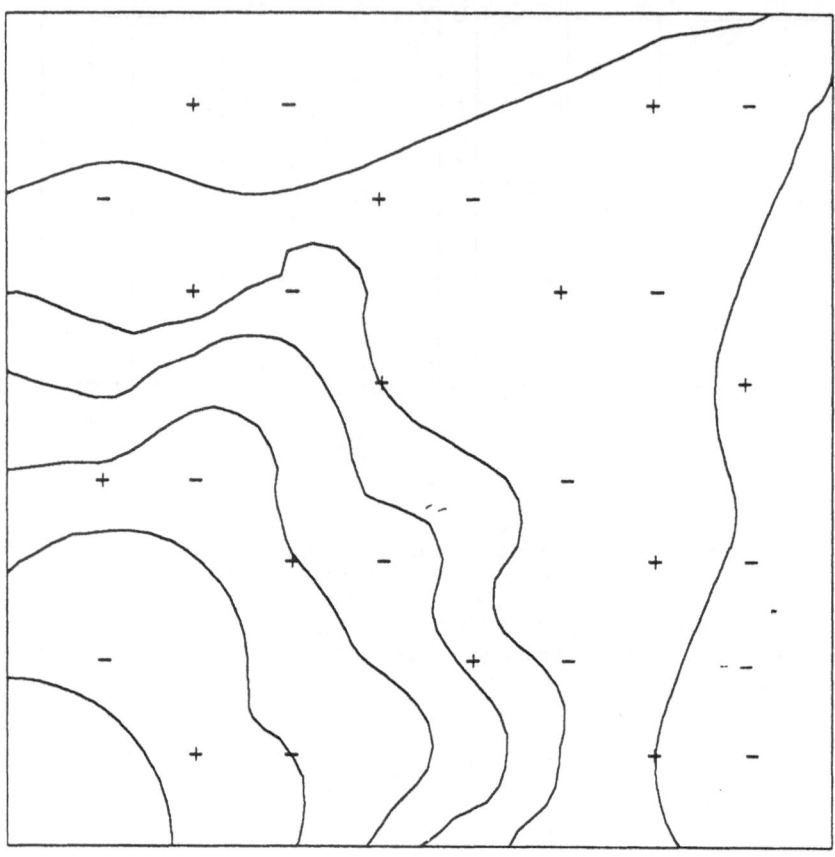

Fig. 3.1 Growth of fingers for immiscible displacement in a heterogeneous reservoir: the temporal evolution of the oil-water discontinuity. The + (−) signs correspond to local maxima (minima) of the porosity field. The viscosity ratio of the water to oil is 1 : 10 .

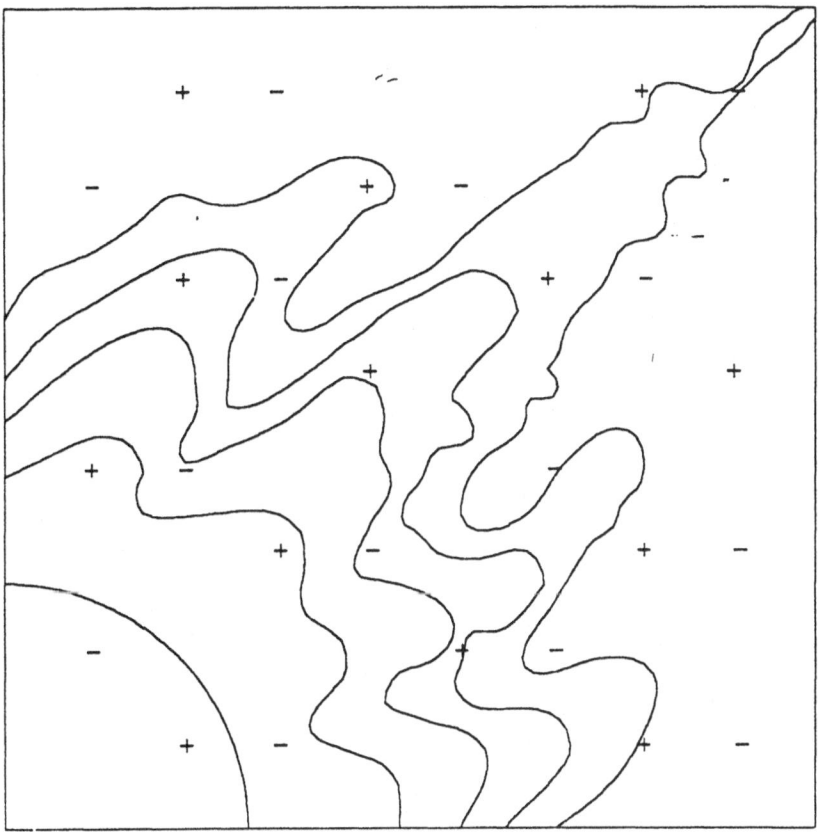

Fig. 3.2 Growth of fingers for immiscible displacement in a heterogeneous reservoir: the temporal evolution of the oil-water discontinuity. The + (−) signs correspond to local maxima (minima) of the permeability field. The viscosity ratio of the water to oil is 1 : 10 .

A Theory of Tension at a Miscible Displacement Front

by

H. Ted Davis
Department of Chemical Engineering and Materials Science
Unversity of Minnesota
Minneapolis, MN 55455

Introduction

When two miscible fluids and placed in contact they will immediately begin to mix diffusively (and convectively if their densities are such as to drive convection) across the concentration front formed at the zone of contact. Although no interface will form at the concentration front, the composition inhomogeneities can give rise to pressure anisotropies and therefore to tension at the mixing zone between the contacted fluids. Diffusive mixing will continuously broaden the mixing zone and reduce the pressure anisotropy and the associated tension. The purpose of this short paper is to examine with the aid of a molecular theory of inhomogeneous fluid the magnitude and rate of reduction of the tension by diffusive mixing of the zone of contact of miscible fluids. The results found here suggest that instabilities in miscible frontal displacement may be similar to those in ultralow tension immiscible frontal displacement, with the added caveat that in the miscible process the tension decreases continuously in time.

Theory of Tension Caused by Fluid Inhomogeneity

Consider an inhomogeneous fluid whose component densities n_α vary in the x-direction only, i.e., $n_\alpha = n_\alpha (x)$. Assume that the fluid particles interact via pair additive centrally, symmetric forces. Because of the density inhomogeneities, the normal component, P_N, transverse component P_T of the fluid differ, and the difference $P_N - P_T$ gives rise to fluid tension in the transverse direction.

Suppose that far ahead of and far behind the position $x = 0$, the fluid is homogeneous (as illustrated in Fig. 1). There the pressure is isotropic, and so $P_N = P_T$. If we define the tension of the inhomogeneous region of the fluid by

$$\gamma = \int_{-\infty}^{\infty} (P_N - P_T) \, dx , \qquad (1)$$

we find from the Irving-Kirkwood pressure tensor that[1,2]

$$\gamma = -\sum_{\alpha,\beta} \frac{1}{2} \int_{-\infty}^{\infty} \int \frac{s_x^2 - s_y^2}{s} \frac{du_{\alpha\beta}}{ds} (s) \, n_\alpha (x) n_\beta (x + s_x) \, g_{\alpha\beta} (s; x, s_x) \, d^3s dx \, , \quad (2)$$

where s_x and s_y are x and y-components of the separation s of a pair of particles, $u_{\alpha\beta}(s)$ is pair potential between particles of components α and β and $g_{\alpha\beta}$ is the pair correlation function.

To simplify the analysis in this paper we assume that the concentration gradients are small enough to approximate $n_\beta(x + s_x)$ by

$$n_\beta (x + s_x) \propto n_\beta(x) + s_x \frac{dn_\beta}{dx} + \frac{s_x^2}{2} \frac{d^2 n_\beta}{dx^2} \quad (3)$$

in Eq. (2). We also assume that the pair correlation function depends only on the particle separation s. These assumptions result in the van der Waals-Korteweg approximation to the pressure tensor[1,3,4]. The resulting formula for γ is

$$\gamma = \sum_{\alpha,\beta} c_{\alpha\beta} \int_{-\infty}^{\infty} \frac{dn_\alpha}{dx} \frac{dn_\beta}{dx} \, dx \, , \quad (4)$$

where

$$c_{\alpha\beta} \equiv \frac{2\pi}{15} \int s^3 \frac{du_{\alpha\beta}}{ds}(s) \, g_{\alpha\beta}(s) \, d^3s \, . \quad (5)$$

Our final simplification is to restrict our analysis to a two component regular solution in which the densities of components 1 and 2 are $n_1 = \phi n$ and $n_2 = (1 - \phi)n$. ϕ is the mole fraction of component 1 and n is the total density, which is constant in a regular solution. For this case, Eq (4) reduces to

$$\gamma = c \int_{-\infty}^{\infty} \left[\frac{d\phi}{dx} \right]^2 dx \, , \quad (6)$$

where

$$c = n^2[c_{11} + c_{22} - 2c_{12}] \quad (7)$$

If the fluid phases beyond and behind the inhomogeneous zones are coexisting phases, then upon contacting the two phases an equilibrium interfacial profile $\phi(x)$ is quickly formed and γ gives the tension of the interface between coexisting phases. If the two fluids are miscible in all proportions, then the profile ϕ will broaden by diffusion and γ will decrease towards zero. However, although γ becomes small as ϕ broadens, it is not strictly zero until ϕ is uniform. *In this sense, there is tension between miscible fluids at different concentrations.*

Reduction of Tension between Miscible Fluids by Diffusion

Fick's law of one-dimensional diffusion in a binary regular solution is

$$\frac{\partial \phi}{\partial t} = D \frac{\partial^2 \phi}{\partial x^2} ,$$ (8)

where D is the diffusion coefficient. We consider the case for which

$$\begin{aligned} \phi &\to \phi_b , & x &\to -\infty \text{ (if} \\ \phi &\to \phi_a , & x &\to +\infty \end{aligned}$$ (9)

where ϕ_b and ϕ_a are uniform mole fractions far behind and far ahead of the inhomogeneous region. If initially we suppose that $\phi = \phi_b$, x <0 and $\phi = \phi_a$, x >0, then the solution to Eq. (8) is

$$\phi = \phi_b + \frac{(\phi_a - \phi_b)}{\sqrt{2\pi}} \int_{-\infty}^{x/2\sqrt{Dt}} e^{-y^2} dy .$$ (10)

Using Eq. (10) to evaluate Eq. (6), we find

$$\gamma = \frac{c\ (\phi_a - \phi_b)^2}{2\sqrt{2\pi Dt}} .$$ (11)

A value of c typical of a hydrocarbon mixture is 10^{-5} dyn.

In Table 1 we list for a miscible system values of γ as a function of time for various values of the diffusivity.

Table 1. Tension γ of a planar front of a miscible fluid as a function of time t of diffusive spreading of the front[*].

Time t(s)	Mixing Zone Width $\sqrt{Dt}$(cm)			Tension γ(dyn/cm)		
	$D = 10^{-5}$cm²/s	$D = 10^{-7}$cm²/s	$D = 10^{-9}$cm²/s	$D = 10^{-5}$cm²/s	$D = 10^{-7}$cm²/s	$D = 10^{-9}$cm²/s
1	3.1×10^{-3}	3.1×10^{-4}	3.1×10^{-5}	6.3×10^{-4}	6.3×10^{-3}	6.3×10^{-2}
10	10^{-2}	10^{-3}	10^{-4}	2.0×10^{-4}	2.0×10^{-3}	2.0×10^{-2}
10^2	3.1×10^{-2}	3.1×10^{-2}	3.1×10^{-3}	6.3×10^{-5}	6.3×10^{-4}	6.3×10^{-3}
10^3	10^{-1}	10^{-2}	10^{-4}	2.0×10^{-5}	2.0×10^{-4}	2.0×10^{-3}
4×10^3	2×10^{-1}	2×10^{-2}	2×10^{-3}	1.0×10^{-5}	1.0×10^{-4}	1.0×10^{-3}

[*] The value c $(\phi_a - \phi_b)^2/2\sqrt{2\pi} = 2 \times 10^{-6}$ dyn was used.

From the entries in this table it follows that the tension of a diffusive mixing zone between miscible fluids, while small, is nevertheless not zero.

Discussion

The work reported here was inspsired by the stability analysis presented by L. Schwartz in this same volume. The results of the simple analysis perhaps offer insight into the theory of stability and fingering of miscible fronts. Since γ is not identically zero, but rather approaches zero as the front widens due to diffusive mixing, it appears reasonable to think of the fingering of a miscible front as the fingering of an immiscible front (non-zero tension) of continuously decreasing tension. According to the Chouke-Taylor-Saffman linear stability theory[7,8] of immiscible displacement the wavelength λ_{cr} of the smallest unstable fingering mode of a planar front moving in the x-direction through a rectangle cross-section of height b in the z-direction and width L in the y-direction is given by

$$\lambda_{cr} = \frac{\pi}{\sqrt{3Ca'}} \, , \tag{12}$$

where

$$Ca' = v\mu L^2/\gamma b^2 \, . \tag{13}$$

v is the velocity of the front, γ the interfacial tension, and μ the viscosity of the displacing fluid (fluid behind the interface). Similarly, they find for the wavelength λ_{max} of the fastest growing finger

$$\lambda_{max} = \frac{\pi}{\sqrt{Ca'}} \, . \tag{14}$$

If we presume that the Chouke-Taylor-Suffman analysis can be used to estimate the fastest growing fingers as a function of tension in a miscible flood, we conclude from Eq. (14) that the fastest growing fingers, and therefore the scale of fingering, goes as $t^{-1/4}$, i.e.,

$$\frac{\lambda_{max}(t_2)}{\lambda_{max}(t_1)} = \left(\frac{t_1}{t_2}\right)^{1/4} \, . \tag{15}$$

Thus, a prediction of the theory presented here is that the scale of fingering of a miscible front should become increasingly finer with time or with distance x from a datum position. There should be a fine structure that appears as a factor of 1/3 with each factor of 8 increase in time or displacement distance.

As a particular example, let us assume v $= 1$ cm/s, $\mu = 10^{-2}$ poise, L/b $= 15$, $\gamma = 10^{-3}$ dyn/cm. Then Ca' $= 2250$ and

$$\frac{\lambda_{max}}{L} = 0.0662 \qquad (16)$$

If $L = 30$ cm, then $\lambda_{max} = 2.0$ cm. At an instant 81 times later the theory predicts the $\lambda_{max} = 0.66$ cm in the same (long) cell. It should be noted that as the width of the mixing zone approaches λ_{max} the mechanism given here breaks down.

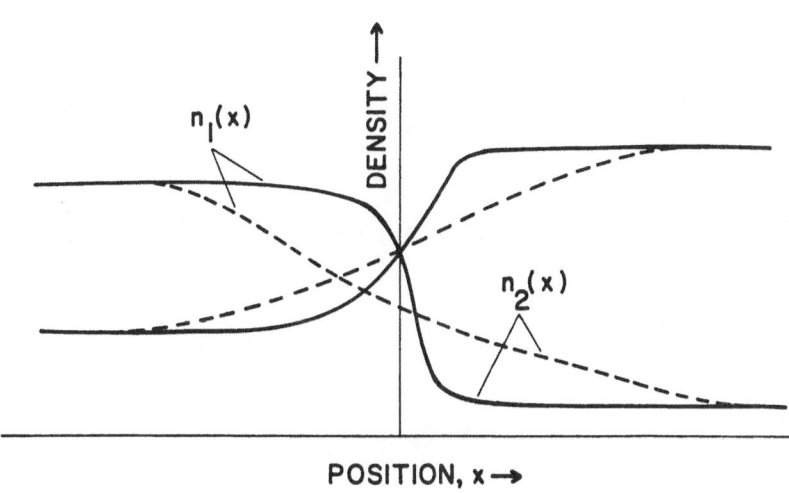

Figure 1. Density profiles between homogeneous fluid regions of different composition. If the fluids are miscible, then the inhomogeneous region will broaden indefinitely due to diffusion. Time t_2 is longer than time t_1. Solid curves are for t_1 and dashed curves are for t_2.

References Cited

1. Davis, H. T. and Scriven, L. E., Advances in Chemical Physics.

2. Irving, J. H. and Kirkwood, J. G., J. Chem. Phys. **18**, 817 (1950).

3. van der Waals, J. D. and Kohnstamm, P., *Lehrbuch der Thermodynamik*, Vol. 1, Maas and van Suchtlen, Leipzig, 1908.

4. Korteweg, D. J., Archives Neerl. Sci. Exacts Nat. **6**, 1 (1904).

5. Prausnitz, J. M., Lichtenthaler, R. N, and de Azevedo, E. G., *Molecular Thermodynamics of Fluid-Phase Equilibria*, 2nd ed. (Prentice-Hall 1986), p. 293.

6. Schwartz, L. W., see article elsewhere in this volume. DeGregoria, A. J. and Schwartz, L. W., J. FLuid Mech. **164**, 383 (1986).

7. Saffman, P. G. and Taylor, G. I., Proc. R. Soc. Lond. **A245**, 312 (1958).

8. Chouke, R. L., van Meurs, P. and van der Poel, C., Transactions AIME **216**, 188 (1959).

A RANDOM BOUNDARY VALUE PROBLEM MODELING SPATIAL VARIABILITY IN POROUS MEDIA FLOW*

EDMUND DIKOW† AND ULRICH HORNUNG‡

Abstract. A steady-state flow problem through a porous 3D-domain is considered. Under the assumption that the permeability (or hydraulic conductivity for water-flow) is a random field a random elliptic boundary value problem has to be solved. A mathematical methodology to prove existence and uniqueness is presented. In addition, the statistical moments of the total flux through the domain - considered as a random variable - are calculated in terms of deterministic boundary value problems.

1. Introduction. In the literature dealing with flow through porous media the assumption adopted by most authors is that Darcy's law holds

$$(1.1) \qquad \overrightarrow{q} = -K\nabla u.$$

Here, the symbols have the following meaning:

$$\overrightarrow{q} = \text{ Darcy's velocity },$$
$$K = \text{ permeability or conductivity},$$
$$u = \text{ pressure potential }.$$

In the absence of sources or sinks one obtains for an incompressible fluid in a saturated rigid medium the continuity equation

$$(1.2) \qquad \nabla \cdot \overrightarrow{q} = 0.$$

Hence, together with (1.1) the elliptic differential equation

$$(1.3) \qquad \nabla \cdot (K\nabla u) = 0$$

comes up which has to be augmented by appropriate boundary conditions to form a boundary value problem in a domain in space. If the permeability K is a known function of the space variable x, the usual theory of linear elliptic equations applies.

In the engineering literature it has been pointed out for many years that the permeability K is a rapidly oscillating function in space, see, e.g., Warren/Price (1961) or Bakr et al. (1978). These spatial variations made the approach obvious of assuming that K is a random field. Another aspect is that in practical applications there is always only a limited number of measurements of K available; and these measurements contain errors. Therefore, it is natural to say that not a function K is known but an estimate of the moments of a spatial stochastic process can be given. This way of thinking about spatially distributed data has been well known in the geosciences; and the statistical methods developed there have been introduced to hydrology in recent years, see,

---

*This work was partly suppported by the Deutsche Forschungsgemeinschaft (Grant # Ho 782/7-1) and the Institute for Mathematics and its Applications, Minneapolis, Minnesota with funds provided by the National Science Foundation.

†Fak. f. Inf. UniBwM, Werner-Heisenberg-Wey 39, D-8014 Neubiberg, West Germany

‡*SCHI*, P.O. Box 1222, D-8014 Neubiberg, West Germany

e.g., Delhomme (1978). There are many research papers dealing with the question how to estimate K and its statistical properties.

Once the assumption has been made that K is not a function of the space variable x alone but is simultaneously a random variable, the question comes up how to set up a framework in which equation (1.3) can be treated as a random partial differential equation.

In hydrology, the Fourier-transform has been applied to equation (1.3), see, e.g., Bakr (1976). This means, the differential equation is considered in all space $\mathbf{R}^n$. It turns out that for stationary random fields K the spectral density can be used to obtain some insight into the structure of possible solutions. The drawbacks of this approach are that precise mathematical statements about existence, uniqueness, stability, etc. are not available, and that it is not applicable to problems in general finite domains with prescribed boundary conditions.

This paper answers the question of which impact the information about the statistical properties of the conducitivty K has onto the solution of the differential equation (1.3). The main results are that, firstly, random boundary value problems can be solved in a precise mathematical sense, and that, secondly, relevant functionals of the solution, such as the total flux Q through a given domain, and their moments, i.e. expectation and variance, can be calculated.

The formulas presented here to approximate the moments of Q have well established error estimates. They are given in terms of solutions of deterministic boundary value problems only. Thus, in principle, it is possible to calculate these quantities without using Monte-Carlo-techniques. One can therefore hope that in the near future it will become possible to apply the results of this paper to random boundary problems and thus to make the Monte-Carlo-approach obsolete in this context.

2. The Random Boundary Value Problem. The first objective is to define precisely the mathematical problem to be solved.

2.1 ASSUMPTIONS. (a) Let $G \subset \mathbf{R}^n$ be a bounded domain the boundary of which is split into parts:

$$\partial G = S \cup T \cup \Gamma, \quad \Gamma = \partial S \cap \partial T,$$

where

$$S \text{ and } T \text{ are } C^{2,\alpha} - \text{ smooth}$$

and

$$u_0 \in C^{1,\alpha}(\overline{G})$$

with vanishing normal derivative on T, i.e.,

$$\partial_\nu u_0 = 0 \quad \text{on} \quad T.$$

(b) Let $(\Omega, \mathfrak{A}, P)$ be a probability space with countably generated σ-algebra $\mathfrak{A}$ and probability measure P on Ω. The expectation operator is denoted by

$$E \cdots = \int_\Omega \ldots dP$$

(c) For all $x \in G$ let

$$K(x, \cdot) : \Omega \to \mathbf{R}$$

be a random variable, i.e. a measurable function, such that

$$1/\mu \le K \le \mu$$

holds $P - a.e.$ with some constant $\mu > 0$.

2.2 NOTATIONS. We use the function spaces

$$\mathfrak{H} = L^2(\Omega, \mathfrak{A}, P),$$

$$C_{0,S}^1(G, \mathfrak{H}) = \{u \in C^1(G, \mathfrak{H}) : u \text{ has compact support in } \overline{G} \setminus S\},$$
$$W_{0,S}^{1,2}(G, \mathfrak{H}) = \text{ closure of } C_{0,S}^1(G, \mathfrak{H}) W^{1,2}(G, \mathfrak{H}) \text{ in the space } W^{1,2}(G, \mathfrak{H}).$$

2.3. DEFINITION. A weak solution of the random boundary value problem (RBVP)

$$\begin{cases} \nabla(K\nabla u) = 0 & \text{in } G \\ u = u_0 & \text{on } S \\ \partial_\nu u = 0 & \text{on } T \end{cases}$$

is a function $u \in W^{1,2}(G, \mathfrak{H})$ such that

(a) $$u - u_0 \in W_{0,S}^{1,2}(G, \mathfrak{H})$$

and the equation

(b) $$\int_G EK\nabla u \cdot \nabla\varphi dx = 0$$

holds for all $\varphi \in W_{0,S}^{1,2}(G, \mathfrak{H})$. This definition goes back to work by Papanicolaou/Varadhan (1982); using the Lax-Milgram lemma one obtains easily the following result.

THEOREM 1. *There is a unique weak solution of the RBVP (2.3).*

By the usual method one also may prove stability, i.e., continuous dependence on the data given.

3. The Random Total Flux. First we define what the random total flux means.

3.1 ASSUMPTION. Let $S_1 \subset S, S_1 \ne \phi, S_1 \ne S$ with dist $(S_1, S \setminus S_1) > 0$ be given, and $\rho \in C^\infty(\mathbf{R}^n)$ with

$$\rho = \begin{cases} 1 & \text{on } S_1 \\ 0 & \text{on } S \setminus S_1. \end{cases}$$

The following statement is almost obvious; it serves as a motivation for the definition to follow.

3.2 LEMMA. *If $\overrightarrow{q} \in C^1(\overline{G})$ is a vector field with*

$$\begin{cases} \nabla \cdot \overrightarrow{q} = 0 & \text{in } G \\ \overrightarrow{q} \cdot \nu = 0 & \text{on } T \end{cases}$$

then

$$\int_{S_1} \vec{q} \cdot \nu dS = \int_G \vec{q} \cdot \nabla \rho dx$$

where this quantity is independent of the special choice of ρ.

3.3 DEFINITION. If u is a weak solution of (2.3) then we define the total flux through the part S_1 of the boundary as

$$Q = \int_G K\nabla u \cdot \nabla \rho dx;$$

we write shortly

$$Q = \int_{S_1} K\partial_\nu udS.$$

3.4 LEMMA. *For the weak solution u of the RBVP (2.3) the quantity Q is an element of $\mathfrak{H}$.*

3.5 ASSUMPTIONS.
(a) $n = 2$ or $n = 3$.
(b) The permeability K can be written as

$$K(x) = K_0(x)\exp(\xi(x)),$$

where K_0 is deterministic and ξ is random.
(c) The function K_0 satisfies

$$K_0 \in C^\alpha(\overline{G})$$

with uniform α-Hölder norm and

$$K_0(x) \geq \lambda_0 > 0$$

uniformly on $\overline{G}$.
(d) For all $x \in G$ the function

$$\xi(x, \cdot) : \Omega \to \mathbf{R}$$

is $\mathfrak{A}$-measurable, i.e., it is a random variable, such that

$$-\Lambda \leq \xi(x) \leq \Lambda \qquad \text{P-a.e.}$$

with a constant Λ independent of x.
(e) The mapping ξ, considered as

$$\xi : G \to L^6(\Omega, \mathfrak{A}, P),$$

is uniformly α-Hölder continuous.
(f) For all $x \in \overline{G}$

$$E\xi(x) = 0$$

holds.
(g) There is an α-Hölder-continuous function $R : \mathbf{R}^n \to \mathbf{R}$ such that

$$Cov(\xi(x), \xi(y)) = R(x - y)$$

holds for all $x, y \in \overline{G}$.

Under these assumptions one obtains the following lemma and theorem; here the definitions (2.3) and (3.3) have to be modifed appropriately.

3.6 LEMMA. *(a) There is a unique weak solution $v_0 \in C^{1,\alpha}(\overline{G})$ of the BVP*

$$\begin{cases} \nabla(K_0 \nabla v_0) = 0 & \text{in } G \\ v_0 = u_0 & \text{on } S \\ \partial_\nu v_0 = 0 & \text{on } T \end{cases}$$

(b) There is a unique weak solution $v_1 \in C^{1,\alpha}(\overline{G}, \mathfrak{H})$ of the RBVP

$$\begin{cases} \nabla(K_0 \nabla v_1) = -\nabla(K_0 \xi \nabla v_0) & \text{in } G \\ v_1 = 0 & \text{on } S \\ \partial_\nu v_1 = 0 & \text{on } T \end{cases}$$

(c) There is a unique weak solution $v_2 \in C^{1,\alpha}(\overline{G}, L^3(\Omega, \mathfrak{A}, P))$ of the RBVP

$$\begin{cases} \nabla(K_0 \nabla v_2) = -\nabla(K_0(\frac{1}{2}\xi^2 \nabla v_0 + \xi \nabla v_1) & \text{in } G \\ v_2 = 0 & \text{on } S \\ \partial_\nu v_2 = 0 & \text{on } T \end{cases}$$

THEOREM 2. *The random total flux is*

$$Q(\xi) = Q_0 + Q'_0(\xi) + \frac{1}{2}Q''_0(\xi, \xi) + r$$

where

$$Q_0 = \int_{S_1} K_0 \partial_\nu v_0 dS$$

is a constant,

$$Q'_0(\xi) = \int_{S_1} K_0(\xi \partial_\nu v_0 + \partial_\nu v_1) dS$$

is a linear functional of ξ, and

$$Q''_0(\xi, \xi) = \int_{S_1} K_0(\xi^2 \partial_\nu v_0 + 2\xi \partial_\nu v_1 + 2\partial_\nu v_2) dS$$

is a quadratic functional of ξ; the remainder $r \in \mathfrak{H}$ satisfies

$$\|r\|_{\mathfrak{H}} \leq C \exp(3\Lambda)|\xi|^3_{0,\alpha,G,\mathfrak{H}}$$

In the latter estimate C depends only on $n, \alpha, \lambda_0, u_0, S$ and G; $|\cdot|_{0,\alpha,G,\mathfrak{H}}$ is the α-Hölder norm of (3.5e).

4. The Moments of the Total Flux.

4.1 NOTATION. Let $\sigma^2 = R(0)$.

4.2 LEMMA. (a) For each $y \in \overline{G}$ there is a unique weak solution $v_y \in C^{1,\alpha}(\overline{G})$ of the BVP

$$\begin{cases} \nabla(K_0 \nabla v_y) = -\nabla(R(\cdot - y)\nabla v_0) & \text{in } G \\ v_y = 0 & \text{on } S \\ \partial_\nu v_y = 0 & \text{on } T. \end{cases}$$

The vector field $f = \nabla v_x(x)$ is in $C^\alpha(\overline{G})$.

(b). There is a unique weak solution $w \in C^{1,\alpha}(\overline{G})$ of the BVP

$$\begin{cases} \nabla(K_0 \nabla w) = -\nabla(f + \frac{1}{2}\sigma^2 \nabla v_0) & \text{in } G \\ w = 0 & \text{on } S \\ \partial_w w = 0 & \text{on } T. \end{cases}$$

THEOREM 3. The expectations of the terms in Theorem 2 are

$$EQ_0'(\xi) = 0$$

and

$$EQ_0''(\xi, \xi) = \int_{S_1} K_0(\sigma^2 \partial_\nu v_0 + 2f \cdot \nu + 2\partial_\nu w) dS.$$

The latter formula can, at least in principle, be evaluated numerically. The expectation of the remainder satisfies a similar estimate as in theorem 2 given for r itself.

4.3 LEMMA. For each $y \in \overline{G}$ and $1 \leq j \leq n$ there is a unique weak solution $V_{y,j} \in C^{1,\alpha}(\overline{G})$ of the BVP

$$\begin{cases} \nabla(K_0 \nabla V) = -\nabla(K_0 \overrightarrow{h}) & \text{in } G \\ V = 0 & \text{on } S \\ \partial_\nu V = 0 & \text{on } T, \end{cases}$$

where $\overrightarrow{h}_{y,j}$ is defined as

$$\overrightarrow{h}_{y,j}(x) = \partial_j v_x(y)\nabla v_0(x).$$

4.4 NOTATION. We use the function

$$g^{i,j}(x,y) = \partial_i V_{y,j}(x).$$

THEOREM 4. The variance of $Q_0'(\xi)$ in Theorem 2 is given by

$$\begin{aligned} \operatorname{Var} Q_0'(\xi) = \int_{S_1} \int_{S_1} K_0(x) K_0(y) (R(x-y)\partial_\nu v_0(x)\partial_\nu v_0(y) \\ + \partial_\nu v_0(x)\nu(y) \cdot \nabla v_x(y) + \partial_\nu v_0(y)\nu(x) \cdot \nabla v_y(x) \\ + \sum_{i,j} \nu^i(x)\nu^j(y) g^{i,j}(x,y)) dS_x dS_y. \end{aligned}$$

5. Perspectives. The proofs of the lemmas and theorems of §3 and 4 are given in Dikow (1986). An application of these ideas to problems in rectangular domains is discussed in Dikow (1987); there Fourier-series are used to simplify the numerical calculations.

It should be pointed out here that the approach presented in this paper works on a macro-scale. A very important open question related to porous media flow is how to derive Darcy's law from more basic physical assumptions in a medium with stochastic geometry on a micro-scale. Relevant works in this direction are the use of homogenization for periodic structures, see Tartar (1980), and flow outside of small balls which are randomly positioned, see Rubinstein (1986).

Another aspect is to apply the technique presented here to miscible displacement problems and to deal with hydrodynamic dispersion. There are papers on this subject, see, e.g., Dagan (1984), Gelhar (1984), and Tompson/Gray (1986). Their results have still to be put onto a sound mathematical basis.

Acknowledgement. The authors thank Kaye A. Smith for her careful typing of the manuscript

REFERENCES

[1] A.A. BAKR, *Stochastic analysis of the effects of spatial variations of hydraulic conductivity on groundwater flow.*, Thesis, New Mexico Institute of Mining and Technology, Socorro, New Mexico (1976).

[2] A.A. BAKR, L.W. GELHAR, A.L. GUTJAHR, J.R. MACMILLAN, *Stochastic analysis of spatial variability in subsurface flows, 1, comparison of one-and three-dimensional flows.*, Water Resources Research, 14 (1978), pp. 263–271.

[3] G. DAGAN, *Solute transport in heterogeneous porous formations*, Journal of Fluid Mechanics, 145 (1984), pp. 151–177.

[4] J.P. DELHOMME, *Kriging in the hydro sciences*, Advances in Water Resources, 1 (1978), pp. 251–266.

[5] E. DIKOW, *Einfluss von räumlicher Variabilität auf Strömungen durch poröse Medien*, Thesis, Universität der Bundeswehr München (1986).

[6] E. DIKOW, *Stochastic analysis of groundwater flow in a bounded domain by spectral methods*, To appear.

[7] L.W. GELHAR, *Stochastic analysis of flow in heterogeneous porous media In: J. Bear,, M.Y. Corapcioglu (Eds.); Fundamentals of Transport Phenomena in Porous Media*, Dordrecht (1984).

[8] G.C. PAPANICOLAOU, S.R.S. VARADHAN, *Boundary value problems with rapidly oscillating random coefficients.*, Colloquia mathematica societatis János Bolyai. North Holland, Amsterdam (1982), pp. 835–873.

[9] J. RUBINSTEIN, *On the macroscopic description of slow viscous flow past a random array of spheres*, Journal of Statistical Physics, 44 (1986), pp. 849–863.

[10] L. TARTAR, *Incompressible fluid flow in a porous medium - convergence of the homogenization process.*, In E. Sanchez-Palencia (Ed.); *Non-homogeneous Media and Vibration Theory*, Lecture Notes in Physics, Springer, Berlin, 127 (1980), pp. 368–377.

[11] A. TOMPSON, W. GRAY, *A second-order approach for the modeling of dispersive transport in porous media*, Water Resources Research, 22 (1986), pp. 591–614.

[12] Y.E. WARREN, H.S. PRICE, *Flow in heterogeneous porous media*, Society of Petroleum Engineers Journal, 1 (1961), pp. 153–169.

NUMERICAL SIMULATION OF IMMISCIBLE FLOW IN POROUS MEDIA BASED ON COMBINING THE METHOD OF CHARACTERISTICS WITH MIXED FINITE ELEMENT PROCEDURES

Jim Douglas, Jr.

Department of Mathematics, University of Chicago, Chicago, IL 60637;
Institute for Mathematics and its Applications, University of Minnesota, Minneapolis, MN 55455

Yuan Yirang

University of Shandong, Jinan, Shandong, People's Republic of China;
Institute for Mathematics and its Applications, University of Minnesota, Minneapolis, MN 55455

Abstract

Two-phase, incompressible, immiscible flow in a porous medium is governed by a system of nonlinear partial differential equations with the dependent variables chosen to represent a pressure and a saturation. The pressure and the associated fluid velocity will be approximated by a mixed finite element method, and the saturation by a finite element method based on a variant of the method of characteristics. Two such schemes are presented and analyzed for convergence for flow in a bounded domain.

1. Introduction

The immiscible displacement of one incompressible fluid by another in a porous medium is described by a system of two second order partial differential equations [4], [14]. One, called the pressure equation, is formally of elliptic type, and the other, the saturation equation, is formally parabolic. The mathematical model of the flow is an initial-boundary problem for this system. The form of the coefficients in the saturation equation is such that a fluid velocity ("Darcy velocity") is required from the pressure equation, but not the pressure itself. As a consequence, it is appropriate to choose a numerical method for the pressure equation that approximates this fluid velocity directly, and a mixed finite element method will be selected. The saturation equation is easily seen from physical considerations to be convection-dominated for modelling flow in domains the size of petroleum reservoirs; thus, a numerical method designed to treat this type of parabolic problem should be chosen, and a method based on handling the transport by a modification of the method of characteristics [10] will be picked for the saturation equation. We shall consider two such schemes in this paper.

A short history of the development of the techniques to be studied herein is as follows. Finite difference and finite element versions of the method to be applied to the saturation equation were considered by Douglas and Russell [10] for a scalar parabolic problem. Russell [16], [17] applied the technique to the concentration equation in a miscible displacement problem for a periodic flow, with the corresponding pressure equation being treated by a standard Galerkin method. A series of papers has been written by various combinations of Douglas, Ewing, Russell, and Wheeler [5], [7], [8], [11], [12], [16], [17], [18] on the periodic miscible displacement problem, with the result that a method very similar to the ones to be considered here is thought to be very effective. A finite difference version of the method has been applied by Douglas [6] to the immiscible problem in a periodic setting.

The mathematical model of the immiscible displacement process will be presented in Section 2. In Section 3 the two numerical schemes will be described. Some projections that will be used in the convergence analysis will be introduced in Section 4, and some technical lemmas will be derived. The two schemes will be analyzed in Sections 5 and 6. Throughout, the symbols C and ϵ will denote, respectively, a generic constant and a generic small constant. Lebesgue and Sobolev spaces will be used with their standard norms.

2. The Mathematical Model

Let Ω be a bounded domain in R^2. The equations describing two-phase, incompressible, immiscible displacement in a gravity-free environment, as given in [14], are

$$\partial(\phi s_o)/\partial t - \nabla \cdot (kk_{ro}(s)\mu_o^{-1}\nabla p_o) = q_o, \ x\epsilon\Omega, \ t\epsilon J, \tag{2.1a}$$

$$\partial(\phi s_w)/\partial t - \nabla \cdot (kk_{rw}(s)\mu_w^{-1}\nabla p_w) = q_w, \ x\epsilon\Omega, \ t\epsilon J, \tag{2.1b}$$

where the subscripts "o" and "w" refer to oil and water, respectively, and s_i denotes the saturation, p_i the pressure, k_{ri} the relative permeability, μ_i the viscosity, and q_i the external volumetric flow rate, each with respect to the ith phase, and J the time interval $[0,T]$; the functions $\phi(x)$ and $k(x)$ are the porosity and absolute permeability, respectively, of the medium. The water and the oil will be assumed to fill the void space in the medium:

$$s_o + s_w = 1. \tag{2.2}$$

Thus, one saturation can be eliminated trivially; let $s = s_0 = 1-s_w$. The pressures in the two phases are related through the capillary pressure $p_c = p_0 - p_w$; we shall assume, as is normal, that p_c is a function of saturation, $p_c = p_c(s)$. Let $\lambda(s) = k_{ro}\mu_o^{-1} + k_{rw}\mu_w^{-1}$ represent a total mobility of the two-phase fluid, and define relative mobilities by $\lambda_i(s) = k_{ri}(s)/\mu_i\lambda(s)$, $i = o,w$.

We prefer to treat the problem in terms of Chavent's global pressure [4],

$$p = \{p_0 + p_w + \int_0^{p_c} [\lambda_0(p_c^{-1}(\xi))-\lambda_w(p_c^{-1}(\xi))]d\xi\}/2. \tag{2.3}$$

Set $q = q_0 + q_w$ and $u = -k(x)\lambda(s)\nabla p$. Then, it follows that

$$u + k(x)\lambda(s)\nabla p = 0, \quad \nabla \cdot u = q, \tag{2.4a}$$

$$\phi \partial s/\partial t + \lambda_0'(s)u \cdot \nabla s - \nabla \cdot (k(\lambda\lambda_0\lambda_w p_c')(s)\nabla s) = \begin{cases} -\lambda_0 q, & q > 0, \\ \\ 0, & q < 0. \end{cases} \tag{2.4b}$$

We suppose the periphery of the reservoir to be impermeable, so that

$$u \cdot \nu = 0, \quad x \in \partial\Omega, \; t \in J, \tag{2.5}$$

where ν is the outer normal to $\partial\Omega$. Compatibility to the incompressibility of the fluids requires that

$$\int_\Omega q(x,t)dx = 0, \quad t \in J. \tag{2.6}$$

The boundary condition (2.5) carries over to the condition

$$\partial s/\partial \nu = 0, \quad x \in \partial\Omega, \; t \in J, \tag{2.7}$$

for the saturation equation. Finally, it is necessary to specify the initial saturation

$$s(x,0) = s^0(x), \quad x \in \Omega. \tag{2.8}$$

3. Formulation of the Approximation Schemes

Let

$$d(x,s) = k(x)(\lambda\lambda_0\lambda_w p_c')(s).$$

Denote the intersection of the supports of λ_0 and λ_w by $[s_{o,res}, 1-s_{cw}]$, and assume that s lies strictly between the end points of this interval; i. e., that s ϵ $[s_{o,res}+\delta, 1-s_{cw}-\delta]$. Then, there exists a positive constant d_0 such that

$$d_0 \leq d(x,s) , \quad x\epsilon\Omega. \tag{3.1}$$

Let $b(s) = \lambda_0'(s)$ and $a(x,s) = k(x)\lambda(s)$. It is a consequence of the nature of the relative permeability functions that there exists a_0 such that

$$0 < a_0 < a(x,s). \tag{3.2}$$

We generalize the right-hand side of the saturation equation slightly and write the equation in the form

$$\phi\partial s/\partial t + b(s)u\cdot\nabla s - \nabla\cdot(d(x,s)\nabla s) = g(x,t,s), \qquad x\epsilon\Omega, \, t\epsilon J. \tag{3.3}$$

The flow is essentially along the characteristic associated with the transport $\phi s_t + b(s)u\cdot\nabla s$; consequently, it is appropriate to introduce differentiation in this characteristic direction. Let

$$\psi(x,s,u) = [\phi(x)^2 + b(s)^2|u|^2]^{1/2}, \tag{3.4a}$$
$$\partial/\partial\tau = \psi^{-1}\{\phi\partial/\partial t + b(s)u\cdot\nabla\}, \tag{3.4b}$$

and note that the direction τ is dependent on space and the saturation and fluid velocity, which vary in space and time. It follows easily that the saturation equation can be written in the form

$$\psi\partial s/\partial\tau - \nabla\cdot(d(s)\nabla s) = g(s), \, x\epsilon\Omega, \, t\epsilon J. \tag{3.5}$$

We shall seek a solution of (3.5) by seeking a map $s:J\to H^1(\Omega)$ such that

$$(\psi\partial s/\partial\tau,z) + (d(s)\nabla s,\nabla z) = (g(s),z), \qquad z\epsilon H^1(\Omega), \, t\epsilon J. \tag{3.6}$$

with $s(x,0) = s^0(x)$.

Since the pressure appears in the saturation equation only through its velocity field u, it is appropriate to rewrite the pressure equation in the form of a first order system:

$$u + a(s)\nabla p = 0, \quad x\epsilon\Omega, \, t\epsilon J, \tag{3.7a}$$
$$\nabla\cdot u = q, \quad x\epsilon\Omega, \, t\epsilon J. \tag{3.7b}$$

In order to formulate (3.6) as a saddle point problem, let $H(div,\Omega)$ consist of vectors $v\epsilon L^2(\Omega)^2$ such that $\nabla\cdot v\epsilon L^2(\Omega)$, and set

$$V = \{v \in H(div,\Omega): v \cdot \nu = 0 \text{ on } \partial\Omega\},$$
$$W = L^2(\Omega)/\{w \equiv \text{constant on } \Omega\};$$

the reduction of $L^2(\Omega)$ by constants removes the ambiguity in the pressure coming from the Neumann boundary condition. Define the bilinear forms

$$A(\theta;u,v) = \sum_{i=1}^{2} (a(\theta)^{-1}u_i,v_i), \quad B(u,w) = -(\nabla \cdot u,w)$$

for u and v in V, w in W, and θ in $L^\infty(\Omega)$. Then, solving the pressure equation is equivalent to finding $\{u,p\} \in V \times W$ for $t \in J$ such that

$$A(S;u,v) + B(v,p) = 0, \qquad v \in V, \qquad (3.8a)$$
$$B(u,w) = -(q,w), \qquad w \in W. \qquad (3.8b)$$

Let us turn to the numerical solution of the problem. Let $h = (h_s, h_p)$, with h_s and h_p being positive and in general different. We shall employ a mixed finite element method for the approximation of the pressure and the Darcy velocity. Let $\tilde{V}_h \times \tilde{W}_h \subset H(div,\Omega) \times L^2(\Omega)$ be a mixed finite element space [2], [3], [15] associated with a quasiregular triangulation or quadrilateralization of Ω such that the elements have diameters bounded by h_p, and set

$$V_h = \{v \in \tilde{V}_h : v \cdot \nu = 0 \text{ on } \partial\Omega\}, \quad W_h = \tilde{W}_h/\{w \equiv \text{constant on } \Omega\}.$$

Denote the norm in $H^{j,r}(\Omega)$ or $H^{j,r}(\Omega)^2$ by $\|\cdot\|_{j,r}$, $0 \leq j$ and $1 \leq r \leq \infty$, and omit the Lebesgue index r for r = 2. Assume the approximation properties, for $v \in V$ and $w \in W$,

$$\inf\{\|v - v_h\|_0 : v_h \in V_h\} \leq C\|v\|_{k+1}h_p^{k+1},$$
$$\inf\{\|\nabla \cdot (v-v_h)\|_0 : v_h \in V_h\} \leq C\|\nabla \cdot v\|_j h_p^j,$$
$$\inf\{\|w - w_h\|_0 : w_h \in W_h\} \leq C\|w\|_j h_p^j,$$

where j is equal to either k or k+1, depending on the selection of the type of the mixed finite element space. Also, let $M_h \subset H^1(\Omega)$ be associated with another quasiregular decomposition of Ω and assume for it the approximation property for $z \in H^1(\Omega)$

$$\inf\{\|z-z_h\|_0 + h_s\|z-z_h\|_1 : z_h \in M_h\} \leq C\|z\|_\ell h_s^\ell.$$

It is necessary to define a bit of notation in order to present the numerical schemes. The time steps for the pressure and saturation variables will be allowed to be different, and it will be convenient to take the first pressure time step smaller than the subsequent ones. Let Δt_s denote the saturation time step, $\Delta t_p{}^0$ the first pressure step, and Δt_p the subsequent ones. Assume that $\Delta t_p/\Delta t_s = j^1$ and $\Delta t_p{}^0/\Delta t_s = j^0$. Set $t^n = n\Delta t_s$, $t_m = \Delta t_p{}^0 + (m-1)\Delta t_p$, $\beta^n = \beta(t^n)$, and $\beta_m = \beta(t_m)$. Let $J^n = (t^{n-1}, t^n)$ and $J_m = (t_{m-1}, t_m)$. Define extrapolation operators as follows:

$$E_1\beta^n = \begin{cases} \beta_0, & t^n \leq t_1, \\ (1+\gamma/j^0)\beta_1 - \gamma\beta_0/j^0, & t_1 < t^n \leq t_2, \quad t^n = t_1 + \gamma\Delta t_s, \\ (1+\gamma/j^1)\beta_m - \gamma\beta_{m-1}/j^1, & t_m < t^n \leq t_{m+1}, \, t^n = t_m + \gamma\Delta t_s; \end{cases}$$

$$E_2\beta^n = \begin{cases} \beta^0, & n = 1, \\ 2\beta^{n-1} - \beta^{n-2}, & n \geq 2; \end{cases}$$

$$\beta_{m+1/2} = \begin{cases} \beta_0, & m = 0, \\ (1+\Delta t_p/2\Delta t_p{}^0)\beta_1 - \Delta t_p\beta_0/2\Delta t_p{}^0, & m = 1, \\ (3\beta_m - \beta_{m-1})/2, & m \geq 2. \end{cases}$$

We shall denote the approximate solution for the saturation by $S^n \epsilon M_h$ at the time level t^n and by S_m at the time level t_m (note that the pressure time levels t_m form a subset of the saturation time levels t^n); the approximate solution for the Darcy velocity and the pressure will be denoted by $\{U_m, P_m\} \epsilon V_h \times W_h$ at t_m. In order to initiate the waterflood, it is necessary to approximate the initial condition s^0 in the space M_h; this we do by an elliptic projection by finding $S^0 = \hat{s}^0 \epsilon M_h$ such that

$$(d(s^0)\nabla\hat{s}^0, \nabla z) + (\hat{s}^0, z) = (d(s^0)\nabla s^0, \nabla z) + (s^0, z), \quad z \epsilon M_h. \tag{3.9}$$

For both schemes that will be defined below the mixed method to be used at the time levels t_m to approximate the pressure equation will take the same form. Assume S_m known, and then find $\{U_m, P_m\} \epsilon V_h \times W_h$, $m \geq 0$, as the solution of

$$A(S_m; U_m, v) + B(v, P_m) = 0, \qquad v \epsilon V_h, \tag{3.10a}$$
$$B(U_m, w) = -(q_m, w), \qquad w \epsilon W_h. \tag{3.10b}$$

The modified method of characteristics procedure [10] for approximating the directional derivative $\psi \partial s/\partial \tau$ is based on backward differencing this term:

$$(\psi \partial s/\partial \tau)^n(x) \approx \phi(x)\{s^n(x) - s^{n-1}(\hat{x}^n)\}/\Delta t_s,$$

where

$$\hat{x}^n = \tilde{x}(x,t^n) = x - \phi(x)^{-1}b(s^n(x))u^n(x)\Delta t_s.$$

Several questions must be addressed in order to implement this concept for our problem. Clearly, s^n and u^n must be replaced by their approximations; however, these values are not known at the time level t^n and it is necessary to use the extrapolations defined above. Set

$$\hat{x}^n = \hat{x}^n(x) = x - \phi(x)^{-1}b(E_2S^n(x))E_1U^n(x)\Delta t_s. \tag{3.11}$$

If $\hat{x}^n(x)\epsilon\Omega$, set $\hat{S}^{n-1}(x) = S^{n-1}(\hat{x}^n(x))$. If not, $\hat{S}^{n-1}$ can be defined by reflection. Let $\hat{y}^n = \hat{y}^n(x)$ be the reflection of $\hat{x}^n(x)$ across $\partial\Omega$ and assume that $\hat{y}^n(x)$ lies in Ω whenever $\hat{x}^n(x)$ fails to do so; if it does not, in a practical reservoir simulation the time step is too large and should be reduced. We shall always assume that $\hat{y}^n(x)$ does fall in Ω when it is required. Set $\hat{S}^{n-1}(x) = S^{n-1}(\hat{y}^n(x))$ when $\hat{x}^n(x)$ lies outside Ω. Note that this definition of $\hat{S}^{n-1}$ would require modification if the homogeneous Neumann boundary condition were replaced by an inhomogeneous one. This definition represents extension that is bounded as a map of $H^3(\Omega)$ to H^3 of a neighborhood of Ω, so its usefulness is limited to linear or quadratic elements; for higher degree elements, use a multipoint extension such as described by Lions and Magenes [13].

We are now ready to define the two procedures for approximating the saturation at time t^n; they differ in the way that the $(d(s)\nabla s,\nabla z)$-term is handled. **Scheme 1** for the waterflood problem consists of the use of (3.10) in combination with the determining of $S^n\epsilon M_h$ as the solution of the equations

$$(\phi\{S^n-\hat{S}^{n-1}\}/\Delta t_s,z) + (d(E_2S^n)\nabla S^n,\nabla z) = (g(E_2S^n),z), \quad z\epsilon M_h, \tag{3.12}$$

for $n\geq1$; S^0 has been defined by (3.9). **Scheme 2** replaces (3.12) by the following equations for $n\geq1$:

$$(\phi\{S^n-\hat{S}^{n-1}\}/\Delta t_s,z) + (d(S_{m+1/2})\nabla S^n,\nabla z) \tag{3.13}$$
$$= (\{d(S_{m+1/2})-d(E_2S^n)\}\nabla E_2S^n,\nabla z) + (g(E_2S^n),z), \quad z\epsilon M_h.$$

For either scheme the solution is computed in the following order: S^0; $\{U_0,P_0\}$; S^1, ... , $S^{j^0}=S_1$; $\{U_1,P_1\}$; S^{j^0+1}, ... , $S^{j^0+j^1}=S_2$; etc.

The computational work associated with Scheme 2 should be less than that required by Scheme 1. If a direct method is used to solve the linear algebraic equations (and note that the algebraic equations are linear) generated at each step, then a factorization must be performed each pressure step for Scheme 2, while one must be made each saturation step for Scheme 1. If conjugate gradient iteration is employed to approximate the solution of (3.12) or (3.13), a new preconditioner in the form of a partial factorization would have to be found less often for Scheme 2 than for Scheme 1.

4. Some Technical Results

The analysis of the convergence of the two schemes will be given under the assumption that the imposed external flow is smoothly distributed. As in all of the discussions of finite element methods for the miscible displacement problem, we follow Ewing and Wheeler [12] and introduce two projections. First, we project the solution $\{u,p\}$ of the pressure equation (2.4a) into the space $V_h \times W_h$; let $\{U^*, P^*\}: J \to V_h \times W_h$ be determined as the solution of

$$A(s;U^*,v) \;+\; B(v,P^*) \qquad\qquad = 0, \qquad v \epsilon V_h, \qquad\qquad (4.1a)$$

$$B(U^*,w) \qquad\qquad\qquad = -(q,w), \quad w \epsilon W_h. \qquad\qquad (4.1b)$$

Optimal order estimates for $u-U^*$ and $p-P^*$ are known for any of the choices of $V_h \times W_h$ mentioned above. In particular, the estimate

$$\|u - U^*\|_0 \leq C\|u\|_{k+1} h_p^{k+1}, \quad t\epsilon J, \qquad\qquad (4.2)$$

is valid for any of these choices [2], [3], [15]. A non-optimal $L^\infty(\Omega)$ estimate can be obtained from this estimate and the quasiregularity assumption on the decomposition; in some cases optimal order L^∞ estimates exist, but they are not needed in our argument.

Next, let $S^*: J \to M_h$ be given by

$$(d(s)\nabla(s-S^*),\nabla z) + (s-S^*,z) = 0, \qquad z\epsilon M_h; \qquad\qquad (4.3)$$

thus,

$$\|s-S^*\|_0 + h_s\|s-S^*\|_1 \leq C\|s\|_\ell h_s^\ell, \quad t\epsilon J. \qquad\qquad (4.4)$$

Optimal order L^∞ estimates exist for $\ell\geq 3$, and the slightly non-optimal estimate of the form $O(\|s\|_{2,\infty} h_s^2 \log(1/h_s))$ holds for $\ell=2$.

The following derivation of estimates for $U-U^*$ and $P-P^*$ is the analogue of a lemma due to Ewing and Wheeler [12]; such a lemma has played a vital role in the analyses of numerical methods for miscible and immiscible problems since. Note that, at time levels t_m,

$$A(S;U-U^*,v) + B(v,P-P^*) = A(s;U^*,v) - A(S;U^*,v), \quad v \epsilon V_h, \tag{4.5a}$$

$$B(U-U^*,w) = 0, \quad w \epsilon W_h, \tag{4.5b}$$

so that the stability theory of Brezzi [1], which is applicable for all of the spaces being admitted for $V_h \times W_h$, implies that

$$\|U - U^*\|_{H(\text{div},\Omega)} + \|P - P^*\|_0 \leq C\|U^*\|_{0,\infty}\|s - S\|_0 \tag{4.6}$$
$$\leq C\|s - S\|_0,$$

since the L^∞ estimates for $u-U^*$ show that U^* is bounded in L^∞.

5. Convergence Analysis for Scheme 1

Let $\xi = S-S^*$ and $\eta = s-S^*$, and let f^{n-1} denote the evaluation of f at the point $(\check{x}^n(x),t^{n-1})$ or its evaluation via reflection, which can be evaluation at $(\check{y}^n(x),t^{n-1})$ or a more complicated one for higher order elements as mentioned above. The reflection process being assumed to extend $H^\ell(\Omega)$ continuously to H^ℓ of a neighborhood of Ω, we can treat $\check{s}^{n-1}$ as if it were evaluated on Ω. Then, a calculation using (3.3), (3.12), and (4.3) leads to the error equation

$$(\phi\{\xi^n-\check{\xi}^{n-1}\}/\Delta t_s,z) + (d(E_2 S^n)\nabla \xi^n, \nabla z) \tag{5.1}$$
$$= (g(E_2 S^n)-g(s^n), z) - (\eta^n, z) - ([d(E_2 S^n)-d(s^n)]\nabla S^{*n}, \nabla z)$$
$$+ ([\phi \partial s^n/\partial t + b(s^n)u^n \cdot \nabla s^n] - \phi\{s^n - \check{s}^{n-1}\}/\Delta t_s, z) + (\phi\{\eta^n - \hat{\eta}^{n-1}\}/\Delta t_s, z)$$
$$= T_1(z) + T_2(z) + T_3(z) + T_4(z) + T_5(z).$$

It would be very desirable to analyze the error using the relation above, which expresses the error in the transport in the (approximate) direction of flow; however, for lack of known argument we must reintroduce the time direction in place of the characteristic direction. Thus, we consider the equation

$$(\phi\{\xi^n-\xi^{n-1}\}/\Delta t_s,z) + (d(E_2 S^n)\nabla \xi^n, \nabla z) = T_1(z) + \cdots + T_6(z), \tag{5.2}$$

where

$$T_6(z) = -(\phi\{\xi^{n-1}-\check{\xi}^{n-1}\}/\Delta t_s,z). \tag{5.3}$$

Now, choose the test function $z=\zeta^n$ and sum over n for n = 1,...,j. If

$$Q_i = Q_i{}^j = \sum_{n=1}^{j} T_i(\zeta^n)\Delta t_s,$$

then

$$\tfrac{1}{2}(\phi\zeta^j,\zeta^j) + \sum_{n=1}^{j} (d(E_2S^n)\nabla\zeta^n,\nabla\zeta^n) \leq Q_1 + \cdots + Q_6. \tag{5.4}$$

We shall estimate the Q_i-terms one at a time, beginning with Q_1. Note that

$$g(E_2S^n)-g(s^n) = [g(E_2S^n)-g(E_2s^n)]+[g(E_2s^n)-g(s^n)]$$

$$\leq C\{|\zeta^{n-1}| + |\zeta^{n-2}| + |\eta^{n-1}| + |\eta^{n-2}| + (\Delta t_s)^2\}$$

for $n\geq2$; for n=1 the terms for index n-2 are omitted and the exponent on Δt_s is reduced to one. Thus,

$$|Q_1| \leq C\sum_{n=1}^{j} \|\zeta^n\|_0^2\Delta t_s + C\{h_s^{2\ell} + (\Delta t_s)^3\}, \tag{5.5}$$

where we are assuming a smooth solution of the differential system. Clearly,

$$|Q_2| \leq C\sum_{n=1}^{j} \|\zeta^n\|_0^2\Delta t_s + Ch_s^{2\ell}. \tag{5.6}$$

The term $\nabla S*^n$ is bounded in L^∞ for smooth solutions; hence,

$$|Q_3| \leq \epsilon\sum_{n=1}^{j}\|\nabla\zeta^n\|_0^2\Delta t_s + C\sum_{n=1}^{j}\|\zeta^n\|_0^2\Delta t_s + C\{h_s^{2\ell} + (\Delta t_s)^3\}. \tag{5.7}$$

In the bounding of Q_4, it is convenient to note that

$$\phi\partial s^n/\partial t + b(s^n)u^n\cdot\nabla s^n$$
$$= \phi\partial s^n/\partial t + b(E_2S^n)E_1u^n\cdot\nabla s^n$$
$$+ \{[b(s^n)-b(E_2s^n)] + [b(E_2s^n)-b(E_2S^n)]\}u^n\cdot\nabla s^n$$
$$+ b(E_2S^n)\{[u^n-E_1u^n] + [E_1u^n-E_1U^n]\}\cdot\nabla s^n.$$

Make the induction hypothesis that

$$\|U_m\|_{0,\infty} \leq C_1, \qquad m =0,1,...; \tag{5.8}$$

the constant C_1 can be taken to be any number larger than the corresponding bound on u. Note that the inequality holds for m=0; it will have to be verified later for $m\geq1$. It follows from the assumed smoothness of the solution of the differential system, the boundedness of the reflection operator, and the induction hypothesis (5.8) that

$$\|[\phi \partial s^n/\partial t + b(E_2 S^n)E_1 U^n \cdot \nabla s^n] - \phi(s^n - \hat{s}^{n-1})/\Delta t_s\|_0$$
$$\leq C\|\partial^2 s/\partial \tau(x, E_2 S^n, E_1 U^n)^2\|_{0,\infty} \Delta t_s \leq C\Delta t_s.$$

Next,

$$\|\{[b(s^n) - b(E_2 S^n)] + [b(E_2 s^n) - b(E_2 S^n)]\} u^n \cdot \nabla s^n\|_0$$
$$\leq C\{(\Delta t_s)^2 + \|\xi^{n-1}\|_0 + \|\xi^{n-2}\|_0 + \|\eta^{n-1}\|_0 + \|\eta^{n-2}\|_0\}.$$

Also, for $t^n \geq t_1$,

$$\|b(E_2 S^n)\{(u^n - E_1 u^n) + (E_1 u^n - E_1 U^n)\} \cdot \nabla s^n\|_0$$
$$\leq C\{(\Delta t_p)^2 + \|u_{m-1} - U_{m-1}\|_0 + \|u_{m-2} - U_{m-2}\|_0\}.$$

These bounds must be modified for t^n in the first pressure time step to replace the $(\Delta t_p)^2$-term by Δt_p^0: it then follows from the estimates above and (4.6) that

$$|Q_4| \leq C\{\sum_{n=1}^{j} \|\xi^n\|_0^2 \Delta t_s + \sum_{t_m \leq t^j} \|\xi_m\|_0^2 \Delta t_p\} + C\{(\Delta t_s)^2 + (\Delta t_p^0)^3 + (\Delta t_p)^4 + h_s^{2\ell} + h_p^{2k+2}\}.$$
(5.9)

The terms Q_5 and Q_6 require the most delicate analysis of the six; however, they can be treated very similarly to analogous terms estimated by Russell on pp. 982-983 of [17]. Only small modifications in the argument are needed to take into account the extension of the saturation to a tubular neighborhood of $\Omega \times J$. As in [17], it is necessary to assume that

$$\Delta t_s = o(h_s) \quad \text{and} \quad \Delta t_s = o(h_p) \tag{5.10}$$

and to make the additional induction hypothesis that

$$\sup\{\|S_m\|_{0,\infty}: 0 \leq t_m < t^j\} \leq C_2/h_s. \tag{5.11}$$

Then, it follows from [17] that

$$|Q_5| + |Q_6| \leq \epsilon \sum_{n=1}^{j} \|\nabla \xi^n\|_0^2 \Delta t_s + C \sum_{n=1}^{j} \|\xi^n\|_0^2 \Delta t_s + C\{(\Delta t_s)^2 + h_s^{2\ell} + h_p^{2k+2}\}.$$
(5.12)

Consequently, (5.4)-(5.7), (5.9), and (5.12) lead to the inequality

$$\frac{1}{2}(\phi\xi^j,\xi^j) + (d_0-\epsilon)\sum_{n=1}^{j}\|\nabla\xi^n\|_0^2\Delta t_s \tag{5.13}$$

$$\leq C[\sum_{n=1}^{j}\|\xi^n\|_0^2\Delta t_s + \sum_{t_m\leq t^j}\|\xi_m\|_0^2\Delta t_p] + C\{(\Delta t_s)^2+(\Delta t_p^0)^3+(\Delta t_p)^4+h_s^{2\ell}+h_p^{2k+2}\}.$$

The remainder of the argument to complete the analysis of Scheme 1, including the verification of the validity of the induction hypotheses, is essentially identical to those given in [17]. It then follows that

$$\max_j\|\xi^j\|_0 \leq C\{\Delta t_s + (\Delta t_p^0)^{3/2} + (\Delta t_p)^2 + h_s^\ell + h_p^{k+1}\}. \tag{5.14}$$

Thus,

$$\max_j\|s^n-S^n\|_0 + \max_j\|u^n-U^n\|_0 \leq C\{\Delta t_s+(\Delta t_p^0)^{3/2}+(\Delta t_p)^2+h_s^\ell+h_p^{k+1}\}. \tag{5.15}$$

Theorem 1. The error estimate (5.15) holds for the solution of Scheme 1, provided that the coefficients and solution of the differential system (2.4) are sufficiently smooth and that the constraints (3.1), (3.2), and (5.10) are valid.

An optimal set of choices for the discretization parameters is given by taking the five quantities on the right-hand side of (5.15) approximately equal.

6. Analysis of Scheme 2

The analysis of Scheme 2 requires the bounding of some extra terms coming from the change in the way the diffusion term is approximated. These terms are of exactly the same form as were treated in [8]; thus, a detailed analysis of Scheme 2 is not necessary. The results of Theorem 1 hold for the error for the approximation given by the second scheme, at least if the mixed method employed for the pressure equation is formally second order correct for the Darcy velocity and the discretization parameters are selected in the manner indicated above for optimality.

References

[1]   F. Brezzi, *On the existence, uniqueness and approximation of saddle-point problems arising from Lagrangian multipliers*, R. A. I. R. O., Anal. Numér. 2 (1974), 129-151.

[2]   F. Brezzi, J. Douglas, Jr., M. Fortin, and L. D. Marini, *Efficient rectangular mixed finite elements in two and three space variables*, to appear.

[3]   F. Brezzi, J. Douglas, Jr., and L. D. Marini, *Two families of mixed finite elements for second order elliptic problems*, Numer. Math. 47 (1985), 217-235.

[4]   G. Chavent and J. Jaffré, Mathematical Models and Finite Elements for Reservoir Simulation, North-Holland, Amsterdam, 1986.

[5]   J. Douglas, Jr., *Numerical methods for the flow of miscible fluids in porous media*, Numerical Methods in Coupled Systems, R. W. Lewis, P. Bettess, and E. Hinton, eds., John Wiley and Sons Ltd, London, 1984.

[6]   J. Douglas, Jr., *Finite difference methods for two-phase incompressible flow in porous media*, SIAM J. Numer. Anal. 20 (1983), 681-696.

[7]   J. Douglas, Jr., *Simulation of miscible displacement in porous media by a modified method of characteristics procedure*, Numerical Analysis, Dundee 1981, Lecture Notes in Mathematics 912, Springer-Verlag, Berlin, 1982.

[8]   J. Douglas, Jr., R. E. Ewing, and M. F. Wheeler, *A time-discretization procedure for a mixed finite element approximation of miscible displacement in porous media*, R. A. I. R. O., Anal. Numér. 17 (1983), 249-265.

[9]   J. Douglas, Jr., and J. E. Roberts, *Global estimates for mixed methods for second order elliptic equations*, Math. Comp. 44 (1985), 39-52.

[10] J. Douglas, Jr., and T. F. Russell, *Numerical methods for convection-dominated diffusion problems based on combining the method of characteristics with finite element or finite difference procedures*, SIAM J. Numer. Anal. 19 (1982), 871-885.

[11] R. E. Ewing and T. F. Russell, *Efficient time-stepping methods for miscible displacement problems in porous media*, SIAM J. Numer. Anal. 19 (1982), 1-67.

[12] R. E. Ewing and M. F. Wheeler, *Galerkin methods for miscible displacement problems in porous media*, SIAM J. Numer. Anal. 17 (1980), 351-365.

[13] J. L. Lions and E. Magenes, Problèmes aux limites non homogènes et applications, vol. 1, Dunod, Paris, 1968.

[14] D. W. Peaceman, Fundamentals of Numerical Reservoir Simulation, Elsevier, Amsterdam, 1977.

[15] P. A. Raviart and J. M. Thomas, *A mixed finite element method for 2nd order elliptic problems*, Mathematical Aspects of the Finite Element Method, Lecture Notes in Mathematics 606, Springer-Verlag, Berlin, 1977.

[16] T. F. Russell, *An incompletely iterated characteristic finite element method for a miscible displacement problem*, Thesis, University of Chicago, 1980.

[17] T. F. Russell, *Time stepping along characteristics with incomplete iteration for a Galerkin approximation of miscible displacement in porous media*, SIAM J. Numer. Anal. 22 (1985), 970-1013.

[18] T. F. Russell and M. F. Wheeler, *Finite element and finite difference methods for continous flows in porous media*, The Mathematics of Reservoir Simulation, Frontiers in Applied Mathematics 1, Society for Industrial and Applied Mathematics, Philadelphia, 1983.

ADAPTIVE GRID-REFINEMENT TECHNIQUES FOR TREATING SINGULARITIES, HETEROGENEITIES, AND DISPERSION

Richard E. Ewing*

Departments of Mathematics,
Petroleum Engineering, and Chemical Engineering
University of Wyoming
Laramie, Wyoming 82071

1. Introduction

Simulation of complex fluid flow processes in heterogeneous porous media is an important aspect of petroleum reservoir engineering. Due to the cost of implementation and flow complexities of enhanced recovery procedures, simulation of these methods is especially important. The purpose of this paper is to address techniques for treating the difficulties caused in simulation by heterogeneities in the reservoir, diffusion and dispersion, viscous fingering, and other highly-localized flow properties such as flow around wells. The common difficulty in modeling these various flow regimes is the inability to treat highly varying length scales in large-scale reservoir simulation.

The flow processes of both miscible and immiscible displacement involve convection, or physical transport, of the fluids through a heterogeneous porous medium. The equations used to simulate this flow at a macroscopic level are variations of Darcy's law. Darcy's law has been derived [75,76] via a volume averaging of the Navier-Stokes equations, which govern flow through the porous medium at a microscopic or pore-volume level. The length scale for Navier-Stokes flow (10^{-4}-10^{-3} meters) is very different from the scale required by normal reservoir simulation (10^2-10^3 meters). Reservoirs themselves have scales of heterogeneity ranging from pore-level to field scale. In the standard averaging process for Darcy's law, many important physical phenomena which may eventually govern the macroscopic flow are lost. The further averaging of reservoir and fluid properties necessary to use grid blocks of the size of 10^2-10^3 meters in field-scale simulators further complicates the modeling process. We will discuss certain techniques to try to address these scaling problems.

Many of the enhanced recovery processes are characterized by the chemical and physical interaction of the fluids. Therefore diffusion and dispersion are often critical to the flow processes and must be understood and modeled. Molecular diffusion is typically quite small. However, dispersion, or the mechanical mixing caused by velocity variations and flow through heterogeneous rock, can be extremely important and should be incorporated in some way in our models.

Since the mixing and velocity variations are influenced at all relevant length scales by the heterogeneous properties of the reservoir, much work must be done in volume averaging of terms like porosity and permeability. Statistical methods have shown promise in this area [2,36]. The author is currently considering some statistical techniques to obtain effective permeability tensors for large-scale models of flow through highly fractured media. However, if the fractures or field-scale heterogeneities are sufficiently large, and can be identified, they should be incorporated in the model via special gridding and high permeability variations. The adaptive local grid-refinement techniques presented in this paper can be very valuable in these applications.

* This research was supported in part by U.S. Army Research Office Contract No. DAAG29-84-K-0002, by U.S. Air Force Office of Scientific Research Contract No. AFOSR-85-0117, and by National Science Foundation Grant No. DMS-85-04360.

Section 2 presents the model equations used in field-scale simulators to describe miscible and immiscible displacement and relates these process models with regard to the diffusion/dispersion concepts. In Section 3, the importance of hydrodynamic dispersion and viscous fingering are discussed. Section 4 then describes various adaptive grid refinement techniques which offer promise in resolving important local phenomena in large-scale simulators in an efficient manner.

2. Equations for Miscible and Immiscible Displacement

The miscible displacement of one incompressible fluid by another, completely miscible with the first, in a horizontal porous reservoir $\Omega \subset \mathbb{R}^2$ over a time period $J = [T_0, T_1]$, is given by

$$-\nabla \cdot \left(\frac{k}{\mu} \nabla p\right) \equiv \nabla \cdot \mathbf{u} = q , \quad x \in \Omega , \quad t \in J , \tag{2.1}$$

$$\phi \frac{\partial c}{\partial t} - \nabla \cdot (\mathbf{D} \nabla c - \mathbf{u} c) = q \tilde{c} , \quad x \in \Omega , \quad t \in J , \tag{2.2}$$

where p and $\mathbf{u}$ are the pressure and Darcy velocity of the fluid mixture, ϕ and k are the porosity and the permeability of the medium, μ is the concentration-dependent viscosity of the mixture, c is the concentration of the invading fluid, q is the external rate of flow, and $\tilde{c}$ is the inlet or outlet concentration. The form of the diffusion-dispersion tensor $\mathbf{D}$ is given by

$$\mathbf{D} = \phi(x) \left\{d_m I + |\mathbf{u}| \left(d_l E(\mathbf{u}) + d_t E^\perp(\mathbf{u})\right)\right\} \tag{2.3}$$

where

$$E_{ij}(\mathbf{u}) = \frac{\mathbf{u}_i \mathbf{u}_j}{|\mathbf{u}|^2} , \tag{2.4}$$

$E^\perp = I - E$, d_m is the molecular diffusion coefficient and d_l and d_t are the longitudinal and transverse dispersion coefficients, respectively. In general, $d_l \approx 10 d_t$. Also, the viscosity μ in Equation (2.1) is assumed to be determined by some mixing rule, such as

$$\mu(c) = \mu_o \left[1 + c \left(\left(\frac{\mu_o}{\mu_s}\right)^{\frac{1}{4}} - 1\right)\right]^{-4} \tag{2.5}$$

where μ_o is the viscosity of the resident fluid and μ_s is the viscosity of the invading fluid. In addition to Equations (2.1) and (2.2), initial and no flow boundary conditions are specified. The flow at injection and production wells is modeled in Equations (2.1) and (2.2) via point sources and sinks.

The equations describing two phase, immiscible, incompressible displacement in a horizontal porous medium are given by

$$\phi \frac{\partial S_w}{\partial t} - \nabla \cdot \left(k \frac{k_{rw}}{\mu_w} \nabla p_w\right) = q_w , \quad x \in \Omega , \quad t \in J , \tag{2.6}$$

$$\phi \frac{\partial S_o}{\partial t} - \nabla \cdot \left(k \frac{k_{ro}}{\mu_o} \nabla p_o\right) = q_o , \quad x \in \Omega , \quad t \in J , \tag{2.7}$$

where the subscripts w and o refer to water and oil respectively, S_i is the saturation, p_i is the pressure, k_{ri} is the relative permeability, μ_i is the viscosity, and q_i is the external flow rates, each with respect to the ith phase.

The saturation constraint is given by

$$S_w + S_o = 1 . \tag{2.8}$$

From Equation (2.8), we see that one of the saturations can be eliminated. Let $S = S_w = 1 - S_o$. The pressure between the two phases is described by the capillary pressure

$$p_c(S) = p_o - p_w . \tag{2.9}$$

Note that $\dfrac{dp_c}{dS} \leq 0$.

Although formally, the equations presented in (2.1) and (2.2) seem quite different from those in (2.6) and (2.7), the latter system may be rearranged in a form which very closely resembles the former system. In order to use the same basic simulator in our sample computations to treat both miscible and immiscible displacement, we will follow the ideas of Chavent [13], and briefly discuss a miscible/immiscible flow analogy.

Let

$$\lambda(S) = \frac{k_{ro}}{\mu_o} + \frac{k_{rw}}{\mu_w} \tag{2.10}$$

denote a total mobility of the two-phase fluid and define relative mobilities by

$$\lambda_i(S) = \frac{k_{ri}(S)}{\mu_i \lambda(S)} , \quad i = w, 0 . \tag{2.11}$$

We will try to rearrange the immiscible system using Chavent's pressure variable [13] given by

$$p = \frac{p_o + p_w}{2} + \frac{1}{2} \int_0^{p_c} \left(\overline{\lambda}_o(\eta) - \overline{\lambda}_w(\eta) \right) d\eta . \tag{2.12}$$

Note that $\overline{\lambda}_w + \overline{\lambda}_o = 1$.

Adding Equations (2.6) and (2.7) and performing some simple calculations, we obtain:

$$-\nabla \cdot (k\lambda(S)\nabla p) = q_w + q_0 = q_t , \tag{2.13}$$

$$v_t = -k\lambda(S)\nabla p . \tag{2.14}$$

Taking the difference of Equations (2.6) and (2.7), the following equation is obtained:

$$\phi \frac{\partial S}{\partial t} + \nabla \cdot \left(k\lambda(S)\overline{\lambda}_o \frac{dp_c}{dS} \nabla S \right) + \nabla \cdot \left(\overline{\lambda}_w v_t \right) = w_w . \tag{2.15}$$

The equations presented above describe miscible and immiscible flow in porous media. They can be used to simulate various production strategies in an attempt to understand and hopefully optimize hydrocarbon recovery. However, in order to use these equations effectively, parameters that describe the rock and fluid properties for the particular reservoir application must be input. The relative permeabilities which are nonlinear functions of water saturation can be estimated via laboratory experiments using reservoir cores and resident fluids. Similarly fluid viscosities are relatively easy to obtain. However the permeability tensor k, the porosity ϕ, the capillary pressure curve $p_c(S)$, and the diffusion and dispersion coefficients are effective values that must be obtained from local properties via scaling techniques. In addition, the inaccessibility of the reservoir to measurement of even the local properties increases the difficulties. See [21,27,49] and the references contained therein for a survey of parameter estimation and history matching techniques which have been applied. Section 3 below then describes the modification of estimated viscosity, permeability, or dispersion coefficients to incorporate the important effects of viscous fingering. Much work is needed in this research area.

The accurate modeling of permeabilities which may vary considerably from grid block to grid block caused more difficulties in reservoir simulation. For flow parallel to permeability discontinuities an arithmetic averaging of permeabilities is appropriate. However, for flow across discontinuities, a harmonic averaging of permeabilities is required [1]. With high velocity variations, this averaging prescription is not easy to maintain. However the mixed finite methods presented in [19,20,28-30,32,33,69] insure these averaging rules and to simultaneously approximate the Darcy velocity directly. These techniques have been extremely successful even with fairly large grid blocks. Their combination with local grid refinement around transmissibility changes holds even more potential.

3. Dispersion and Viscous Fingering

The effects of dispersion in various flow processes have been discussed extensively in the literature [11,47,61,62,77,78,83]. Russell and Wheeler [69] and Young [86] have given excellent surveys of the influence of dispersion and attempts to incorporate it in present reservoir simulators. Various terms which effect the length of the dispersive mixing zone include viscosity and velocity variations and reservoir heterogeneity. Much work is needed to quantify these effects and to obtain useful effective dispersion coefficients for field-scale simulators. The dispersion tensor (see Equation (2.3)) has strong velocity dependence [11,22,61,69]. The longitudinal dispersion is often an order of magnitude larger than the transverse dispersion. This variation enhances unstable flow regimes induced by viscosity differences and reservoir heterogeneity.

In both miscible and immiscible displacements, the process of pushing a heavy, viscous oil through a heterogeneous porous medium with a lighter, less viscous fluid can be a very unstable one. If the flow rate is sufficiently high, the interface between the resident petroleum and the invading fluid becomes unstable and tends to form long fingers which grow in length toward the production wells, bypassing much of the hydrocarbon. Once a path consisting of the injected fluid has extended from an injection well to a production well, that production well will henceforth produce primarily the injected fluid which flows more easily due to its lower viscosity and higher mobility. The production of petroleum from that well is then greatly reduced. This phenomenon, termed *viscous fingering*, is well known [3,14,41,43,45,46,48,50,51,55-68,70-74,80,81,84,85] and is a serious problem in hydrocarbon recovery. This problem and different techniques for understanding and modeling it were surveyed in [26].

The equations used to describe fluid flow in porous media are obtained by applying various averaging techniques to the Navier-Stokes equations which describe the macroscopic flow properties. The averaging is done over tubes of varying geometry and tortuosity. Clearly, the Navier-Stokes equations do not describe the flow on a macroscopic level. Analysis must be applied to the Navier-Stokes equations to understand the conditions under which small flow perturbations caused by tortuous flow in the porous media will become unstable and will grow into large fingers which affect the flow on a macroscopic level. The understanding of this instability is crucial, both in attempts to stabilize the flow via polymers, etc., and to model the fingering phenomenon when it cannot be controlled. There are three distinct problems associated with the viscous fingering phenomenon. Knowledge of the conditions causing the onset of the instability is one major goal. Then the understanding of the nonlinear effects which cause certain fingers to grow preferentially and coalesce to form larger fingers is essential. The rate of this finger growth must also be determined. Once the growth of the fingers is understood, from a microscopic to a macroscopic level, statistical methods should be used to incorporate the large-scale effects of fingering into the mathematical models used to simulate the field scale process without trying to treat the individual fingers.

The linear stability condition for multiphase flow was obtained by Chuoke [14] and Saffman and Taylor [71,72] for the Hele-Shaw problem. If mixing or hydrodynamic dispersion is allowed, as in miscible displacement, a somewhat different stability analysis was obtained [35,43,46,62,64]. These conditions predict only the onset of the instability and cannot give information about the finger growth.

Peters and Flock [66] have introduced a "wettability number" into the Chuoke analysis and have correlated experimental data with the number. Using a water-wet and an oil-wet number they obtained much information from dimensional analysis and the linear stability analysis. In this case, the dimensional analysis produces the conclusion that the physics is not complete concerning the wettability phenomenon and more understanding and new equations are needed to explain these effects.

An important example of a study of the stability and finger growth problem is a controlled setting, rather than the highly inhomogeneous porous media flow, is the work on the Hele-Shaw cell. Recently, McLean and Saffman [51] have calculated the finger formed in a Hele-Shaw cell with surface tension and found that the finger was unstable. Observations from experiments contradict this conclusion. Park and Homsy [60] have developed a comprehensive theory of the Hele-Shaw cell which allows wetting of the walls, basically including local 3-D effects, and promises many new results. A better understanding of the Hele-Shaw problem will give insights which should aid in the solution of porous media problems.

Generalizing these concepts to flow in porous media, Jerauld *et al.* [45] use a traveling wave solution of permanent form from which to perturb. In [45], equations for the stability parameters are calculated numerically. In a "middle" parameter range, these parameters behave qualitatively like the Chuoke solution. A similar approach has been taken by Yortsos and Huang [85]. The obvious next step in this line of research is the incorporation of surface tension and the resulting finger growth.

A complete understanding of the fingering phenomenon must include the nonlinear terms that "shape" the fingers. To include nonlinear effects, Outmans [57-59] has used asymptotic methods on the nonlinear equations. His calculations indicate that the nonlinear terms play a significant role in the behavior and growth of fingers. Nayfeh [55] has developed a "weakly nonlinear" theory for fingering in porous media flow.

Another important influence upon the finger growth is the velocity dependence of the dispersion tensor. Once a finger is initiated, flow is preferentially directed along the finger axis. Then since the longitudinal dispersion (in the direction of flow) is considerably larger than the transverse dispersion, the dispersive effects tend to increase finger growth. Therefore, a complete understanding of the viscous fingering phenomena is not possible without a better model for the dispersion tensor and the relative length scales of the mixing zone and the finger size.

When viscous fingering occurs, the finger wavelengths are generally much smaller than the typical computational cells which are used in a simulator, and therefore cannot be represented on a standard numerical grid. In order to correctly model the physics of the fingering phenomenon, one would have to add a prohibitive number of grid points around the moving front. If the fingering effect is not modeled, in general an overly optimistic value of fluid recovery is predicted. Realizing the inability to realistically model the physics of fingering with reasonable grids, a lumped-parameter approach, similar in principle to the averaging of the Navier-Stokes flow equations which was previously discussed, has been tried. The original attempt along these lines was due to Koval [46], utilizing a global Buckley-Leverett approach.

Koval attempted to account for the viscous fingering effect by adjusting the viscosity and/or density of the displacing phase when conditions were suitable for fingering to occur. Two variants of this approach are being used extensively in simulation: that of Koval [46] and that of Todd and Longstaff [79]. In Koval's approach, the effective viscosity for the injected fluid is calculated by

$$\mu_{\text{eff}} = \frac{\mu_o}{\left[0.78 + \left(\frac{\mu_o}{\mu_s}\right)^{\frac{1}{4}}\right]^4} \tag{3.1}$$

where μ_o and μ_s are the viscosities of the oil and the pure solvent, respectively. Comparison with Equation (2.5) shows a similarity of this approach with the mixing rules used in miscible displacement. Todd and Longstaff [79] modify both the viscosity and the density via a mixing parameter $w[0, 1]$ which is empirically matched to flooding data. Often, the results of simulation are quite sensitive to w which can be a function of grid size used.

Other attempts to incorporate the effects of fingering in a simulator without trying to model the physics on its smaller scale are discussed by Gardner and Ypma [35] and by Fishlock and Rodwell [34]. See [12] for a survey of some of this work together with a method for incorporating velocity variation in the modeling of viscous fingering. None of these attempts is completely satisfactory and the author feels this is an important area for future research. Attempts to use statistics of the flow to empirically determine the longitudinal and transverse dispersion coefficients in the dispersion tensor (see Equation (2.3)) to incorporate fingering effects are in progress.

Given the importance of dispersion and viscous fingering in the modeling of most enhanced recovery processes, research must progress in several directions. First, the averaging processes used to change length scales must be improved, perhaps via more statistical techniques, to obtain better effective reservoir coefficients for the macroscopic models presented in the next section. Simultaneously, the effective length scales of dispersion and its effect upon viscous fingering must be better understood. Third, better macroscopic techniques for including the effects (perhaps statistical) of viscous fingering and dispersion must be developed. Finally, since the fluid interactions are so critical to enhanced recovery procedures, we cannot afford to continue to use coarse grids and standard upstream weighting techniques around the fluid interfaces in our simulators, because these methods introduce an artificial diffusion [22] which dominates the physical diffusion/dispersion and smears fronts over coarse grid blocks. This involves higher resolution of the fluid interfaces which, hopefully, can be achieved through adaptive, local-grid refinement techniques presented in this paper.

4. Adaptive Grid-Refinement Techniques

Many time-dependent fluid flow problems involve both large-scale processes and highly localized phenomena that are often critical to the overall chemical and physical behaviors of the flows. For large-scale applications, it is frequently impossible to use a uniform grid which is sufficiently fine to resolve the local phenomena without yielding numbers of unknowns which will overburden even the largest of today's supercomputers. Since these local processes are often dynamic, efficient numerical simulation requires the ability to perform dynamic self-adaptive local grid refinement. The need for adaptive techniques has provided the impetus for the development of local grid refinement software tools, some of which are used in day to day applications for small to mid size problems. Software and engineering tools capable of dynamic local-grid refinement need to be developed for large-scale, fluid flow applications. The adaptive grid refinement algorithms must also be closely matched with the architecture features of the new advanced computers to take advantage of possible vector and parallel capabilities.

There are two different types of self-adaptive grid refinement that have been applied in reservoir simulation. The first technique is a truly local grid refinement where an arbitrary level of refinement can be applied at any region or point in space. This technique requires a special multi-linked tree data structure for effective matrix set-up as well as special algorithms for efficient solution. A data structure has been developed that will support truly local refinement and dynamic "unrefinement" in both space and time (see [16,17,23-25]). The special tree structure allows truly local grid refinement and is implemented via an efficient multi-linked list. It was developed at the Mobil Research and Development Corporation Laboratory in Dallas and represents a general labeled tree structure supporting two-dimensional finite element meshes for discretization of partial differential equations. The dynamic multi-linked list representation efficiently allows both placement and removal of local meshes. A local grid analysis triggers the dynamic changes in the trees for adaptivity.

The data structures have proven to be very effective for elliptic or time-independent partial differential equations. However, the complexity of the data structures and the associated solution processes make many of the truly local refinement procedures inefficient for large-scale, time-dependent problems. If different grids are used for each time-step in a large problem, the overhead associated with the data structures and the grid generation can easily dominate the overall computation times. For this reason, alternate techniques which do not require complex data structures or regeneration of the grid at each time-step are desirable.

A technique termed patch refinement may be an attractive alternative to truly local refinement. This method does not require as complex a data structure but does involve ideas of passing information from one grid to another. Berger and Oliger have been using patch refinement techniques for hyperbolic problems using finite different discretizations for some time [4,5]. The idea of a local patch refinement method is to pick a patch that includes most of the critical behavior requiring better resolution, and use a special, possibly uniform, refinement within this patch. If a uniform fine grid is utilized in the patch, very fast solvers, perhaps utilizing vector-based algorithms, can be applied locally in this region using boundary data from the coarse original grid.

The local patch refinement techniques [10,31] have proven to be very effective for obtaining local resolution around fixed singular points such as wells in a reservoir. We will discuss the patch approximation technique first in the context of local refinement around a point like a well and then move to dynamic problems. The major input and output from a reservoir in various production procedures is through wells. Hence it is important to obtain an accurate approximation to the flow nearby.

If fluid flow rates are specified at injection or production wells, the use of Dirac delta functions as point sources and sinks in the mathematical equations has been shown to be a good model for well-flow behavior beyond some minimal distance away from the wells. In this case, the pressure (which determines the flow) grows like $\ln r$ where r is the distance to that well. A different well model, involving specification of a bottom hole pressure as a boundary condition, also gives rise to a logarithmic growth in pressure up to a finite specified pressure.

Because of the singular behavior in the vicinity of wells, accurate pressure approximations require local grid refinement.

Consider a simple example problem. From Equation (2.1) pressure p, of a fluid in a horizontal reservoir $\Omega \subset \mathbb{R}^2$ satisfies

$$-\nabla \cdot \frac{k}{\mu}\nabla p = q \quad \text{in} \quad \Omega \ . \tag{4.1}$$

Assuming no flow boundary conditions, we have

$$\frac{k}{\mu}\frac{\partial p}{\partial \nu} = 0 \quad \text{on} \quad \partial\Omega \ , \tag{4.2}$$

where $\frac{\partial}{\partial \nu}$ is the outward normal on $\partial\Omega$. For the existence of p we assume that the mean value of q is zero and for uniqueness we impose that p have mean value zero. If fluid flow rates at injection and production wells are specified via Dirac delta functions at the N_w wells x_i with associated flow rates q_i, then

$$q = \sum_{i=1}^{N_w} \delta(x - x_i)q_i \ . \tag{4.3}$$

Several techniques which assume radial flow near the well have been used to obtain local properties of p. One such technique involves subtracting out the singular behavior of p around the wells [15,22,30,32]. A radial flow assumption is probably not bad around injection wells but is inadequate for production wells. For production wells, different techniques, such as local grid refinement are often needed. It has been shown [82] that appropriate local grid refinement around these singularities can greatly increase the accuracy throughout the reservoir.

In an involved production strategy, new wells are drilled and old wells are often shut in to produce better sweep efficiency by the injected fluid and increase the hydrocarbon recovery. Thus the need to dynamically turn wells on or off necessitates the ability to add or remove local refinements around the wells of the simulation without regenerating the entire grid. Many of the current solution algorithms used in large-scale simulators are highly vectorized, depend upon a regular matrix structure, and do not allow the dynamic changing of the number of grid points or elements. The ability to have truly local refinement capabilities often necessitates the use of a fairly complex data structure with a fairly large computational overhead in field scale applications. The aim of this section is to discuss new techniques which will greatly reduce overhead requirements, will allow dynamic local grid refinement, and can be incorporated efficiently in existing codes.

We have developed fast solution methods for the approximation of problems requiring mesh refinement. These techniques are related to various domain decomposition methods (cf. [6-9]). High accuracy throughout the computational region is obtained by incorporating local refinements around wells. A composite grid is obtained by superimposing these refinements on a quasi uniform grid on the original domain. Composite grid methods [39,40,42,44] have been presented for applications of this type. These techniques usually have no systematic way of dealing with such questions as interface interpolation, mass conservation, and degree of grid overlap. Also, instead of actually trying to formulate and solve the problem on the composite grid, the problem is considered on the coarse grid and the refined patches are used only to achieve better local resolution. In the methods to be discussed below, the problem is formulated with a composite operator on the composite grid. The techniques are iterative procedures which drive the residual of this composite grid operator to zero.

A new domain decomposition variant is developed to efficiently solve the resulting matrix equations. This involves the development of a preconditioner. This preconditioner is novel in that the task of computing its inverse applied to a vector reduces to the solving of separate matrix systems for the local refinements and the matrix system for the quasi uniform grid on the original domain. Note that this quasi uniform grid overlaps the regions of local refinement and its corresponding matrix problem remains invariant when local refinements are dynamically added or removed. This local refinement technique can be incorporated in existing reservoir codes without extensive modification. Furthermore, if the nodes on the quasi uniform grid are chosen in a regular pattern, highly vectorizable algorithms for the solution of the corresponding matrix system can be developed.

Extension of earlier domain decomposition algorithms [6-9,39,40] to situations involving local refinements are possible. However, these extensions involve the solution of coarse grid problems with the regions corresponding to the refinements removed. Thus the coarse grid problem would change as refinements are added or removed. More complex solution procedures would have to be used to handle this problem since many of the original grid nodes would no longer be in the resulting problem.

We now describe a technique which allows ease of implementation in existing codes. First a composite grid is formed from a coarse uniform grid with superimposed refinements in subregions denoted collectively by Ω_2. The coarse grid remains in the region $\Omega_1 = \Omega/\Omega_2$. The discretization in this example is given via finite element techniques. However, extensions to finite difference methods are progressing well.

Multiplying (4.1) by an arbitrary (sufficiently regular) function ϕ, integrating by parts and using (4.2), we see that the solution p satisfies

$$A(p, \phi) = (f, \phi) \tag{4.4}$$

where

$$A(u, v) = \int_\Omega \frac{k}{\mu} \nabla u \cdot \nabla v \; dx$$

and

$$(u, v) = \int_\Omega uv \; dx \; .$$

The Galerkin approximation to (4.4) is to find a function P in a suitable finite dimensional substace M_h of the Sobolev space $H^1(\Omega)$ such that

$$A(P, \phi) = (f, \phi) \; , \quad \text{for all} \quad \phi \in M_h \; . \tag{4.5}$$

Since the bilinear form $A(\cdot, \cdot)$ corresponds to the composite operator, (4.5) is, in general, difficult to solve for P. Instead we will use a preconditioned iterative method to obtain P. We must then find a comparable form $B(\cdot, \cdot)$ such that, given g, the problem of finding $W \in M_h$ satisfying

$$B(W, \phi) = (g, \phi) \; , \quad \text{for all} \quad \phi \in M_h \; . \tag{4.6}$$

is relatively easy.

As was described in [10], the problem of calculating the action of the inverse of the preconditioner essentially reduces to the solution of discrete mixed problems on the refined subgrids, and discrete Neumann problems on the original grid. Due to the regularity of the mesh geometry, such problems are generally easier to solve than the system resulting from the composite grid discretization.

We first split the bilinear form into parts $A(u,v) = A_1(u,v) + A_2(u,v)$, where

$$A_i(u,v) = \int_{\Omega_i} \frac{k}{\mu} \nabla u \cdot \nabla v \; dx \; . \tag{4.7}$$

Then we decompose any $V \in M_h(\Omega)$ as follows: $V = V_p + V_r$ where V_p equals V on Ω_1, $V_p \in M_h(\Omega_2)$ on Ω_2 and $V_r \in M_h(\Omega)$ satisfies

$$A(V_r, \phi) = 0 \quad \text{for all} \quad \phi \in M_h(\Omega_2) \tag{4.8}$$

Then, as in [10], we see that for $V \in M_h(\Omega)$,

$$A(V,V) = A_1(V,V) + A_2(V_p, V_p) + A_2(V_r, V_r) \; . \tag{4.9}$$

The action of the inverse of (4.9) is not easy to obtain. However, by replacing $A(V,V)$ by

$$B(V,V) = A_1(V,V) + A_2(V_p, V_p) + A_2(V_c, V_c) \; , \tag{4.10}$$

where V_c is determined by the original coarse grid $M_h^c(\Omega)$ and satisfies $V_c = V$ in Ω, and

$$A_2(V_c, \phi) = 0 \quad \text{for all} \quad \phi \in M_h^c(\Omega_2) \; , \tag{4.11}$$

then the action of the inverse of (4.10) is relatively easy to obtain and the form $B(\cdot, \cdot)$ is comparable to the form $A(\cdot, \cdot)$ with comparability constants independent of the grid size h [8-10]. As described in [10], the following algorithm suffices for solving

$$B(W, \phi) = (g, \phi) \quad \text{for all} \quad \phi \in M_h \; , \tag{4.12}$$

given g.

ALGORITHM FOR COMPUTING W [10]:

1. Find U_p by solving mixed problems on the regions Ω_2.

2. Pass the local information to the right hand side of the original problem and compute any solution U_c of the coarse grid problem

$$A(U_c, \phi) = (g, \overline{\phi}) - A_2(U_p, \overline{\phi}) \quad \text{for all} \quad \phi \in M_h^c \tag{4.13}$$

where $\overline{\phi}$ is any function in M_h which equals ϕ on Ω_1.

3. Find U_r on Ω_2 by computing the discrete harmonic extension with respect to the refinement subspaces.

4. Compute $\overline{U}$, the mean value of $U = U_p + U_r$. Set $W = U - \overline{U}$.

Since this algorithm only requires two separate solutions of mixed problems on the subregions (each subregion possibly being considered via a different parallel processor) and one solution on the original, uniform coarse grid, it is relatively easy to perform. Similarly, no complex data structure is required and the algorithm can be implemented in existing large-scale codes without severely disrupting the solution process. Promising numerical results for the algorithm appeared in [10]. Work is underway to extend these results to more general reservoir simulation problems.

For time-dependent problems, often there is much information which can be used from preceding time-steps to help drive our adaptivity process. In parabolic problems, where the solution changes smoothly in time, the "optimal" grid used at the previous time-step should be a very good approximation to the desired grid at the next time step. Thus beginning with a new coarse grid at each time step and using the elliptic techniques of error estimators to drive the local refinement would be wasteful. For small parabolic problems, when the grid is changing very slowly in time, a much better technique would be to take the grid from the last time-step, apply a grid analysis to determine where new grid is needed and where grid is no longer needed, and then change only the grid that indicates need for change. For large time-dependent problems, iterative solution processes are much more efficient than direct solution techniques. For problems with fairly smoothly changing solutions, the same preconditioner can generally be used for several time steps, because the matrices change smoothly, greatly saving in computational effort. If the size of the grid and hence the number of unknowns is constantly changing, clearly the preconditioner must be changed. Similarly, as mentioned earlier, changing the number of unknowns greatly hinders vectorization techniques. Therefore, a considerably more efficient alternative to constantly changing the grid is to use a larger refined area within which the action is maintained for several time steps and to move the patch less frequently, after several steps.

For hyperbolic or transport-dominated parabolic partial differential equations arising in fluid flow problems, sharp fluid interfaces move along characteristic or near-characteristic directions. The computed fluid velocities determine both the local speed and direction of the regions where local refinement will be needed at the upcoming time steps. This information can be utilized in the adaptive method to move the local refinement with the front. We are currently experimenting with using the computed fluid velocities to move the patch grids in quantum jumps. In these techniques, great care must be taken to preserve mass balance when grid is removed and the flow properties must be averaged and described on the new coarser grid.

A technique is being implemented which has some of the better properties of both the fully local grid refinement and the moving finite element (MFE) techniques [18,37,38,52-54] where the grid is moved continuously to follow the dynamics of the flow. The grid is first separated into various macro-cells. The multi-linked data structure developed to treat this double layer of refinement (macro-cells with interior micro-cells) will be able to handle this construction with only minor complexity. The refinement interior to each macro-cell can be implemented in any of a number of ways: (a) fully local refinement requiring the complexity of a more complex data structure and tree traveling algorithms during the solution; (b) uniform local refinement as in the patch techniques; and (c) local MFE methods. We are experimenting with combinations of (b) and (c), neither of which requires a local tree structure. If the macro-cells are suffciently small, the local movement of the nodes in the MFE method should not create the long, thin triangles or rectangles with bad approximation properties. Care must be taken in transferring the information from one macro-cell to the next when a front passes the cell boundary; this is an area of research focus. One idea is to create a variable intermediate macro-cell which moves in quantum jumps with the front to keep the severe local refinement away from a macro-cell boundary. Coupling these ideas with the patch preconditioning techniques described earlier to give flexibility to the solution technique on each macro-cell appears very attractive. Progress will be described in a later report.

REFERENCES

[1] K. Aziz and A. Settari, *Petroleum Reservoir Simulation*, Applied Science Publishers, 1979.

[2] A.A. Baker, L.W. Gelhar, A.L. Gutjahr, and J.R. Macmillan, "Stochastic analysis of spatial variability in subsurface flows, I, Comparison of one- and three-dimensional flows", *Water*

Resour. Res., 14 (2) (1978), pp. 263-271.

[3] A.L. Benham and R.W. Olson, "A model study of viscous fingering", *Soc. Pet. Eng. J.*, (June 1963), pp. 138-144.

[4] M.J. Berger, "Data structures for adaptive mesh refinement", in *Adaptive Computational Methods for Partial Differential Equations*, I. Babuska, J. Chandra, and J.E. Flaherty, eds., SIAM, Philadelphia (1983) pp. 237-251.

[5] M.J. Berger and J. Oliger, "Adaptive mesh refinement for hyperbolic partial differential equations", Man. NA-83-02 (Computer Science Department, Stanford University, 1983).

[6] P.E. Bjorstad and O.B. Widlund, "Iterative methods for the solution of elliptic problems on regions partitioned into substructures", preprint.

[7] P.E. Bjorstad and O.B. Widlund, "Solving elliptic problems on regions partitioned into substructures", *Elliptic Problem Solvers II*, G. Birkhoff and A. Schoenstadt, eds., Academic Press, 1984, pp. 245-256.

[8] J.H. Bramble, J.E. Pasciak, and A.H. Schatz, "An iterative method for elliptic problems on regions partitioned into substructures", *Math. Comput.*, pp. 361-370.

[9] J.H. Bramble, J.E. Pasciak, and A.H. Schatz, "The construction of preconditioners for elliptic problems by substructuring, I", *Math. Comput.*, to appear.

[10] J.H. Bramble, R.E. Ewing, J.E. Pasciak, and A.H. Schatz, "A preconditioning technique for the effcient solution of problems with local grid refinement", *Computer Methods in Applied Mechanics and Engineering*, to appear.

[11] W.E. Brigham, P.W. Reed, and J.N. Dew, "Experiments on mixing during miscible displacement in porous media", *Soc. Pet. Eng. J.*, 1 (1961), pp. 1-8.

[12] E.S. Carlson, "Velocity distributions and overall mobility: their roles in the initiation and propagation of viscous fingering in first contact miscible systems", *Ph.D. Thesis*, University of Wyoming, 1986.

[13] G. Chavent, "A new formulation of diphasic incompressible flows in porous media", *Lecture Notes in Mathematics*, No. 503, Springer-Verlag (1976).

[14] R.L. Chuoke, P. Van Meurs, and C. Van Der Poel, "The instability of slow, immiscible, viscous, liquid-liquid displacements in permeable media", *Trans. AIME*, 216 (1959), pp. 188-194.

[15] B.L. Darlow, R.E. Ewing, and M.F. Wheeler, "Mixed finite element methods for miscible displacement in porous media", *Proc. Sixth SPE Symposium on Reservoir Simulation*, New Orleans, 1982, pp. 137-146; and *Soc. Pet. Eng. J.*, 4 (1984), pp. 391-398.

[16] J.C. Diaz, R.E. Ewing, R.W. Jones, A.E. McDonald, L.M. Uhler, and D.V. von Rosenberg, "Self-adaptive local grid refinement for time-dependent, two-dimensional simulation", in *Finite Elements in Fluids*, Vol. VI, Wiley, New York (1985), pp. 279-290.

[17] J.C. Diaz and R.E. Ewing, "Potential of HEP-like MIMD architectures in self-adaptive local grid refinement for accurate simulation of physical processes", *Proc. Workshop on Parallel Processing Using the HEP*, Norman, Oklahoma, 1985, pp. 209-226.

[18] M.J. Djomehri and K. Miller, "A moving finite element code for general systems of PDE's in 2-D", PAM-57, Center for Pure Appl. Math., University of California, Berkely, CA, (1981) pp. 1-49.

[19] J. Douglas, Jr., R.E. Ewing, and M.F. Wheeler, "The approximation of the pressure by a mixed method in the simulation of miscible displacement", *R.A.I.R.O. Analyse Numerique*, 17 (1983), pp. 17-33.

[20] J. Douglas, Jr., R.E. Ewing, and M.F. Wheeler, "A time-discretization procedure for a mixed finite element approximation of miscible displacement in porous media", *R.A.I.R.O. Analyse Numerique*, 17 (1983), pp. 249-265.

[21] R.E. Ewing, "Determination of coefficients in reservoir simulation", *Numerical Treatment of Inverse Problems for Differential and Integral Equations*, P. Deuflhardt and E. Hairer, eds., Birkhauser, Berlin, 1982, pp. 206-226.

[22] R.E. Ewing, "Problems arising in the modeling of processes for hydrocarbon recovery", in *The Mathematics of Reservoir Simulation*, R.E. Ewing, ed., Frontiers in Applied Mathematics, Vol. 1, SIAM, Philadelphia, 1983, pp. 3-34.

[23] R.E. Ewing, "Adaptive mesh refinement in large-scale fluid flow simulation", *Accuracy Estimates and Adaptivity for Finite Elements*, Ch. 16 (I. Babuska, O. Zienkiewicz, and E. Oliveira, eds.), John Wiley and Sons, New York, 1986, pp. 299-314.

[24] R.E. Ewing, "Efficient adaptive procedures for fluid flow applications", *Comp. Meth. Appl. Mech. Eng.*, 55 (1986), pp. 89-103.

[25] R.E. Ewing, "Adaptive grid refinement methods for time-dependent flow problems", *Comm. Appl. Num. Meth.*, to appear.

[26] R.E. Ewing and J.H. George, "Viscous fingering in hydrocarbon recovery processes", *Mathematical Methods in Energy Research*, K.I. Gross, ed., SIAM, Philadelphia, 1984, pp. 194-213.

[27] R.E. Ewing and J.H. George, "Identification and control of distributed parameters in porous media flow", *Distributed Parameter Systems*, Lecture Notes in Control and Information Sciences (M. Thoma, ed.), Springer-Verlag, 70, May 1985, pp. 145-161.

[28] R.E. Ewing and R.F. Heinemann, "Incorporation of mixed finite element methods in compositional simulation for reduction of numerical dispersion", *Proc. Seventh SPE Symposium on Reservoir Simulation*, SPE No. 12267, San Francisco, November 15-18, 1983, pp. 341-347.

[29] R.E. Ewing and R.F. Heinemann, "Mixed finite element approximation of phase velocities in compositional reservoir simulation", *Computer Meth. Appl. Mech. Eng.*, R.E. Ewing, ed., 47 (1984), pp. 161-176.

[30] R.E. Ewing, J.V. Koebbe, R. Gonzalez, and M.F. Wheeler, "Mixed finite element methods for accurate fluid velocities", *Finite Elements in Fluids*, Vol. 4, John Wiley, (1985) pp. 233-249.

[31] R.E. Ewing, S. McCormick, and J. Thomas, "The fast composite grid method for solving differential boundary-value problems", *Proc. Fifth ASCE Specialty Conference*, Laramie, Wyoming, 1984, pp. 1453-1456.

[32] R.E. Ewing, T.F. Russell, and M.F. Wheeler, "Simulation of miscible displacement using mixed methods and a modified method of characteristics", *Proc. Seventh SPE Symposium on Reservoir Simulation*, SPE No. 12241, San Francisco, November 15-18, 1983, pp. 71-82.

[33] R.E. Ewing, T.F. Russell, and M.F. Wheeler, "Convergence analysis of an approximation of miscible displacement in porous media by mixed finite elements and a modified method of characteristics", *Computer Meth. Appl. Mech. Eng.*, R.E. Ewing, ed., 47 (1984), pp. 73-92.

[34] T.P. Fishlock and W.R. Rodwell, "Improvements in the numerical simulation of carbon dioxide displacement", European Paris Conference, 1983.

[35] J.W. Gardner and J.G.J. Ypma, "Investigation of phase behavior-macroscopic bypassing interaction in CO_2 flooding", SPE 10686.

[36] L.W. Gelhar and C.L. Axness, "Three-dimensional stochastic analysis of macro-dispersion in aquifers", *Water Resour. Res.*, 19 (1) (1983), pp. 161-180.

[37] R. Gelinas, S. Doss and K. Miller, "The moving finite element method: application to general equations with multiple large gradients", *J. Comput. Phys.*, 40 (1981) pp. 202-249.

[38] R. Gelinas, S. Doss, P. Vajk, J. Djomehri and K. Miller, "Moving finite elements in 2-D: fluid dynamics examples", in: Proceedings Tenth IMACS Congress, Montreal, 1982.

[39] G.H. Golub and D. Meyers, "The use of preconditioning over irregular regions", Proceedings of the Sixth International Conference on Computing Methods in Science and Engineering, (to appear).

[40] W. Hackbusch, "Domain decomposition techniques", in: Error Asymptotics and Defect Correction (H.J. Stetter and K. Bohmer, eds.), a special issue of Computing, 1984.

[41] J.M. Hagoort, "Displacement stability of water drives in water-wet connate-water-bearing reservoirs", *Soc. Pet. Eng. J.*, (February 1974), pp. 63-74.

[42] A. Harten and J. Hyman, "Self-adjusting grid methods for one-dimension hyperbolic conservation laws", *J. Comp. Phys.*, to appear.

[43] J.P. Heller, "Onset of instability patterns between miscible fluids in porous media", *J. Appl. Physics*, 37 (1966), pp. 1566-1579.

[44] J. Hyman, "The numerical solution of time dependent PDE's on an adaptive mesh", Los Alamos Scientific Lab Rep. No. NAUR-3701.

[45] G.R. Jerauld, H.T. Davis, and L.E. Scriven, "Frontal structure and stability in immiscible displacement", SPE/DOE Fourth Symposium on Enhanced Oil Recovery 2 (1984), pp. 135-144.

[46] E.J. Koval, "A method for predicting the performance of unstable miscible displacement in heterogeneous media", *Soc. Pet. Eng. J.*, (June 1963), pp. 145-154.

[47] L.W. Lake and G.J. Hirasaki, "Taylor's dispersion in stratified porous media", *Soc. Pet. Eng. J.*, 21 (1981), pp. 459-468.

[48] R.G. Larson, H.T. Davis, and L.E. Scriven, "Displacement of residual nonwetting fluid from porous media", *Chem. Eng. Sci.*, 36 (1981), pp. 75-85.

[49] T. Lin and R.E. Ewing, "Parameter estimation for distributed systems arising in fluid flow problems via time series methods", *Proceedings of Conference on "Inverse Problems"*, Oberwolfach, West Germany, Birkhauser, Berlin (to appear).

[50] J.C. Melrose, "Interfacial phenomena as related to oil recovery mechanisms", *Can. J. Chem. Eng.*, 48 (1970), pp. 638-644.

[51] J.W. McLean and P.G. Saffman, "The effect of surface tension on the shape of fingers in a Hele-Shaw cell", *J. Fluid Mech.*, 102 (1981), pp. 455-469.

[52] K. Miller and R. Miller, "Moving finite elements, I", *SIAM J. Numer. Anal.*, 18 (1981) pp. 1019-1032.

[53] K. Miller, "Moving finite elements, II", *SIAM J. Numer. Anal.*, 18 (1981) pp. 1033-1057.

[54] K. Miller, "Moving node finite element methods", in *Accurate Estimates and Adativity for Finite Elements*, I. Babuska, O.C. Zienkiewicz, and E. Arantes e Oliveira, eds. (Wiley, New York, 1985).

[55] A.H. Nayfeh, "Stability of liquid interfaces in porous media", *Physics of Fluids*, 15 (10) (1972), pp. 1751-1754.

[56] S.G. Oh and J.C. Slattery, "Interfacial tension required for significant displacement of residual oil", *Soc. Pet. Eng. J.*, 19 (1979), pp. 83.

[57] H.D. Outmans, "Transient interfaces during immiscible liquid-liquid displacement in porous media", *Soc. Pet. Eng. J.*, (June 1962), pp. 156-164.

[58] H.D. Outmans, "Nonlinear theory for frontal stability and viscous fingering in porous media", *Soc. Pet. Eng. J.*, (June 1962), pp. 165-176.

[59] H.D. Outmans, "On unique solutions for steady-state fingering in a porous medium", *J. Geophysical Research*, 68 (1963), pp. 5735-5737.

[60] C.W. Park and G.M. Homsy, "Two-phase displacement in Hele-Shaw cells: theory", *J. Fluid Mech.*, to appear.

[61] T.K. Perkins and O.C. Johnston, "A review of diffusion and dispersion in porous media", *Soc. Pet. Eng. J.*, 3 (1963), pp. 70-84.

[62] T.K. Perkins, O.C. Johnston, and R.N. Hoffman, "Mechanics of viscous fingering in miscible systems", *Soc. Pet. Eng. J.*, (December 1965), pp. 301-317.

[63] T.K. Perkins and O.C. Johnston, "A study of immiscible fingering in linear models", *Soc. Pet. Eng. J.*, (March 1969), pp. 39-46.

[64] R.L. Perrine and A.M. Gay, "Unstable miscible flow in heterogeneous systems", *Soc. Pet. Eng. J.*, (September 1966), pp. 228-238.

[65] R.L. Perrine, "Stability theory and its uses to optimize solvent recovery of oil", *Soc. Pet. Eng. J.*, (March 1961), pp. 9-16.

[66] E.J. Peters and D.L. Flock, "The onset of instability during two-phase immiscible displacement in porous media", *Soc. Pet. Eng. J.*, (April 1981), pp. 249-258.

[67] E. Pitts, "Penetration of fluid into a Hele-Shaw cell: the Saffman-Taylor experiment", *J. Fluid Mech.*, 97 (1980), pp. 53-64.

[68] H.H. Rachford, Jr., "Instability in water flooding oil from water-wet porous media containing connate water", *Soc. Pet. Eng. J.*, (June 1964), pp. 133-148.

[69] T.F. Russell and M.F. Wheeler, "Finite element and finite difference methods for continuous flows in porous media", in *The Mathematics of Reservoir Simulation*, (R.E. Ewing, ed.) *Frontiers in Applied Mathematics*, SIAM, Philadelphia, 1983.

[70] P.G. Saffman, "Fingering in porous media", Lecture Notes in Physics *154*, R. Burridge, ed., 1982, pp. 208-215.

[71] P.G. Saffman and G. Taylor, "The penetration of a fluid into a porous medium or Hele-Shaw cell containing a more viscous liquid", *Proc. Roy. Soc. A*, 245 (1958), pp. 312-329.

[72] P.G. Saffman, "Exact solutions for the growth of fingers from a flat interface between two fluids in a porous medium or Hele-Shaw cell", *Quart. J. Mech. and Appl. Math.*, 12 (1959), pp. 146-150.

[73] A.E. Scheidegger, "General spectral theory for the onset instabilities in displacement processes in porous media", *Geof. Pura Appl.*, 47 (1960), pp. 41.

[74] A.E. Scheidegger, "Growth of instabilities on displacement fronts in porous media", *Physics of Fluids*, 3 (1960), pp. 94.

[75] J.C. Slattery, "Single-phase flow through porous media", *AIChE J.*, 15 (1969), pp. 866-872.

[76] J.C. Slattery, "Two-phase flow through porous media", *AIChE J.*, 16 (1970), pp. 345-352.

[77] G.I. Taylor, "Dispersion of soluble matter in solvent flowing slowly through a tube", *Proc. Roy. Soc.*, A219 (1953), pp. 183-203.

[78] G.I. Taylor, "Conditions under which dispersion of a solute in a stream of solvent can be used to measure molecular diffusion", *Proc. Roy. Soc.*, A225 (1954), pp. 473-477.

[79] M.R. Todd and W.J. Longstaff, "The development, testing and application of a numerical simulator for predicting miscible flood performance", *Jour. Pet. Tech.*, 253 (1972), pp. 874-882.

[80] P. Van Meurs and C. Van Der Poel, "A theoretical description of water-drive processes involving viscous fingering", *Trans. AIME*, 213 (1958), pp. 103-112.

[81] J.M. Vanden-Broeck, "Fingers in a Hele-Shaw cell with surface tension", *Physics of Fluids*, 26 (8) (1983).

[82] D.U. von Rosenberg, "Local mesh refinement for finite difference methods", Paper SPE 10974, presented at 1982 SPE Annual Technical Conference and Exhibition, New Orleans, LA.

[83] J.E. Warren and F.F. Skiba, "Macroscopic dispersion", *Soc. Pet. Eng. J.*, 4 (1964), pp. 215-230.

[84] R.A. Wooding, "The stability of an interface between miscible fluids in a porous medium", *Zeit. Fur Ang. Math. Und Physik*, 13 (1962), pp. 255-265.

[85] Y.C. Yortsos and A.B. Huang, "Linear stability of immiscible displacement including continuously changing mobility and capillary effects", SPE/DOE Fourth Symposium on Enhanced Oil Recovery 2 (1984), pp. 145-162.

[86] L.C. Young, "A study of spatial approximations for simulating fluid displacements in petroleum reservoirs", *Comp. Meth. in Appl. Mech. Eng.*, 47 (1984), pp. 3-46.

AN ANALYTIC MODEL FOR CHANGE OF TYPE
IN THREE-PHASE FLOW

Barbara Lee Keyfitz

Department of Mathematics
University of Houston
Houston, Texas   77004

ABSTRACT   Some commonly used models for three-phase oil-
reservoir flow, such as Stone's model, may exhibit change
of type: there exists a compact region in phase space in
which the system is not hyperbolic. A simple model exemplifying
this phenomenon is presented here, for which the structure of
shocks and simple waves can be completely described analytically.
The system discussed here differs in significant ways from
change-of-type examples which have been previously found, and
seems to be more closely related to the three-phase flow problem.
Stability and well-posedness for this model problem are described.

INTRODUCTION

This paper begins a mathematical investigation into some systems of
equations that change type. The present work was motivated by numerical
results on a system of equations used to model three-phase porous-medium
flow. Those results, and an explanation of the model, can be found in the
recent paper of Bell, Trangenstein and Shubin [1]. The standard assumptions
of Darcy's law for the phases (oil, gas and water) and homogeneity of the
medium are made, as well as neglect of gravitational and compressibility
effects. The system of parabolic equations that results, which couples the
phase velocities to the elliptic pressure equation, can be further simplified
for the purpose of studying flows that are predominately hyperbolic in
character (for example, sharp fronts which propagate like hyperbolic waves
with relatively small dispersion). This is done by ignoring capillary
pressure, which amounts to setting the dispersion terms to zero. The further
simplification of considering one-dimensional flows allows decoupling of the
pressure equation. These approximations are reasonable, not because the
neglected effects are unimportant, but because the solutions of the full
system of equations will incorporate the properties of the simplified system,
and so the latter must be understood before the question of how well the full

system models the actual dynamics of the flow can be answered. Bell, Trangenstein and Shubin [1] examine the pair of equations comprising the simplified system:

$$\frac{\partial}{\partial t}\left[\begin{array}{c} s_w \\ s_g \end{array}\right] + \frac{\partial}{\partial x}\left[\begin{array}{c} f_w \\ f_g \end{array}\right] = 0 \qquad (1)$$

where s_i, $i = o,w,g$ are relative phase saturations of oil, water and gas respectively, and $f_i = \lambda_i/\lambda_t$ for $\lambda_i = k_i/\mu_i$ where k_i is the relative permeability, μ_i the viscosity and $\lambda_t = \Sigma\lambda_i$. Note that $s_o = 1-s_w-s_g$ and that $k_i = k_i(s)$. The system (1) can actually exhibit change of type. What is meant by this is that the Jacobian matrix A in the system

$$\frac{\partial}{\partial t}\left[\begin{array}{c} s_w \\ s_g \end{array}\right] + A(s_w,s_g) \frac{\partial}{\partial x}\left[\begin{array}{c} s_w \\ s_g \end{array}\right] = 0 \qquad (2)$$

obtained by differentiating the flux functions in (1) may have complex conjugate eigenvalues for some values of s_w and s_g. When Bell, Trangenstein and Shubin studied (1) using Stone's model, a standard semi-empirical functional form for the three-phase relative permeabilities, they found complex eigenvalues for A in a small open set in the interior of the saturation triangle, $0 \leqslant s_w, s_g \leqslant 1$. It is known that many models share this property. At first blush, such a system must be ill-posed in the classical mathematical sense: linearizing about any state in the elliptic region will result in a system displaying the well-known Hadamard instability – exponential amplification of high-frequency modes. Though this effect can be modified by adding enough dispersion, it will still be present in convection-dominated flows, and, as explained above, must be investigated mathematically if one is to have any confidence in the model.

Until recently, the standard conclusion was that change of type was prima facie evidence of ill-posedness, that Stone's model had some deficiency, and that other models for three-phase relative permeabilities must be sought. But recent studies of phase transitions in fluids [10] and of dynamic phase boundaries in elasticity [3] have also produced simple model equations that change type; work initiated by Joseph [4] on visco-elastic fluids shows that change of type may occur even when constitutive relations are derived from thermodynamic principles and not merely modelled empirically, as are the three-phase permeability functions. At the same time, current mathematical work on the continuum mechanics problems shows that some kind of stable behavior can be expected. For a summary of this work see [8], also [2,7,10].

While none of the results could be used to prove the validity of Stone's model, or any other model which exhibits change of type, it seems relevant to the applications to describe the behavior of different systems which have this property.

In this paper, I will present a system of model equations which exhibits change of type in the sense described above: the matrix A has complex eigenvalues in a region of phase space, which we will call the elliptic region. The region is not compact, as in the Stone's model studied by Bell, Trangenstein and Shubin; in fact, my model is not derived from the three-phase flow equations by any approximation or simplification process. The motivation for examining it was the following: all the mixed-type models studied so far have certain properties in common which preclude their having Riemann solutions of the type displayed in [1]. The system I will describe here has solutions much like those in [1]. In the rest of the paper, I present the model, describe some of the structure of nonlinear wave solutions and explain what aspects of the model account for its differences from other change of type models. To what extent such a system is "well-posed" and whether it actually occurs in Stone's or other relative permeability models for three-phase flow will be left for future discussion.

MODELLING CHANGE OF TYPE

The simplest model equations for studying nonlinear wave structure are those for which the characteristic speeds, λ_i, which are the eigenvalues of A in (2), depend only on a single variable. Pairs of conservation laws of the form

$$u_t + (f(u) - v)_x = 0$$
$$v_t + g(u)_x = 0 \qquad (3)$$

have this property; here

$$A = \begin{bmatrix} f' & -1 \\ g' & 0 \end{bmatrix} \qquad (4)$$

and

$$\lambda_{1,2} = \frac{1}{2} \left[f'(u) \mp \sqrt{(f'(u))^2 - 4g'(u)} \right] \qquad (5)$$

It is conventional to study the Riemann problem for systems like (3): this is the special initial-value problem with data

$$U(x,0) = U_0 (x) = \begin{cases} U_L, & x < 0 \\ U_R, & x \geqslant 0 \end{cases} \qquad (6)$$

where U_L and U_R are constant. Here, and elsewhere, $U = (u,v)$. At an even more basic level, one can study the structure of the centered waves which comprise the solution of the Riemann problem. One can identify an important class of systems. Systems of "nonlinear wave equation" type, such as the so-called p-system in elasticity and the equations of isentropic Lagrangian gas dynamics, are special cases of (3) with $f \equiv 0$ and $g' < 0$. In [6] we considered the (mildly degenerate) possibility $g' \leqslant 0$, and, more generally, $(f')^2 - 4g' \geqslant 0$ and $g' \leqslant 0$; at the values of u where equality holds, the system (3) has a "parabolic degeneracy". The mixed type systems that occur in the van der Waals model of phase transitions in fluids [10] and in modelling elastic/plastic phase transitions [3] are also of this type; here one still has $f \equiv 0$, but now g is no longer monotonic. In these examples, g looks, qualitatively, like a cubic function of u; the finite interval on which $g' > 0$ forms the "elliptic strip" in phase space. In our study of (3), [6], we observed that an important property of nonlinear wave equations was that $\lambda_1 = -\lambda_2$ or, more generally, that λ_1' and λ_2' have opposite signs. (We called this property "opposite variation"; in [5] we noted its usefulness in proving global existence of solutions to the Riemann problem for strictly hyperbolic equations. Without this property, in fact, systems like (3), even if strictly hyperbolic, may not have global solutions.) One might also note in passing that the quadratic systems studied by Shearer et al. [9], and the extension to mixed type systems due to Holden [2], although they are not of type (3), all have the property that $\lambda_1 = -\lambda_2$ and so $\nabla\lambda_1 = -\nabla\lambda_2$; thus they are all of nonlinear wave equation (opposite variation) type. The significance of this becomes clear when one studies the elementary waves near Γ, the boundary of the elliptic region, E. When the eigenvalues of A in (4) change from real to complex as u varies, then A has an eigenvector deficiency at the point where $\lambda_1 = \lambda_2$ (this is what leads us to speak of a parabolic degeneracy). We define the <u>rarefaction curves</u>, R_i, of (3) as the integral curves in the u-v plane of the right eigenvectors of A. Two points on a rarefaction curve correspond to states that can be joined by a centered rarefaction wave if λ_i is monotonic on the segment of curve between them. The shock curves, S_i, are also close to the R_i for weak shocks. Hence at a point on Γ, for a system like (3), all four curves are tangent to the unique right eigenvector of A, and the direction of variation of λ_1 and λ_2 along this eigenvector

direction determines the detailed structure of the shock and rarefaction waves
- for example whether shocks of one or both families can penetrate the
elliptic region (clearly, the rarefactions can never do so, since there are no
real integral curves when the eigenvalues are complex).

By convention, one chooses an orientation for the right eigenvectors,
$\vec{r}_i$, so that $\nabla\lambda_i.\vec{r}_i > 0$. (This supposes that we are not at a point of
degeneracy where $\nabla\lambda_i.\vec{r}_i = 0$.) If opposite variation holds near Γ, then
$\vec{r}_1 = \vec{r}$ and $\vec{r}_2 = -\vec{r}$ necessarily point in exactly opposite directions there.
This situation is apparently not the case in the example studied by Bell,
Trangenstein and Shubin in [1], to judge from their sketch of the direction
fields. There is further evidence for this: in all solutions calculated to
the Riemann problem for opposite variation examples - Shearer's [7] and
Holden's [2], for example - there are no shocks near Γ which look like the
"new" shock found in [1] which traverses the elliptic region from a point on
its boundary. We will describe this shock in more detail below.

This is all the more remarkable because one would expect as one crosses
from H to E in a direction tangent to $\vec{r}$, along a curve parameterized by
μ, say, that to lowest order the eigenvalues would behave like $\pm\sqrt{\mu}$, and thus
would display opposite variation, at least if $\vec{r}$ is transversal to Γ.
Indeed, the sketch of the direction fields in [1], and numerical
calculations I have done on that model, indicate that opposite variation does
hold on one side of the tear-drop shaped elliptic region, same variation on
the other; at the two points where the type of variation changes, $\vec{r}$ is
tangent to Γ, and a more degenerate situation obtains. It is my conjecture
that it is the existence of these two "sides" of Γ that is responsible for
the rather tidy solution to the Riemann problem found by Bell, Trangestein
and Shubin. That is, it is the existence of same and opposite variation on
two parts of Γ, rather than the compactness of E, that allows us to solve
the Riemann problem for initial states in the elliptic region with a pattern
of only four or five states joined by shock and rarefaction waves all of which
are almost of classical "entropy" type. This contrasts with the solutions
found by Holden and Shearer in their examples, where if both U_L and U_R are
in the elliptic region, then there must appear in the solution shocks of
"undercompressive" or "stationary" type, which explicitly violate the Lax
entropy condition on the number of characteristic speeds which enter and leave
the shock. In the model problem I shall describe, I use the term "classical
entropy type" to refer to a shock that violates the Lax condition only in the

sense that strong inequalities are replaced by weak inequalities. This occurs in my model in a new way, which has not been observed before (except in the numerical simulation of [1]). In the next section, I discuss this new kind of shock, and show that it has a viscosity profile, of a somewhat unorthodox type.

My model, then, was devised to illustrate a new type of wave behavior. It is a simple example of a mixed type equation in which the boundary Γ has two components, on one of which the system displays opposite variation, with same variation appearing on the other. By studying this model, I hope to understand some different possibilities for mixed-type equations which might arise in applications. The model system may be added to the growing repertory of prototypes of mixed type flow. I hope to be able to use it to shed light on well-posedness for mixed type systems, and also to apply it in physical models, such as (1) using Stone's model, as part of a global solution of the Riemann problem there. The actual system studied was of the type (3), with f and g chosen to give

$$u_t + (u^2 - v)_x = 0$$

$$v_t + \left[\tfrac{1}{3}u^3 + \tfrac{1}{4}u^4 - \tfrac{1}{5}u^5\right]_x = 0 \tag{7}$$

We note that the eigenvalues (from (5)) are

$$\lambda_{1,2} = u \mp \sqrt{u^3(u-1)} \; ; \tag{8}$$

they are complex in the strip $0 < u < 1$. Near $u = 0$, the system displays same variation; for $u > 1$, opposite variation. The eigenvalues are sketched in Figure 1; we mention that genuine nonlinearity in the faster family fails at a value of u approximately equal to $-.3$. This complicates the wave structure for the solution of the Riemann problem, but has nothing to do with either the reservoir model or with change of type. It could have been avoided by choosing a slightly more complicated function for g in (3). The rarefaction curves through a point (u,v) are obtained by integrating

$$\frac{du}{dv} = u \pm u \sqrt{u^2 - u} \tag{9}$$

either by numerical quadrature or by deriving a straightforward but unenlightening formula.

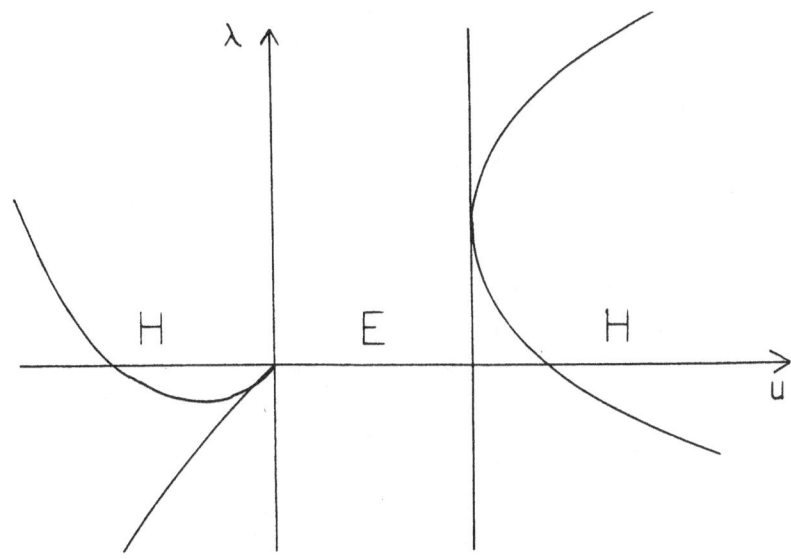

Figure 1

A shock may join a pair of points which satisfy the
Rankine-Hugoniot relation

$$s[U] = [F] \tag{10}$$

where $[U] = U_+ - U_-$ (these refer to the two sides of the discontinuity); the
formulas for the curves and speeds in this case are given by

$$[v] = \frac{1}{2} \left[[f] \pm \sqrt{[f]^2 - 4[g][u]} \right] \tag{11}$$

$$s = \frac{1}{[u]} \left[[f] - [v] \right]. \tag{12}$$

Note that the expression which appears under the square root in (11) is
quartic in u_+ for a fixed value of u_- (modulo a factor of $(u_+ - u_-)^2$),
and hence that it is possible for the Hugoniot locus through a point
$U_- = (u_-, v_-)$ to have detached branches and loops. This occurs in the model
considered in [1], though there the calculation of the shock curves can be
done only numerically. The admissibility of a shock on a detached branch
which satisfies the Lax entropy condition is not completely clear; the Lax
criterion is local, and its generalization, by Liu, does not apply when there
is not a continuum of Hugoniot solutions between the two states. This
situation arose in the mixed type model considered by Shearer in [7]. In
addition, if U_- is in the elliptic region, there may be real solutions of
(11), but all real states U_+ will be remote from U_-; in fact, they will
be in the hyperbolic region. Such shocks, again, were used to solve the

Riemann problem in [7]. In Figure 2 we indicate how a sequence of shocks and rarefactions combines to solve the Riemann problem for a pair of states U_L and U_R in E. Qualitatively, this diagram looks like the solution in [1].

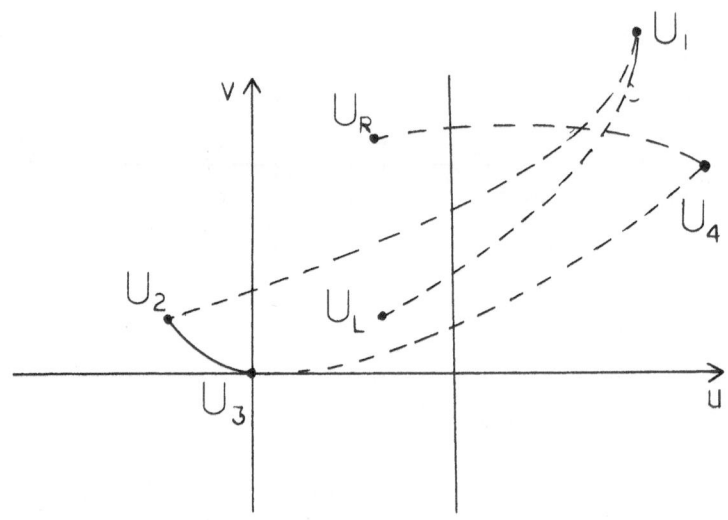

Figure 2

One difference is that U_L and U_R are not very close together; in our model, bringing U_L and U_R closer would move the intermediate state U_2 farther to the left and would make the wave pattern joining U_2 and U_3 more complicated because of the failure of genuine nonlinearity. The basic structure would probably remain the same, though. It is as follows: U_1 is a 1-shock on $H(U_L)$; there is a one-parameter family of such states. The point U_2 is the Buckley-Leverett point, uniquely defined for states U_1 in some range, at which the speed of the shock is equal to the characteristic speed at U_2, the state on the right. The state U_3, on the boundary of E, is also uniquely defined for each U_2, as the furthest point to which U_2 can be joined by a rarefaction wave of the second family. The head speed of this wave is zero, the common value of λ_1 and λ_2 on the left boundary of E. The state U_4 is also unique; it is a point on $H(U_3)$ at which the shock speed is also zero. This is the "new" shock, noted in [1]; we devote the next section to a discussion of it. Finally, from U_4 there is a one-parameter family of states U_R in the elliptic region on $H(U_4)$, to which U_4 can be joined by a fast shock. Thus, for each U_L, there is an open region of E consisting of states U_R to which U_L can be joined uniquely by the sequence of shocks and rarefaction waves indicated. Solution of the Riemann problem for all states, and global questions of uniqueness will be studied in future work. We now turn our attention to the "new" shock.

THE 'NEW' SHOCK IN THE MODEL

It is a simple matter to verify, in (11), that if U_3 is on the left boundary of E, say $U_3 = (0,0)$, without loss of generality, then there is a detached branch of $H(U_3)$ which exists for all u larger than the positive root, u_*, of $12u^2 - 15u - 5 = 0$, which is about 1.5. In fact, for all $u > u_*$, there are two points on $H(U_3)$. For one of these, the shock speed, s_2, is always positive; this can never satisfy the Lax entropy condition since the two speeds on the left are both zero. The other speed is zero at the positive root, u_4, of $12u^2 - 15u - 20 = 0$, and is negative for all larger values. Thus the point we seek is the point on $H(U_3)$ with $v_4 = u_4{}^2 \sim 4.24$.

In what sense is the zero speed discontinuity joining U_3 and U_4 an admissible shock? An attempt to solve the Riemann problem for these states numerically, using a first-order Lax-Friedrichs scheme, provided lukewarm support: a monotone "shock profile" appeared, but the shock was very smeared, its exact speed was impossible to guess, and, even after a large number of iterations, there were slow oscillations about the state U_4 on the right.

In this section we shall present more convincing evidence: there exists a viscous shock profile joining U_3 and U_4 when a scalar artificial viscosity matrix, εU_{xx}, is added to the right hand side of (7). In the standard way, we week a solution to the resulting system in the form

$$U = U \left[\frac{x-st}{\varepsilon} \right] = U(\xi) \tag{13}$$

which satisfies the boundary conditions

$$U(-\infty) = U_3, \ U(+\infty) = U_4 \tag{14}$$

and for which $s=o$. The usual procedure of integrating once and using the boundary conditions produces the system

$$\left. \begin{array}{l} u' = u^2 - v \\[2mm] u' = \frac{1}{3}u^3 + \frac{1}{4}u^4 - \frac{1}{5}u^5 . \end{array} \right\} \tag{15}$$

Here $' = d/d\xi$.

The state U_4 is a saddle for this system, while the origin, corresponding
to U_3, is a degenerate critical point, whose linearization is

$$\begin{bmatrix} 0 & -1 \\ 0 & 0 \end{bmatrix}. \tag{16}$$

In the neighborhood of the origin, this system has the same structure as a
system which occurs in an asymptotic analysis of a strictly hyperbolic, same
direction case; it is sketched in Figure 3. To exhibit a trajectory joining
U_3 to U_4, it is sufficient to find a negatively invariant region. The

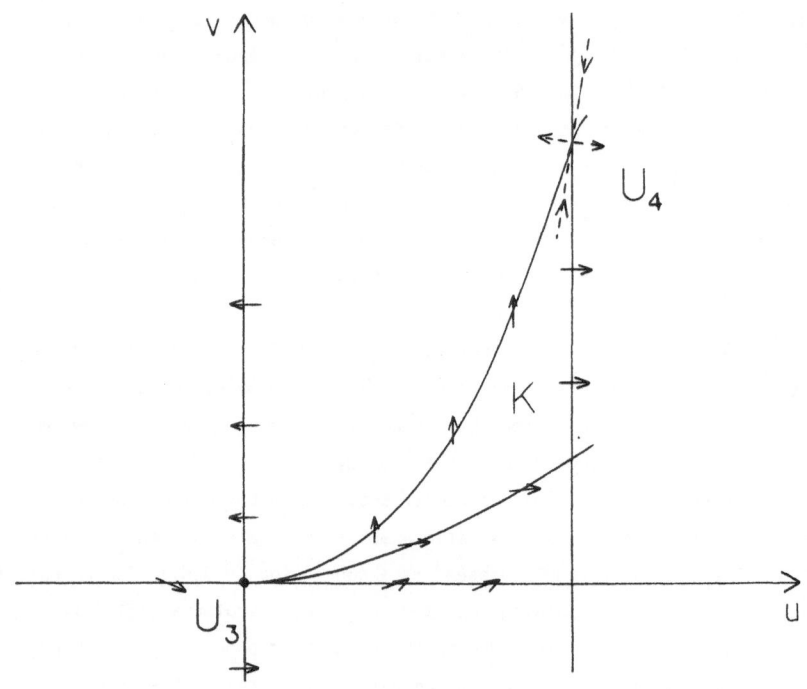

Figure 3

curvilinear triangle formed by

$$v = u^2$$
$$u = u_4$$
$$v = u^2/\sqrt{6}$$

is such a region, as a simple calculation shows.

Thus there is a viscosity profile joining U_3 and U_4. Because of the
degeneracy of the vectorfield at the origin, the approach of the trajectory to

U_3 as $\xi \rightarrow -\infty$ will not be exponential, but polynomial. This may account for the difficulty in obtaining a sharp profile numerically.

CONCLUSIONS

We have introduced and discussed a simple system of conservation laws which changes type. Because of the nature of the variation of the wave speeds along the rarefaction curves, especially on the two segments of the boundary of the elliptic region, the Riemann problem for this system admits a solution, in terms of classical centered rarefaction waves and entropy shocks, for some initial states in the elliptic region. In this it differs from the model equations displaying change of type which have been solved up to now. Moreover, the solution exhibited here resembles solutions found by careful numerical integration of equations for three-phase flow which employ Stone's model for the three-phase relative permeabilities.

The example we have considered here does not display the unstable behavior one might expect from equations that change type; a stable wave pattern that depends continuously on the data seems to be established. Further study of this and related examples seems justified. It also seems worthwhile to study Stone's model in more detail to see whether it displays similar variation of the characteristic speeds near the elliptic region.

ACKNOWLEDGEMENTS

The research reported here was partially funded by the AFOSR under the grant number 86-0088. In addition, I am grateful to the IMA, whose invitation to the Workshop on Oil Recovery stimulated much of this investigation. I benefited from discussions with Michael Shearer, John Trangenstein, John Bell, Jim Greenberg and George Papanicolaou.

REFERENCES

[1] J.B. Bell, J.A. Trangenstein and G.R. Shubin, Conservation laws of mixed type describing three-phase flow in porous media, SIAM Jour. Appl. Math. 46 (1986) 1000-1023.

[2] H. Holden, On the Riemann problem for a prototype of a mixed type conservation law, Comm. Pure Appl. Math., 40 (1987) 229-264.

[3] R.D. James, The propagation of phase boundaries in elastic bars, Arch. Rat. Mech. Anal. 73 (1980) 125-158.

[4] D.D. Joseph and J.C. Saut, Change of type and loss of evolution in the

flow of visoelastic fluids, Jour. Non-Newtonian Fluid Mech. 20
(1986) 117-141.

[5] B.L. Keyfitz and H.C. Kranzer, Existence and uniqueness of entropy
solutions to the Riemann problem for hyperbolic systems of two nonlinear
conservation laws, Jour. Diff. Eqns., 27 (1978) 444-476.

[6] B.L. Keyfitz and H.C. Kranzer, The Riemann problem for a class of
hyperbolic conservation laws exhibiting a parabolic degeneracy, Jour.
Diff. Eqns. 47 (1983) 35-65.

[7] M. Shearer, The Riemann problem for a class of conservation laws of
mixed type, Jour. Diff. Eqns. 46 (1982) 426-443.

[8] M. Shearer, Loss of strict hyperbolicity of the Buckley-Leverett
equation for three phase flow in a porous medium, preprint.

[9] M. Shearer, D.G. Schaeffer, D. Marchesin and P.L. Paes-Leme, Solution
of the Riemann problem for a prototype 2x2 system of non-strictly
hyperbolic conservation laws, Arch. Rat. Mech. Anal. 97 (1987) 299-320.

[10] M. Slemrod, Admissibility criteria for propagating phase boundaries in
a van der Waals fluid, Arch. Rat. Mech. Anal. 81 (1983) 301-315.

VISCOUS FINGERING AND PROBABILISTIC SIMULATION

Michael J. King

Standard Oil Research & Development
4440 Warrensville Center Road
Cleveland, Ohio, USA 44128-2837

ABSTRACT

The problem of viscous fingering of fluids flowing through porous media is studied using a probabilistic simulation method.   Both miscible and immiscible floods are examined. Immiscible flow shows the stabilizing effect of multiphase competition for the pore space (relative permeability).   Unstable miscible flood is shown to undergo a dynamic stabilization through a crossover from fractal behavior to a compact pattern, as a function of mobility ratio.

1. Introduction

The unstable flow of fluid through porous media (viscous fingering) is an important aspect of assessing the efficiency of many petroleum recovery processes. Until recently, such assessment was hampered by the inability of existing numerical simulation techniques to adequately model unstable flow [1,2].   In this presentation we will argue that the essential physical feature which is missing from traditional approaches is the fluctuation of flow inherent in porous media.   We describe a probabilistic simulation technique which includes fluctuations in a simple manner, and use the method to describe the flow of both immiscible and miscible fluids.

In the next section we describe the antecedents to this work, both experimental and theoretical.   The experiments are primarily those of Habermann [3].   Paterson [4] first described flow through porous media in the context of diffusion limited aggregation (DLA). To present the probabilistic interpretation of the differential equations we make a detour to one dimension (Section 3), where viscous fingering cannot, of course, occur.   In sections 4 and 5, results for two dimensional miscible and immiscible floods are presented, respectively. Scaling of miscible flood is examined in detail in Section 6.

2. Experimental and Theoretical Antecedents

Much of the recent theoretical work on viscous fingering is based on the experiments of Blackwell et. al. [5] and of Habermann [3]. Blackwell examined unstable miscible flood in linear unconsolidated sand packs for a variety of aspect ratios. The work of Habermann was for quarter five spot consolidated sand packs at fluid mobility ratios, M, ranging from 0.037 to 131. The underlying fluid dynamics is seen to be the same for both sets of data. For M<1, the flood is stable and smooth concentration profiles form. As soon as M>1, a convective instability leads to the breakup of fronts, which become jagged. These "jags" form fingers which themselves may split and form additional fingers later in a flood. In both geometries, at sufficiently high mobility ratios, the flood is eventually dominated by a single finger. This is particularly evident in the quarter five spot where the pressure draw down of the producer augments the growth of fingers.

For unstable flood, the minimum width of a finger seems to be set by physical dispersion. This is especially evident in the photo of reference 3 where the fluid concentrations are seen to vary continuously at finger edges. Additional examples are given in the review article of Homsy [6]. However, because the finger edge length scale is much smaller than the developed finger, it is approximated by a sharp edge in the experimental sketches. This same approximation will be made in the theoretical work.

The theoretical antecedent for a probabilistic approach was the application of DLA [7] to two fluid displacements in porous media by Paterson [4]. The latter successfully simulated miscible fluid floods for zero and infinite mobility ratios. The breakthrough recovery calculated on a 100 x 100 grid agrees remarkably well with the experimental results of [3,5] extrapolated to M→∞.

Diffusion limited aggregation uses an explicit random walk to "solve" Laplace's equation for the pressure field external to a growing cluster. The solution technique relies upon the relation between Brownian motion and potential theory [8], which may be described as follows. In a region of homogeneous mobility (or concentration) the pressure field satisfies $\nabla^2 p=0$. Consider the discrete form of this equation on a square lattice for a probability field, ϕ:

$$\phi_{ij} = \frac{1}{4} \sum_{i'j'} \phi_{i'j'} \tag{1}$$

where the sum is over the four sites (i',j') immediately adjacent to (i,j). Associate a random walk with (1) by introducing a sense of sequence:

$$\phi_{ij}^{n+1} = \frac{1}{4} \sum_{i'j'} \phi_{i'j'}^n \tag{2}$$

where ϕ_{ij}^n is the probability of being at site (i,j) at step n of the walk. Equation (2) is the Master equation for the propagation of probability.

The initial condition for the walk is that a single particle start at infinity. By equation (2) the total probability of being on the domain is conserved, $\Sigma\phi_{ij}^n=1$. The boundary condition for the walk is trapping on sites adjacent to the cluster, $\phi_{ij}^{n+1}=\phi_{ij}^n$. which ensures that in the long time limit (n→∞) all probability has collapsed to the traps. In fact, the probability that a trap be filled is proportional to the velocity of the fluid interface at that location [9]. Since the trapping boundary completely shields the cluster, $\phi=0$ within the cluster. This corresponds to constant pressure within the cluster for the fluid flow equations. Alternatively, a random walk within the cluster to a trapping edge corresponds to constant pressure outside of the cluster. These are the M→∞ and M→0 limits of the full set of equations, described below.

In a computer simulation of DLA, the random walker moves on the lattice until encountering a trap. The cluster is then extended, providing new traps. The next walk from infinity is subject to the new boundary condition. This method of "solving" Laplace's equation differs from the Brownian motion method [8] in one important aspect. Since there is only a single random walk for each boundary condition, the fluctuations in the solution are large. In particular, the flux is represented as one at one site and zero everywhere else. To properly utilize Brownian motion, an ensemble of walkers must be used, without modifying the boundary shape. The ensemble average of the flux will rigorously solve Laplace's equation. The DLA algorithim ensures that the average flux is correctly obtained, even though only a single sample is taken.

The "shot noise" due to the sparse sampling introduces local perturbations into the flow. For unstable flood the local perturbations nucleate large scale fluctuations while for stable flow the local perturbations decay. The continual presence of fluctuations implicitly performs a (non-linear) stability analysis of the growing cluster.

3. One Dimensional Waterflood

The probabilistic interpretation of a differential equation is much more general than the DLA algorithim. To introduce the approach, consider the flow of immiscible fluids in one dimension. The evolution equations for (normalized) water saturation are

$$\frac{\partial s}{\partial t} + \frac{\partial fw}{\partial x} = 0 \tag{3a}$$

$$fw = F(s) \tag{3b}$$

where fw is the fractional water flow in the absence of capillarity and x and t are dimensionless distance and time. We will solve (3) subject to the initial and boundary conditions for the waterflood of an oil saturated rock.

$$t = 0; \quad s = 0 \tag{4}$$

$$x = 0; \quad fw = 1 \tag{5}$$

Example profiles will use quadratic mobility functions

$$\lambda_w(s) = Ms^2 \tag{6a}$$

$$\lambda_o(s) = (1-s)^2 \tag{6b}$$

$$\lambda(s) \equiv \lambda_w(s) + \lambda_o(s) = Ms^2 + (1-s)^2 \tag{7}$$

leading to

$$F(s) = Ms^2 / [Ms^2 + (1-s)^2]. \tag{8}$$

The mobilities are normalized to that of oil, $\lambda_o=1$.

The traditional interpretation of $\frac{\partial s}{\partial t}(x)$ is the rate of change of s at x. For the probabilistic approach, $\frac{\partial s}{\partial t}$ dx is interpreted as the probability density for growth in saturation at x. Choice of an event as a saturation change Δs, leads to a (differential) time change $dt=\Delta s$ dx. To show that $\frac{\partial s}{\partial t}$ dx may be treated as a probability density notice that

$$\frac{\partial s}{\partial t} = -\frac{\partial f_w}{\partial x} \geq 0 \qquad \text{since} \qquad \frac{\partial s}{\partial x} < 0 \tag{9a}$$

and

$$\int_0^\infty \frac{\partial s}{\partial t} \, dx = 1 \tag{9b}$$

by mass balance. To generate a realization of the evolving saturation profile, we choose a point at random according to the cumulant. For a walk counter to the fluid flow, the cumulant is

$$f_w = \int_x^\infty \frac{\partial s}{\partial t} \, dx \tag{10}$$

Therefore, we may treat fw as the probability that the flux is water. When the flux changes from water to oil, in a realization, the water saturation will evolve.

As an example, consider the evolution for $\Delta s=1/4$, with a spatial discretization, Δx. The first three time steps are shown in Figure 1, with a cumulant given by (8) with M=2.5. For the initial time step the random number chosen is R=0.535. By (5), (10), water flux enters the system at x=0 and flows until fw drops below R. Initially, s=0 and this occurs in the first grid block. At step 2, R=0.973, which is above the fractional flow in block one. For step 3, R=0.660 with corresponding saturation 0.469, and the injected flux flows to the second grid block.

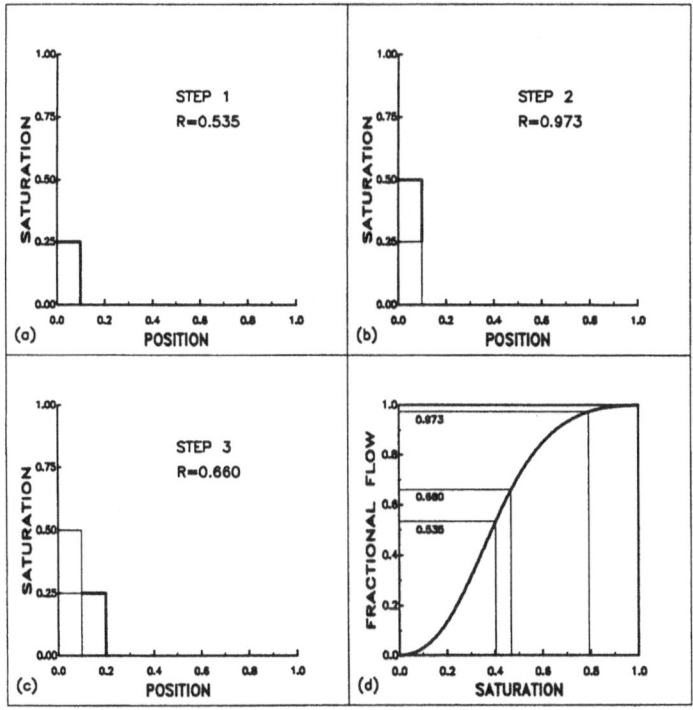

Figure 1.   Probabilistic waterflood.   Time steps 1-3. (a), (b), (c) with cumulant (d) for M=2.5.   Spatial units of Δx.

A more serious calculation is given in Figure 2 with four curves.   The first two curves show the effect of Δs refinement.   Since Δt=Δs•Δx, this is also a time step reduction.   Contrast with the finite difference solution indicates that as Δs→0, the saturation profile converges to the one point upstream solution.   This will be proven, below.   However, even a value of Δs=0.1 gives a good representation of the saturation profile.   The analytical solution is due to Buckley and Leverett [10]:

$$x = t \, F'(s), \quad s \geq s* \tag{11a}$$

$$x = t \, F'(s*), \quad s \leq s* \tag{11b}$$

where $s*$ satisfies

$$F'(s*) = F(s*)/s*. \tag{12}$$

The shock saturation, $s*$, is chosen such that the convective speed $F'(s*)$ is equal to the shock speed $F(s*)/s*$, ensuring mass conservation.   With mobilities (6), one obtains $s* = 1/\sqrt{1+M}.$

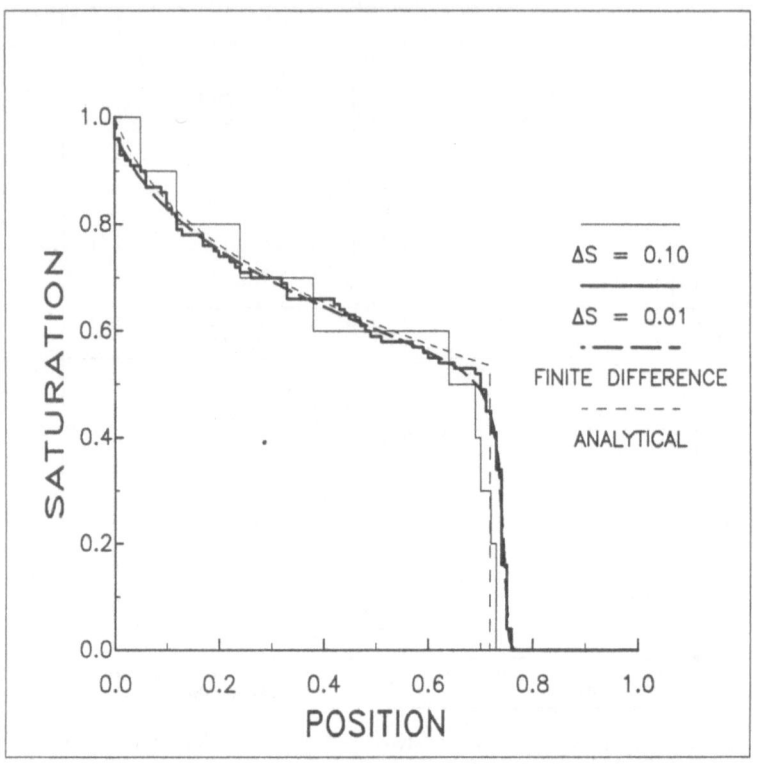

Figure 2.   Saturation profiles, Δs=0.10 and Δs=0.01 with Δx=0.01.   Included are the one point upstream and analytical solutions.

For finite Δs, the profile is represented by a series of saturation steps.   The profile may be written as a sum of a base state (analytical or finite difference solution) plus a saturation perturbation.   This is the initial condition for the next evolution step, providing an implicit non-linear stability analysis.

To prove the ensemble convergence of the probabilistic method, consider the evolution of saturation step edge x_j.

$$x_j \equiv \frac{1}{\Delta s} \int_{s_{j-1}}^{s_j} x(s,t) \ ds.$$

(13)

from (3),

$$\frac{dx_j}{dt} = (F_j - F_{j-1})/\Delta s$$

where

$$F_j \equiv F(s_j) = F(j \cdot \Delta s)$$

and therefore

$$dx_j = (F_j - F_{j-1}) \, dx. \tag{14}$$

If we define as an event $x_j \rightarrow x_j + dx$ with probability P_j then the expected change in x_j is

$$<dx_j> = P_j \, dx$$

from which we obtain

$$P_j = F_j - F_{j-1} . \tag{15}$$

Here, $< >$ denotes an ensemble average. The expected variance for a single realization is

$$\sigma_j = \langle (\tfrac{dx}{dx}j)^2 - \langle \tfrac{dx}{dx}j \rangle^2 \rangle = P_j \, (1-P_j)$$

With N independent realizations,

$$\sigma_j = P_j(1-P_j)/N$$

leading to a fractional uncertainty in x_j of

$$\sqrt{\sigma_j} \; / \; \langle \tfrac{dx}{dx}j \rangle = \sqrt{\tfrac{1}{N} \tfrac{1-P_j}{P_j}} \tag{16}$$

As $N \rightarrow \infty$ one obtains ensemble convergence to the one point upstream finite difference solution.

A stronger result may be proven since we are in one dimension. Consider N time steps of a single realization. One expects

$$<dx_j> = N \, P_j \, dx$$

and

$$\sigma_j = N \, P_j \, (1-P_j).$$

The fractional uncertainty varies as $1/\sqrt{N}$, just as before.

$$\sqrt{\sigma_j} \; / \; \langle \tfrac{dx}{dx}j \rangle = \sqrt{\tfrac{1}{N} \tfrac{1-P_j}{P_j}}$$

Therefore, time evolution provides convergence just as did an ensemble average. This proves ergodicity for the one dimensional flow of water and oil.

4. Two Dimensional Miscible Flood

To return to the topic of viscous fingering, examine two dimensional miscible flood without diffusion. The concentration evolution equations are

$$\frac{\partial c}{\partial t} + \nabla \cdot (\vec{v}\, c) = 0 \tag{17}$$

$$\nabla \cdot \vec{v} = 0 \tag{18}$$

$$\vec{v} = -\lambda(c)\, \nabla p \, . \tag{19}$$

We have suppressed the fluid sources and sinks on the right hand side of (17) and (18). Since we are examining c=1 flooding c=0 and there is no diffusion, the mobility function only enters through the values at c=0,1: $\lambda(0) = 1$, $\lambda(1) = M$.

The mobility ratio, M, determines the stability of the concentration profile. To see this, perform the linearized stability analysis of a planar interface in an infinite domain. One finds that a perturbation grows as $e^{\omega t}$,

$$\omega = \left(\frac{M-1}{M+1}\right) k \tag{20}$$

where k is the wave number. For M>1, small wavelengths (large k) can grow in an unbounded manner, since the small length scale physics of diffusion has been neglected. In the absence of diffusion, the interface evolution is mathematically ill-posed: small variations in initial conditions lead to solutions which diverge exponentially from each other. For M<1, these same perturbations decay exponentially. Physical diffusion is not needed to provide a well-posed problem.

To implement the probabilistic interpretation of (17), it is convenient to transform from the Eulerian equations for c(x,y,t) to Lagrangian equations of motion for $\xi(\varsigma,c,t)$ where ξ and ς are local coordinates constructed with respect to a concentration contour (Figure 3). It is also convenient to introduce the stream function ψ defined by

$$\vec{v} = \nabla \times (\psi \hat{z}) \tag{21}$$

and appropriate boundary conditions. The stream function construction ensures that (18) is satisfied identically. Velocities are normalized so that ψ satisfies $0 \leq \psi \leq 1$.

Equations (17), (21) provide the Lagrangian equations for concentration motion:

$$\frac{\partial \xi}{\partial t}(\varsigma,c,t) = \vec{v} \cdot \hat{n} = \frac{\partial \psi}{\partial \varsigma} \tag{22}$$

where (by equation (19)), ψ must solve

$$\nabla \cdot (\lambda^{-1}(c)\nabla\psi) = 0. \tag{23}$$

Since the initial concentration profile is sharp (c=0,1, only), and there is no physical diffusion, the concentration contours reduce to a single interface. In general this simplification will not be available and separate concentration contours must be followed.

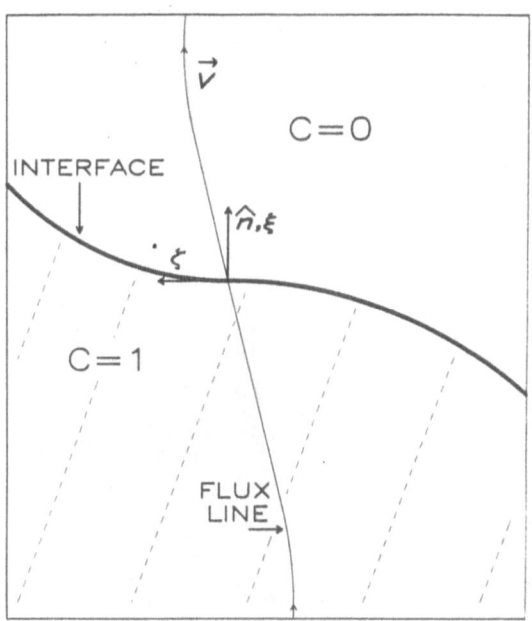

Figure 3. Lagrangian coordinates for concentration evolution (no diffusion).

As in the previous section we obtain the probabilistic approach by interpreting a rate of change as a probability. On the concentration interface:

$$\frac{\partial \xi}{\partial t} \, d\varsigma = \frac{\partial \psi}{\partial \varsigma} \, d\varsigma = d\psi \tag{24}$$

Therefore, we choose a flux line with probability density $\vec{v}\cdot\hat{n} \, d\varsigma$ and corresponding cumulant ψ. Evolution occurs when the flux line intersects the concentration interface. To ensure that the interface remains sharp, we require that concentration will change by an amount $\Delta c = 1.0$. This completely eradicates smearing of concentration profiles and corresponding numerical diffusion. Typical results are shown in Figure 4 for M=1000 and M=0.1, both on 100x100 grids. Clearly, the probabilistic approach has introduced fluctuations on the scale of the spatial grid. For M<1 these fluctuations have decayed, providing an overall stable flood, while for M>1 the small scale fluctuations nucleate large scale fingers.

The solution of equation (23) is the most CPU intensive aspect of the problem. Unlike DLA, the fluxes may not be "solved" by a single random walk, because the two domains do not decouple. To show this, consider a boundary integral solution to (23) for a quarter five spot (Figure 5). Introduce a Green function $w(\vec{x}, \vec{x}_o)$ which solves Poisson's equation for a point charge in a conducting box, D.

$$\nabla^2 w + \delta^2(x - x_o) = 0 \tag{25a}$$

$$w = 0 \qquad \text{on } \partial D \tag{25b}$$

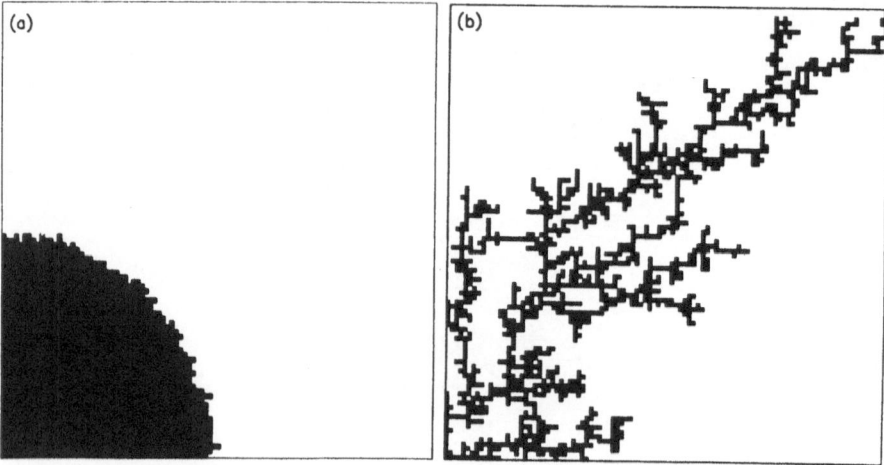

Figure 4. Miscible flood at equal injected volumes for (a) M=0.1 (b) M=1000 on 100 x 100 grids.

Integrating (23) with w gives

$$0 = \iint_D \nabla \cdot (-\lambda^{-1} \nabla \psi) \, w$$

$$= \oint_{\partial D} \lambda^{-1} \psi \frac{\partial w}{\partial n} + \iint_D \lambda^{-1} \psi \, \delta(x - x_o) + \iint_D \psi \, \nabla \lambda^{-1} \cdot \nabla w$$

The first term is known since it only references ψ on the boundary of D. The second term is proportional to ψ at x_o and the third term becomes a line integral since $\nabla \lambda^{-1}$ is a line delta function at the concentration interface. For x_o away from the interface, we obtain an equation for $\psi(x_o)$ in terms of ψ on ∂D and the interface. Choosing x_o on the interface gives an integral equation for ψ at the interface:

$$0 = \oint_{\partial D} \lambda^{-1} \psi \frac{\partial w}{\partial n} + \frac{1}{2\pi} \left(a_1 + \frac{a_2}{M} \right) \psi + \left(1 - \frac{1}{M} \right) \int \psi \frac{\partial w}{\partial n}_1$$

where a_1, a_2 are the angles subtended by the subdomains D_1, D_2 at x_0, and the line integral is along the concentration interface. Splitting up the interface line integral, and the contour integral, allows the equation to be written on each subdomain:

$$0 = \left\{ \oint_{\partial D_1} \psi \frac{\partial w}{\partial n} + \frac{a_1}{2\pi} \psi \right\} + \frac{1}{M} \cdot \left\{ \oint_{\partial D_2} \psi \frac{\partial w}{\partial n} + \frac{a_2}{2\pi} \psi \right\} \tag{26}$$

In all integrals, $\hat{n}$ is outwardly directed from the subdomain.

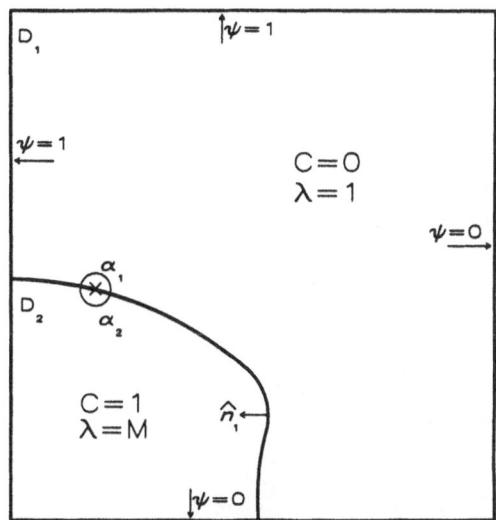

Figure 5.   Quarter five spot stream line boundary conditions.

It is evident from (26) that the problem simplifies as $M \to \infty$ or as $M \to 0$. Since ψ is bounded and w is independent of M, each term in brackets is bounded. Therefore, in the DLA limit ($M \to \infty$) one obtains a random walk over domain D_1.

$$0 = \oint_{\partial D_1} \psi \frac{\partial w}{\partial n} + \frac{a_1}{2\pi} \psi \quad , \quad M \to \infty$$

while as $M \to 0$ one obtains a walk over D_2:

$$0 = \oint_{\partial D_2} \psi \frac{\partial w}{\partial n} + \frac{a_2}{2\pi} \psi \quad , \quad M \to 0$$

Only in these limits do the subdomains decouple, allowing a single random walk to estimate the value of ψ on the concentration interface. These are the limits examined by Paterson [4].

In one dimensional waterflood we found that statistical convergence was obtained by averaging many choices for a single step, or by examining a single realization after a large number of steps. Further, we found that the evolved ensemble average was identical to the ensemble average of realizations -- the system was ergodic. For two dimensional miscible flood, and M<1, the realizations satisfy ergodicity and are cylindrically symmetric (Figure 4a). However, for M>1 (Figure 4b) ergodicity fails. The evolved ensemble average remains symmetric while the average of realizations remains fingered. This is not a failure of the probabilistic method but instead a consequence of the physical instability of the flood.

5. Two Dimensional Waterflood

The viscous fingering of a waterflood combines the convective instability of a miscible flood with the additional effects of relative permeability. The evolution of water saturation is described by a non-linear convection equation:

$$\frac{\partial s}{\partial t} + \nabla \cdot (F(s) \vec{v}) = 0 \qquad (27)$$

where $\vec{v}$ is determined by (18), (19) (with $\lambda(c)$ replaced by $\lambda(s)$). For quadratic relative permeabilities, $\lambda(s)$, $F(s)$ are given by (7), (8). Introducing local coordinates ξ, ς, along a saturation contour allows (27) to be written as

$$\frac{\partial s}{\partial t} = -\vec{v} \cdot \nabla F = - \frac{\partial \psi}{\partial \varsigma} \frac{\partial F}{\partial \xi}$$

from which we obtain a spatial probability density of

$$\frac{\partial s}{\partial t} \; dxdy = dF \; d\psi. \qquad (28)$$

Here $dF(s)$ is the probability density that a saturation level be chosen for evolution while $d\psi$ chooses a flux line, as in the miscible case.

A single time step proceeds by choosing a point at random in the (F, ψ) unit square and determining the

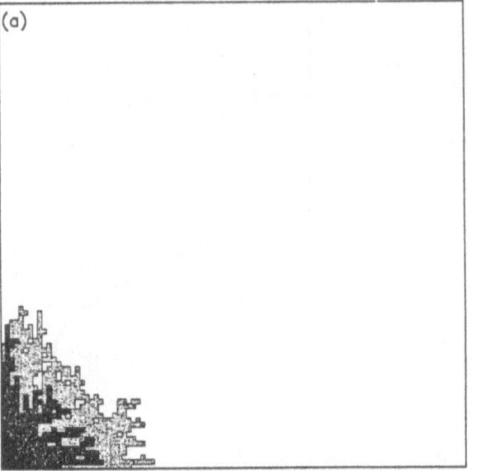

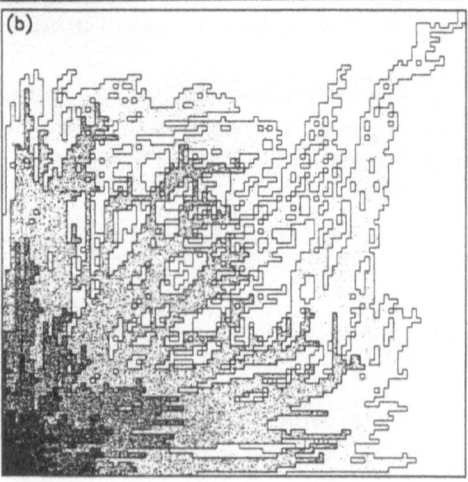

Figure 6.   Immiscible flood at equal injected volumes for (a) M=10, $\Delta s=1/3$ (b) M=100, $\Delta s=1/10$, on 100 x 100 grids.

corresponding location on the spatial (x,y) domain. With a spatial discretization, Δx, Δy, and a choice of saturation interval Δs, we obtain a time step of $\Delta t = \Delta x \Delta y \Delta s$. In practice the choice of Δs is made to minimize numerical dispersion at the waterflood shock front by choosing $\Delta s \simeq s^*$. This prevents shocks from smearing and maintains frontal structure. Two typical results are shown in Figure 6 - for $M=10$ ($\Delta s=1/3$) and $M=100$ ($\Delta s=1/10$).

Waterflood is significantly more stable than the corresponding miscible flood, because of the saturation gradients due to relative permeability (Figure 2). Following Hagoort [11], we note that the mobility at the saturation shock,

$$M^* \equiv \lambda(s*) \ , \tag{29}$$

is generally less than the endpoint mobility ratio, M. For example, quadratic relative permeabilities give $M^*=2(1-1/\sqrt{1+M}\,)$. As $M \to \infty$, one finds $M^* \to 2$. When $M^* > 1$, the dominant instability is driven by M^*, not M. For $M>1$ but $M^* < 1$, numerical studies [12] have shown that saturations for which $\lambda(s)$ is greater then one, evolve in an unstable manner, although the early growth is no longer exponential.

6. Scaling

The major distinction between waterflood and miscible flood is the degree of instability. Even for highly adverse mobility ratios, a non-fingered saturation profile is not a bad approximation to the fingered flood. This is because the instabilities, when they occur, grow rather slowly due to relative permeability effects. In contrast, miscible flood is highly fingered, with a strong convective singularity. To probe the singularity, we examine the integrated response of the system as a function of spatial mesh refinement. For a measure, we examine breakthrough recovery, measured in pore volumes. This is the fraction of area covered by the injected fluid when it is first produced. Because of the singularity evident in equation (2), the maximum finger growth rate is expected to vary as $\sim 1/\Delta x$. An unstable finger may collapse to the grid size. Therefore, coarse grids imply fatter fingers, which displace more fluid than thin fingers. Recovery is expected to decrease as the number of blocks in the NxN grid increases. Figure 7a shows two scaling curves: $M=0.1$ and $M=1000$. These are log-log plots of breakthrough recovery as a function of domain size. The straight line behavior of the $M=1000$ curve indicates fractal behavior. The best fit straight line has a slope of -.34 with correspoinding fractal dimension 1.66. This is in good agreement with the expected dimension [9] of $D=5/3$, although, because of the small range in N, the significance of this match is not clear. For $M=0.1$, the breakthrough recovery does not depend upon grid size, as is expected for a compact object.

To examine the crossover from fractal ($M \to \infty$) to compact ($M<1$), we determined scaling curves for a range of mobility ratios. We found two regimes for unstable flood,

M>1. On all plots, the recovery decreases as the grid is refined, with a scaling which appears independent of M. However, when a certain number of grid blocks, N*, is reached, the recovery stabilizes. The value of N* is an increasing function of M.

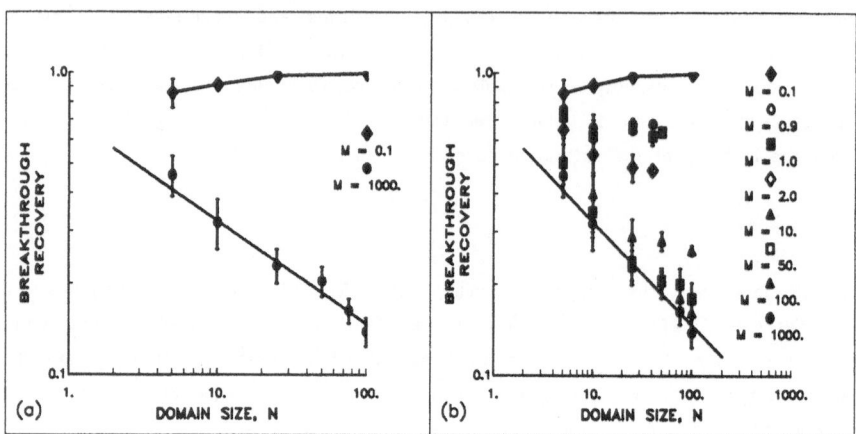

Figure 7.   Scaling results for (a) M=0.1, M=1000 (b) M=0.1, 0.9, 1.0, 2., 10., 50., 100., 1000.

Crossover is due to a flux relaxation in response to the fluctuations introduced by the probabilistic algorithim. From the linearized stability analysis, we recognize that a concentration fluctuation of size Δ, leads to a flux response of size $\Delta \cdot (M+1)/(M-1)$. The amplification factor $A \equiv (M+1)/(M-1)$ is close to unity for large M. However, after n steps we expect an amplification of

$$A^n = (1 + \frac{2}{M+1})^n = 1 + \frac{2}{M-1} n + \ldots \qquad (30)$$

In the fractal scaling regime, the number of steps to breakthrough is related to grid size by fractal dimension $n \sim N^D$. We expect this scaling to break down when the second term in the binomial expansion in (30) is comparable to the first:

$$N^* \sim (\frac{M-1}{2})^{1/D} \qquad (31)$$

For infinite M (DLA) this crossover will never be seen, while for M<1, no scaling region arises. The dependence of N* on M is consistent with the data of Figure 7b.

7. Discussion

We have developed a probabilistic interpretation of the differential equations describing miscible and immiscible floods. The resulting probabilistic algorithim has been used to study the stability of these floods under perturbations. Although individual runs do vary in detail, depending upon the sequence of random numbers, overall behavior and integrated response are well determined. In our experience, all of the cases we have examined converge under an ensemble average (see the error bars on Figure 7, for example). Dynamically stable floods converge faster than do the unstable ones.

If the underlying flow pattern is singular, then the results scale under grid refinement instead of remaining fixed. However, we have identified a novel stabilization mechanism due to the continual presence of fluctuations. Not only do fluctuations initiate fingers, but they may stabilize them as well.

The presence of fluctuations is important when studying non-linear dynamical systems since the partial differential equations, together with initial and boundary conditions, may not be sufficient to uniquely specify a solution. However, physical solutions are also stable under perturbations. Physically, flow through porous media is continually subject to fluctuations [13], due to very small scale heterogeneities, even if the rock can be treated as homogeneous. The dynamic stabilization we have seen may occur in actual fluid flow, and may be associated with the large value of effective dispersion observed in flow through porous media.

In addition to the work we have performed using a general probabilistic framework [14-16], other research groups [17,18] have utilized a single interface algorithim to study miscible flood. Current work has extended the probabilistic approach to include physical dispersion and capillarity, in order to study the small length scale physics [19].

REFERENCES

1. H.H. Rachford, Trans AIME 231, 133 (1964).
2. R.M. Giordano, S.J. Salter and K.K. Mohanty, Soc. Pet. Eng. preprint #14365 (1985).
3. B. Habermann, Trans AIME 219, 264 (1960).
4. L. Paterson, Phys. Rev. Lett. 52, 1621 (1984).
5. R.J. Blackwell, J.R. Rayne and W.M. Terry, Trans AIME 216, 1 (1959).
6. G.M. Homsy, Ann. Rev. Fluid Mech. 19, 271 (1980).
7. T.A. Witten, Jr. and L.M. Sander, Phys. Rev. Lett. 47, 1400 (1981).
8. R. Hersh and R.J. Griego, Sci. Am. 220, 67 (1969).
9. L. Turkevich and H. Scher, Phys. Rev. Lett. 55, 1026 (1985).
10. S.E. Buckley and M.C. Leverett, Trans AIME 146, 107 (1942).
11. J. Hagoort, Soc. Pet. Eng. J., 63 (Feb. 1974).
12. M.J. King, W.B. Lindquist and L. Reyna, Soc. Pet. Eng. preprint #13593 (1985), unpublished.
13. W.G. Laidlaw, R. Maier, Chem. Eng. Sci. 40, 1689 (1985).
14. M.J. King, H. Scher, Soc. Pet. Eng. preprint #14366 (1985), to appear SPE Res. Eng.
15. M.J. King, H. Scher, Bull. Am. Phys. Soc. 31, 673 (1986).
16. M.J. King, H. Scher, Phys. Rev. A 35, 929 (1987).

17. J.D. Sherwood and J.Nittmann, J. Phys. $\underline{47}$, 15 (1986).

18. A.J. DeGregoria, Phys. Fluids $\underline{28}$, 2933 (1985).

19. M.J. King, H. Scher, Bull. Am. Phys. Soc. $\underline{32}$, 586 (1987).

ACKNOWLEDGEMENTS

The work reported here was performed in collaboration with Harvey Scher. We thank Standard Oil Research and Development for permission to publish this work and Dr. Mary Wheeler for organizing the Oil Recovery Symposium.

A MARRIAGE OF GEOLOGY AND RESERVOIR ENGINEERING

Larry W. Lake

Department of Petroleum Engineering
The University of Texas
Austin, Texas 78712

Introduction

Conditional simulation is an interesting and practical way of combining geology and reservoir engineering. This marriage is now possible because of advancements in the size, speed and efficiency of modern computing, and because of a new generation of geologic classification based on quantitative interpretations. This paper gives an over-view of conditional simulations as applied to fluid flow in petroleum reservoirs. Most of the results discussed herein are from The University of Texas program on reservoir characterization. Readers interested in more details or in secondary literature citations should consult the references.

A characterization procedure - A plausible procedure for accomplishing a geological/engineering prediction might be the following:

1.   Generate a statistical description (means, trends, variances and correlations) of the reservoir flow field. Doing this requires a rather massive amount of data; primary sources are well data, outcrop analogues, seismic profiling and "type" functions based on stratification types and depositional environment. In an ideal case, there should be such a statistical description for every input variable for the reservoir simulator.

2.   Generate a synthetic flow field consisting of one or more stochastic variables. There are a number of techniques to convert raw statistics to a collection of point values which consist of a single "realization" of the flow field.

3.   Condition the flow field to be consistent with deterministic information from isolated point or line observations. Points at which values are actually measured should not be treated as random. Such conditioning will allow information from well tests, specific core observations, and seismic data to be entered into the procedure. Also, the flow field in step two must be globally conditioned to make sure the statistics derived therefrom are the same as those desired.

4.   Lay down a simulation grid on the conditioned stochastic flow field. All practical simulation procedures require assigning a discrete static volume (grid block) to each point in the flow field. The number (and hence the size) of these volumes are usually determined by the available computation facilities, although in some cases geologic phenomena will indicate an "best" grid size and geometry.

5.   In even the most favorable case the minimum granularity of the simulation (smallest possible grid size) will be larger than the smallest discernable geologic

event. If such events are important to the fluid flow, and we make this presumption here, they must be added to the simulation by defining appropriate averaged or lumped "pseudo" properties. The latter should reflect the detail within the block as well as the block geometry and size. Typical properties which must be averaged in this fashion are porosity, absolute permeability, relative permeability, capillary pressure, and dispersivity. We deal with permeability and dispersivity in more depth below.

6.   Perform a fluid flow simulation using the tabular input from step five. Just as the flow field itself, the results of this simulation are a single realization. Repeating steps 2-6 will lead to a distribution of results from which one may estimate the risk inherent in the flow process in the subject environment.

The above procedure requires a level of technology and computer commitment which is currently quite laborious; however, several aspects are becoming better defined with active research. The procedure is clearly desirable from two standpoints: it provides a direct mechanism for incorporating detailed geologic observation (steps 1 and 2) and it provides a means for assessing the effects of the geologic uncertainties on the prediction (step 6).

This paper touches on all aspect of the above procedure, though it is steps 1 and 5 (in reverse order) that occupy the most attention.

Permeability Adjustment

Shales, very fine grain clastic material, are probably the most significant barrier to flow in natural media. They occur in two forms: segregated and dispersed. Dispersed shales affect fluid transmisivity significantly, particularly if chemical changes occur, but their effect is usually incorporated in the intrinsic permeability of the medium itself. Segregated shales are more complex. If they are of great lateral extent, their presence can be easily detected and incorporated into simulation; these are "deterministic" shales. If their extent is smaller than the well spacing, their position (and indeed sometimes even their presence) is unknown; these are "stochastic" shales. Our first example deals with stochastic shales.

Figure 1 shows a synthetic cross-section containing randomly distributed, stochastic shales. In this profile, the shale positions are random and uncorrelated, the shale width distribution from cores, and the length distribution from outcrop studies according to environment type. These sources are a blend of direct observation, geologic inference and statistics. The procedure that develops such a profile adds shales to the cross-section (with correction for overlap) until a pre-specified and known ratio of sand to shale is reached. This constitutes the global conditioning step in the shale generation. In an actual application the shales would be adjusted to conform to the actual observations at observations at the wells from where the thickness statistics were derived (local conditioning); this conditioning was not done in Fig. 1.

As elaborate as the presentation in Fig. 1 is, it is of little use to simulation unless the shale distribution can be translated into effective block properties. To do this we explicitly simulated single-phase flow in media with

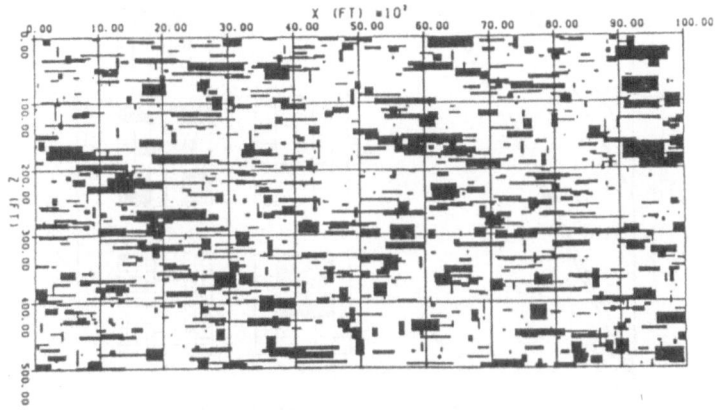

Figure 1. A single realization of a synthetic reservoir cross-section containing sand (light) and shales (dark). The lengths, thicknesses and center coordinates of the shales are independent, random events taken from known distribution functions. Shales are accumulated until a pre-specified global degree of shale area/total area (f_s = 0.24 in this case) is attained. The light boxes correspond to grid blocks.

several specific shale distributions. Figure 2 shows the results of this work expressed as a ratio of effective permeability to intrinsic "sand" permeability versus aspect ratio of the system. To be sure, the specific shale arrange strongly affects the permeability reduction; however, certain patterns emerge in the limits of large or small aspect ratios. For large aspect ratios, as are invariably encountered for horizontal permeability in numerical simulation, the permeability reduction is slight and independent of aspect ratio, being mainly attributable to the reduction in cross-sectional area caused by the shales. For small aspect ratios, as in vertical permeability, the reduction is severe and varies as the first power of aspect ratio. In this limit, the reduction is because of the increased tortuosity of the flow paths. Both extremes can be accurately calculated from the specific distribution using analytic formulas; hence, we can now associate effective permeabilities with each of the grid blocks in Fig. 1.

Dispersivity Estimates

Permeability deals with average properties of the flow paths in a simulation block; dispersivity deals with the variability of the flow paths. Our second illustration of condition simulation deals with estimating dispersivity.

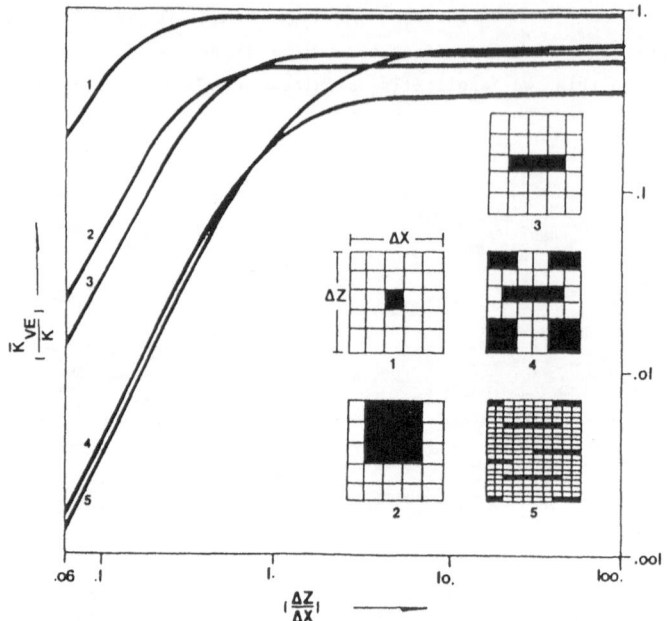

Figure 2. Permeability reduction because of the presence of non-communicating shales. When the aspect ratio ($\Delta z/\Delta x$) is small, as in the case of vertical permeability, the reduction is pronounced. When it is large, as for horizontal permeability, the reduction is small. Both extremes can be dealt with analytically.

Simple theory – The concentration C of a miscible agent of unit injected concentration displacing a fluid of zero concentration in a one-dimensional, semi-infinite medium is given by the familiar error function solution:

$$C = \frac{1}{2}\left\{1 - erf\left(\frac{x - t}{2\sqrt{t/N_{Pe}}}\right)\right\} \tag{1}$$

where N_{Pe} is the Peclet number, a dimensionless ratio of convective to dispersive transport

$$N_{Pe} = \frac{uL}{\phi K_{\ell}} \tag{2}$$

See the Nomenclature for the definition of other symbols. The term K_{ℓ} is the longitudinal (for parallel to bulk fluid flow) dispersion coefficient. A great deal of experimental evidence, mainly in laboratory corefloods, suggests that K_{ℓ} takes the form

$$K_{\ell} = \frac{D_o}{\phi F} + \alpha_{\ell} v^{\beta} \tag{3}$$

where D_o is the effective binary diffusion coefficient between the displacing and displaced fluid. α_ℓ is the longitudinal dispersivity, a characteristic length of the medium and the prime subject of this section. The exponent β is determined by experiment to be about 1-1.2; in what follows we take it to be exactly one. If diffusion is neglected, then, Eq. (3) serves as a definition for the diffusivity. Note that the Peclet number now becomes

$$N_{Pe} = \frac{L}{\alpha_\ell} \qquad (4)$$

because $v = u/\phi$. In much of what follows we will use the inverse Peclet number (α_ℓ/L) rather than N_{Pe}.

Mixing scales - A stochastic simulation will generate $C_s(t,x,y)$ for a unit inlet concentration in a two-dimensional flow field. The goal of this generation is to deduce the behavior of the dispersivity and relate these back to the statistical properties of the field; however, there are two ways to do this each manifesting a particular scale of mixing.

Let the operator $E(\cdot)$ denote averaging in the y-direction. Then $E(C_s)$ is not a function of y. We can denote one scale of mixing by minimizing the function

$$S_{ME} = \int_0^\infty (C(t,x;\alpha_{ME}) - E(C_s))^2 dx \qquad (5)$$

where the subscript ME denotes megascopic and the semicolon in the argument for C separates variables from parameters. In actual practice we use a more sophisticated analytic solution than Eq. (1) in the minimization (5) but the idea is the same. Also, the integration in (5) must be truncated to reflect the finite nature of the system, an effect which will be apparent below, and C_s must be corrected for truncation error. The minimization in (5) is through least-squares regression. The process yields a megascopic dispersivity which is, in general, a function of time or mean distance traveled.

If we perform the minimization in (5) on C_s rather than $E(C_s)$ we obtain a dispersivity α_ℓ which is a function of t and y only, the x-dependency having been eliminated by the integration in (5). For this case we define the macroscopic dispersivity as $\alpha_{MA} = E(\alpha_\ell)$. Macroscopic dispersivity is a measure of the local or point mixing taking place in the medium.

Megascopic dispersivity - The megascopic scale is the full-aquifer dispersivity whose value determines the volumetric sweep in numerical simulation blocks. Figure 3 shows the behavior of α_{ME} (expressed as inverse Peclet number) as a function of time for miscible displacements in a two-dimensional stochastic permeability field. The parameter V_{DP} is the Dykstra-Parsons coefficient, a dimensionless measure of the spread of the permeability distribution to which the flow field was conditioned.

$V_{DP} = 0$ corresponds to a homogeneous medium and $V_{DP} = 1$ is infinitely heterogeneous; see the Appendix for more discussion of V_{DP}. In Fig. 3 and elsewhere the results have been corrected for the presence of numerical truncation error in these explicit finite-difference simulations.

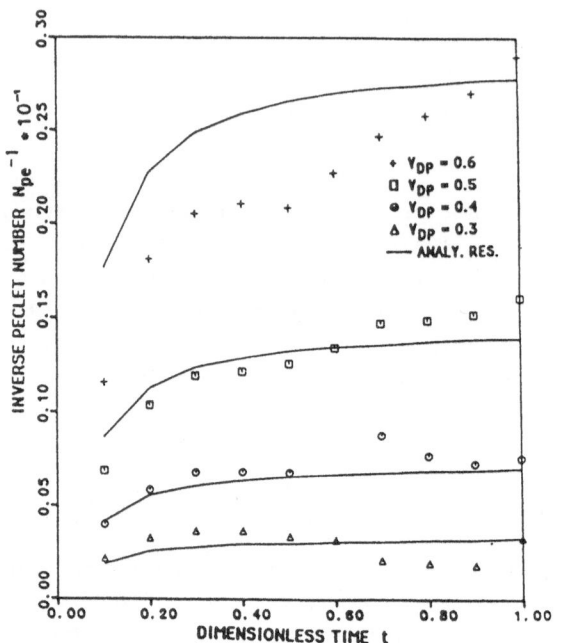

Figure 3. Growth of megascopic dispersivity (points) with throughput or dimensionless travel distance for four different degrees of heterogeneity. Dispersivity generally grows with distance and is larger at larger heterogeneities (as measured with the Dykstra-Parson coefficient V_{DP}). Dispersivities are from a single-phase, diffusion-free simulation in a stochastic permeability generated by the Heller procedure. Solid curves are from Taylor's equation.

As intuitively expected, the higher heterogeneity runs have larger dispersion, but α_{ME} grows with time in general. In fact for small t we see linear growth passing over to near constancy at larger time. Clearly, there are other factors in the permeability distribution besides its spread that account for this behavior. One of these factors is the presence of correlation in the permeability field.

Figure 4 shows a similar plot of α_{ME} versus t except now V_{DP} is being held constant while the dimensionless correlation length λ_D is changed for each run. The correlation length expresses the distance over which the covariance of the permeability distribution falls to near zero. Figure 5 shows the autocorrelation functions (covariance/variance) for the permeability fields used to construct Fig. 4. In general, runs with small λ_D (λ_D is the correlation length divided by the

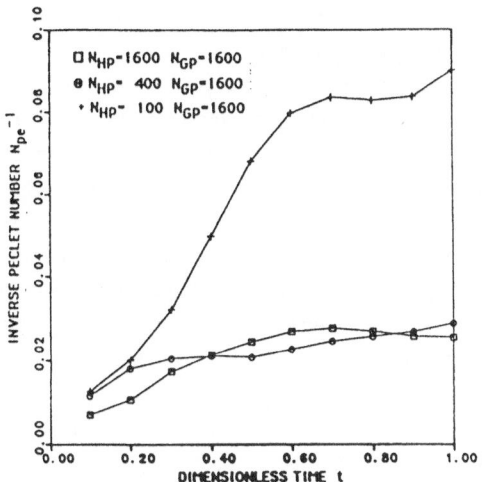

Figure 4. Growth of megascopic dispersivity (points) with throughput or dimensionless travel distance for three different dimensionless correlation lengths. Cases with small correlation lengths generally have constant dispersivities; those with large correlation lengths generally grow, reaching a plateau at large time. Heterogeneity is constant for all cases. Abnormal growth near effluent end ($x_D = 1$) here and in Fig. 3 is an artifact from the averaging of the outflow boundary condition in the finite difference solution. Diffusion is zero.

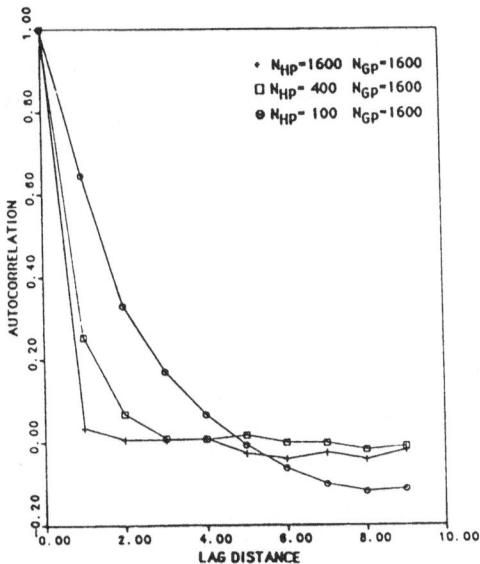

Figure 5. Permeability autocorrelograms for the three realizations whose dispersivity is in Fig. 4. We can change the dimensionless correlation length by varying the ratio of Heller points to grid blocks $\lambda_D = [N_{HP}]^{-\frac{1}{2}}$. The horizontal axis is in grid block units. The curves are only approximately described by an exponential model. In each case, the permeability distribution itself is log-normal.

system length) have relatively constant and small α_{ME}; runs with large λ_D have α_{ME} that is larger and grows significantly with t.

Another way to view correlation length is that it is the distance over which permeability values are roughly the same on average. Thus, it is not surprising that α_{ME} grows with time when λ_D is large because the medium is becoming more layered, a situation for which it is well-known that megascopic dispersivity, as defined here, grows linearly with time. Figures 6 and 7 confirm that this is so. Recall when comparing these two figures that they both have the same degree of heterogeneity V_{DP} and both were generated through stochastic means--the permeability field was not conditioned to be layered. Yet, layering of the concentration distribution $C_s(x,y,t)$ is evident in Fig. 7 and absent in Fig. 6.

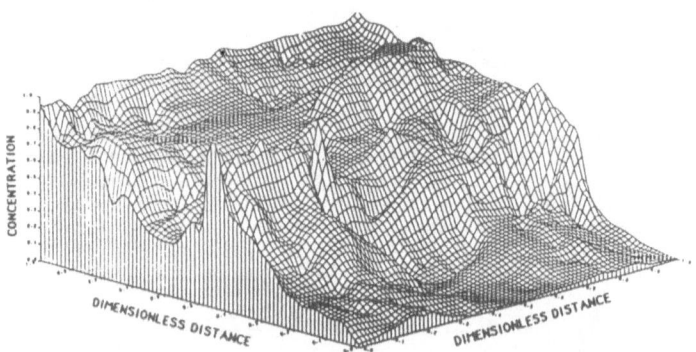

Figure 6. Isometric map of injectant concentration for a case with small correlation lengths and no diffusion. The longitudinal direction (parallel to flow) is x; the transverse direction (perpendicular to flow) is y. Concentration is quite irregular because there are no preferred paths.

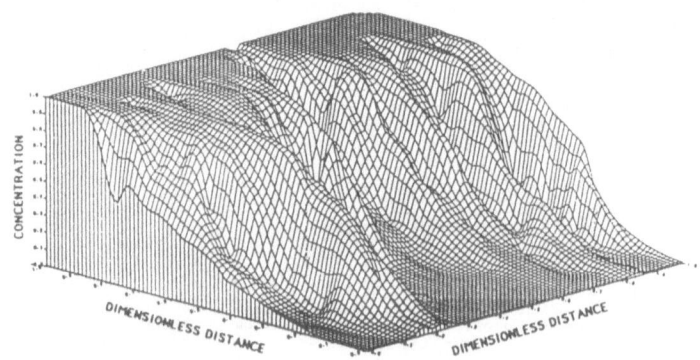

Figure 7. Isometric map of injectant concentration for a case with large correlation lengths and no diffusion. The longitudinal direction (parallel to flow) is x; the transverse direction (perpendicular to flow) is y. Concentration gives the appearance of layering even though there are none present.

Taylor's theory - To explain the behavior of α_{ME} in Figs. 3 and 4 we borrow from G. I. Taylor's theory of continuous movements, one of the pioneering works in the representation of turbulence. Casting Taylor's equation over into our notation and restricting the theory to an exponentially-declining correlation function, as seems appropriate from Fig. 5, yields the following for the dependence of α_{ME} upon t:

$$\frac{\alpha_{ME}}{L} = C_v^2 \; \lambda_D (1 - e^{-t/\lambda_D}) \tag{6}$$

This equation is shown on Fig. 3 as the solid lines. The coefficient of variation C_v and the correlation length λ_D were all derived from the properties of the underlying permeability field--there was no history match. (Strictly speaking, the statistical properties should come from the velocity field, but they appear to be the same within the accuracy of determining C_v and λ_D.)

We accept the quality of agreement between Eq. (6) and the simulated dispersivities in Figs. 3 and 4, and will use it to draw further conclusions below. However, the agreement is not perfect. Some of the reasons for this are the absence of a permeability correlation function that is precisely exponential (Fig. 5), and the lack of multiple realizations in the simulated results. Equation (6) actually expresses the expectation of the dispersivity; thus, it should agree best with the mean of a large number of simulations each with the same statistical properties but with differing flow fields. In fact, the agreement improves by taking such means. In actual practice there will also be uncertanties in estimating V_{DP} (see Appendix).

Equation (6) has two limiting cases which provide a basis for classifying displacements and permeable media. If the correlation length is small Eq. (6) becomes

$$\frac{\alpha_{ME}}{L} = C_v^2 \; \lambda_D \tag{7}$$

which predicts that dispersivity is independent of t (that is, an intrinsic property of the medium) and proportional to the coefficient of variation squared and the dimensionless correlation length. We should call such a medium uniform. When the correlation length is large Eq. (6) becomes

$$\frac{\alpha_{ME}}{L} = C_v^2 \; t \tag{8}$$

a result which says that dispersivity grows in proportion to time or travel distance. Since this consequence occurs when the displacement is channeling, media with λ_D large are layered. In general, the entire Eq. (6) applies; however it is curious that layering predominates at relatively small correlation lengths.

Molecular diffusion adds an additional correlation scale to the behavior of α_{ME}. As Fig. 8 shows, increasing the diffusion hastens the attainment of a constant

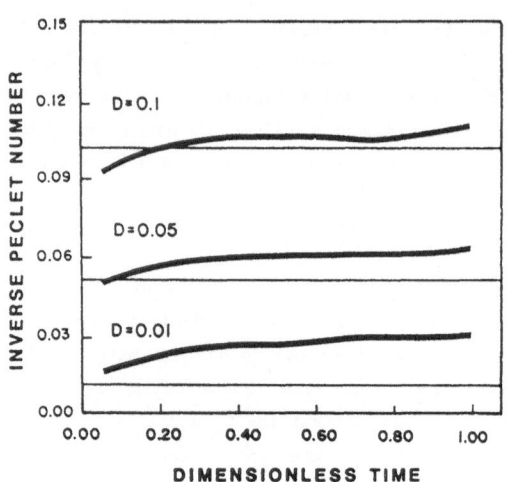

Figure 8. Growth of megascopic dispersivity with distance traveled for cases with varying amounts of diffusion. The dimensionless (input) diffusion coefficient D is (D_0/vL) which is shown in the above as dark horizontal lines. Dispersivity approaches pure diffusion for large D wherein all transverse concentration gradients are zero. Dispersivities being larger than D is caused by velocity variations (as in Figs. 3 and 4) and by the combined effect of such variations with transverse diffusion (Taylor's diffusion).

dispersivity that is always greater than what would be present because of diffusion alone. The result of diffusion is to lessen the transverse concentration gradients and thereby reduce channeling. The increased "apparent" dispersivity is because of the combined effects of isotropic diffusion and the velocity fluctuations. The values of diffusion shown in Fig. 8, however, are generally quite a bit larger than what is evidenced is actual displacements.

Figure 9 summarizes the observations on megascopic dispersivity. All displacements go through regimes of channeling (non-Fickian), transition and diffusive (Fickian) mixing. This behavior is characterized by Taylor's theory in Eq. (6) with the appropriate limiting cases as in Eqs. (7) and (8). The extent to which, if any, dominate a particular displacement depends on the heterogeneity and correlation structure of the permeability field, a subject discussed in the last part of this paper. However, the behavior suggests two important conclusions about how such mixing can be handled in numerical simulators, scaling, and, indeed, in the very nature of permeability heterogeneity.

Lumping mixing effects - The two limiting cases, Eqs. (7) and (8), can be easily captured in heuristic models--that is transport relations whose effect is to mimic the actual behavior. In the case of diffusion mixing, Eq. (7), we simply input the appropriate dispersivity to the simulator. In the case of channeling, as in Eq. (8), such fingering/bypassing models as the Koval, Todd-Longstaff, and Young models are available. However, none of these will mimic the general behavior in Eq. (6) of Fig. 9. To be sure it is possible to use Eq. (6) directly as a heuristic model, but this is impractical since time-dependent parameters would be required. Deducing a practical but efficient way to model the general megascopic mixing case is a very fruitful avenue for future research.

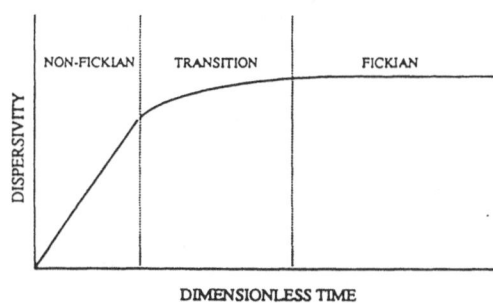

Figure 9. Schematic representation of the growth of megascopic dispersivity with distance traveled. All displacement pass through three regimes (non-Fickian, transition and Fickian) which may be large or small depending on the correlation length and D. Even displacement which are strongly Fickian do not lose the evidence of their non-Fickian beginning.

In the same vein, the general behavior of Fig. 9 is not easy to scale from field to laboratory conditions. To see this imagine that we have a laboratory and a field displacement which follow exactly the same unknown α_{ME} curve. (This does not actually happen, but the additional complexity is superimposed on these remarks.) From the laboratory experiment we know α_{ME} at some small time t_1 and we adjust the heuristic models, Eqs. (7) and (8), to these values. Now let use use the adjusted heuristic models to predict mixing at a much longer time t_2. Since the models do not in general model the true behavior, α_{ME} and the mixing zone size will not be predicted correctly at t_2. Using Eq. (7) will under predict the mixing zone size and Eq. (8) will over predict. Indeed, experience with using Eq. (8) suggests that the "effective" coefficient of variation must be reduced during scaling.

Finally, let use consider the behavior of actual field-measured dispersivities. Figure 10 shows a remarkable plot of α_{ME} versus measurement distance from over 60 datum taken from both laboratory and field measurements. Despite the considerable scatter there is a clear trend of increasing dispersivity with distance. If we assume that the measurement distance is the same as the distance traveled or the horizontal axis in Figs. 3 and 4 then we immediately notice that there is no suggestion of a leveling off even at very large measurement distances. Two contra-

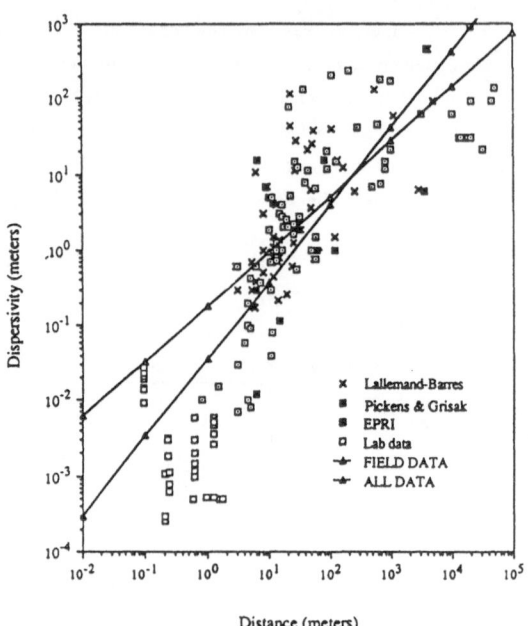

Figure 10. Summary of experimentally measured dispersivities ranging from small (laboratory) to large (field) scale. Solid curves are best fit to all data (slope = 1.13) and to field data only (slope = 0.755). Data includes experiments in a large variety of permeable medium, tracers and experimenters which accounts for much of the scatter, but the trend with increasing dispersivity with distance is evident.

dictory conclusions are possible based on this: measurement distance and travel distance are not synonymous, or the correlation structure of real media do not lead to constant dispersivities at large travel distances. Considering the latter, one could always argue that the correlation distance is actually larger than the horizontal scale of Fig. 10, but this seems unlikely given the very large range encompassed. An alternate possibility is that permeability is distributed as a fractal whence we expect no leveling out regardless of the measurement length. Indeed, the field data in Fig. 10 can be fit with a fractal dimension of 1.12. In the last part of this paper we discuss our attempts to directly measure correlation in naturally-occurring media.

Figure 10 makes apparent the futility of estimating large-scale dispersivities from laboratory experiments: megascopic dispersivities are different between the two by several factors of ten even in the same media and under the same flow conditions. What is more likely to be comparable is the field-scale macroscopic dispersivity and the laboratory-scale megascopic dispersivity, but even these comparisons are tenuous because the laboratory experiments generally are more homogeneous, have smaller length scales and are more influenced by diffusion. We discuss macroscopic dispersivity below, but first we conclude this section by giving results on megascopic dispersivity on the laboratory scale.

Figure 11 shows the results of the simulations, plotted as a normalized dispersion coefficient versus a dimensionless rate, compared to experimental data

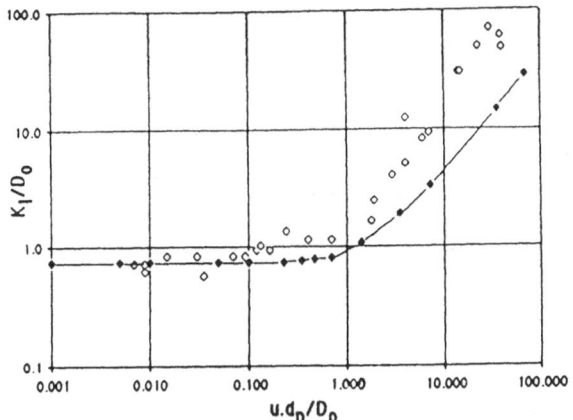

Figure 11. Comparison of calculated macroscopic dispersivities (solid curve) to experimentally-measured megascopic (laboratory scale) dispersivities. d_p is the grain diameter for the packing in the experimental measurements and the correlation length for the calculated curve. Better agreement could be obtained by letting d_p be about five times the correlation length in the calculated curve.

from sandstone cores. The acceptable agreement indicates three conclusions about the nature of dispersion in laboratory experiments:

1. Diffusion and mechanical mixing dominate at low and high rates, respectively. This is consistent with Eq. (2).

2. In the mechanical mixing regime (large velocities), dispersion coefficients become proportional to velocity to the first power. Thus, we are justified in taking $\beta = 1$ in Eq. (2) since a rate-independent α_{ME} agrees with Eq. (7).

3. Correlation lengths are as little as five grain diameters in laboratory experiments. This length is very much smaller than even the smallest experimental dimensions and we, therefore, expect time-invariant dispersivities even in the absence of diffusion. It is possible, in principle, to calculate C_v and λ_D from this observation and Eq. (7).

Macroscopic dispersivity - α_{MA} is of interest in many enhanced oil recovery displacements (such as development of miscibility or generation of optimal salinities) where local mixing controls the success of the displacement. We can also extract α_{MA} from the randomly heterogeneous runs discussed above but our conclusions are much less precise because we have no theoretical model like Eq. (1) for local mixing in two-dimensional flow. But we can compare α_{MA} to α_{ME} under identical circumstances.

Figures 12 and 13 make this comparison for two divergent cases. We see that in general like Eq. (1) α_{MA} grows with time in the same manner as α_{ME}. Its dependence on V_{DP} may be somewhat less. Most importantly, α_{MA} is always smaller than α_{ME}, but the difference decreases as correlation length becomes small. The reduction in the difference with small correlation length is consistent with Eq. (7) for in this limit of a uniform medium sampling one point would be, on average, equivalent to sampling all points at the same cross-section. The difference cannot be made to vanish entirely, however, as α_{ME} always retains its initial channeling component, a feature which is absent in α_{MA}. We are very much in need of a consistent way to estimate α_{MA} from medium properties--an analogue to Eq. (6)--for in this way we could devise a laboratory medium that would truly represent a characteristic degree of mixing in large-scale displacements.

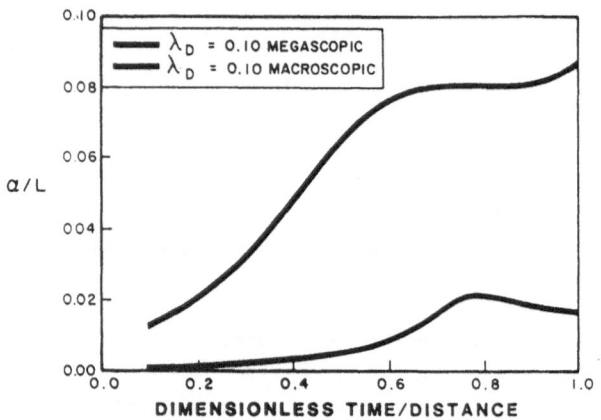

Figure 12. Comparison of megascopic (upper curve) and macroscopic (lower curve) dispersivities from simulations with large correlation length and no diffusion. In all cases, the macroscopic dispersivity is smaller and grows slower.

Geologic Descriptions

The most significant barrier to wide-spread use of conditional simulation is the lack of insight into the statistical properties of real permeable media. While we can and are making great strides into understanding flow in generic media (as was done above) there can be no predictive power unless the stochastic field is realistically representable for specific cases. Stochastic assignments and conditional simulation are probably the only fruitful line of inquiry in making predictions, but tools are still lacking when confronted with realistic media. In this final section we describe efforts to statistically describe a large-scale outcrop.

The subject outcrop was an ancient eolian sand near Glen Canyon dam in northern

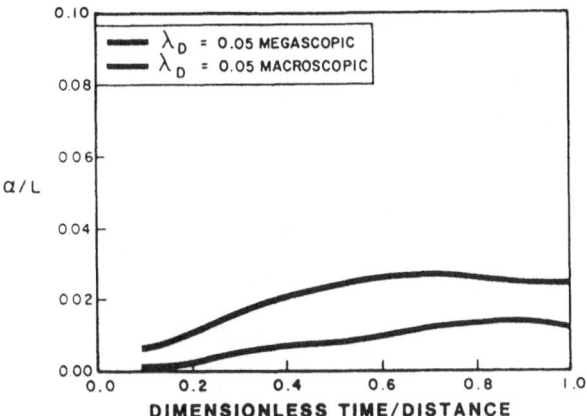

Figure 13. Comparison of megascopic (upper curve) and macroscopic (lower curve) dispersivities from simulations with small correlation length and no diffusion. When correlation length decreases the two dispersivities approach each (compare with Fig. 12) an observation consistent with less channeling when media are uncorrelated.

Arizona. Eolian sands were originally deposited by wind-blown deposition, but their character manifests itself in several scales of heterogeneity as illustrated in Fig. 14. We observed the following stratification types: intradune (sand-free zones between fingers of sand deposition), grainfall (avalanching caused by oversteepening of dune faces) and wind-ripples (reworking of sand grains into undulating patterns caused by low velocity winds). Even with this relatively few number of types the outcrop displayed formidable complexity through the ordering, geometry, orientation and repetition of the patterns.

We sampled the outcrop for permeability using a minipermeameter, a device designed to make a large number of nondestructive permeability measurements in a short time. This device was invaluable inasmuch as very large data sets are required to make statistically meaningful statements. Nearly as important was the sampling schemes. We used three: sampling according to stratification type, sampling along a line (traverses) and sampling along concentric grids. Each yielded particular insights, but it is only the latter that we discuss here.

Figure 15 shows the five concentric grids superimposed on one wall of the outcrop. The grids were intended to all be centered on a single point, but the irregular surface prevented us from doing this exactly. We intended to make the largest grid completely cover the outcrop face and the smallest grid to be such that adjacent samples would be of the same separation distance as in laboratory cores. The largest three grids sampled several beds and the smallest two are entirely within a single bed. Each grid contained more than 50 samples on a regular two-

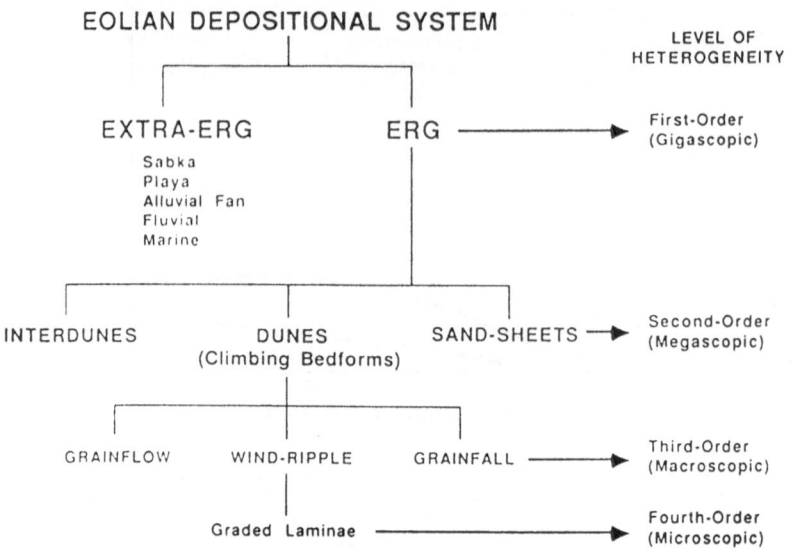

EOLIAN DEPOSITIONAL SYSTEM

LEVEL OF
HETEROGENEITY

EXTRA-ERG   ERG ——————▶   First-Order
  Sabka   (Gigascopic)
  Playa
  Alluvial Fan
  Fluvial
  Marine

INTERDUNES   DUNES   SAND-SHEETS ——▶   Second-Order
           (Climbing Bedforms)   (Megascopic)

GRAINFLOW   WIND-RIPPLE   GRAINFALL ———▶   Third-Order
                                             (Macroscopic)

           Graded Laminae ——————————————▶   Fourth-Order
                                             (Microscopic)

Figure 14. The heirarchy of heterogeneity scales for an eolian sand. Permeability groups by the third—order heterogeneity classifications. Our measurements deal primarily with wind ripple, grainflow and intradune groups.

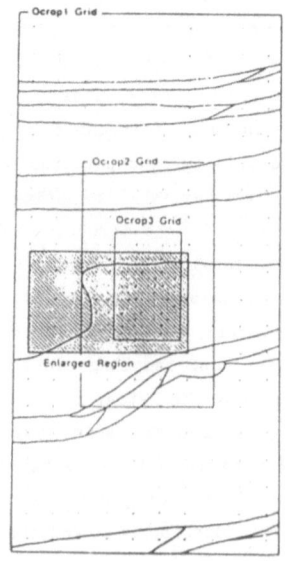

Figure 15. Schematic of the five grids superimposed on the outcrop face of an eolian sand (Page sandstone, Coconino county Arizona). The three largest grids contains more than one bed. Size of the largest grid is about 100 by 180 ft. and the smallest about 15 inches square.

dimensional pattern.

Figure 16 shows the coefficient of variation C_v for each grid plotted as a function of the grid area. C_v is the same for the three largest grids, dips to a minimum at the fourth grid and then returns to a large value at the smallest grid. Statistically, the fourth grid is different from all the others which are themselves

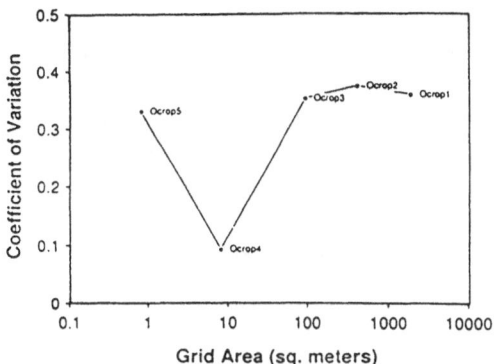

Figure 16. Coefficient of variation for the five grids in Fig. 14. The statistically significant drop for OCROP4 reflects properties being measured entirely within one bed. Inter-septile range shows the same effect although the mean permeability did not vary greatly.

the same. We can explain the dip in the C_v curve of this data by observing that the fourth grid is the first one entirely within one bed. Indeed, histograms of grids one through three show evidence of two or more populations. The increase in the fifth grid is because of very small-scale laminae which are present all over the outcrop, but noticeable only in the statistics of the fifth grid. The mean permeability was the same for all grids.

Whatever the geologic causes, there are several purely statistical inferences to be drawn from Fig. 16 which bear directly on the issue of reservoir simulation. The size of grid four may be a natural choice for the grid block size in a deterministic simulation model. Such a selection would minimize the variation between blocks and may, in fact, make stochastic assignments of secondary importance (thus, reducing the differences between realizations). The variation of the fifth scale would be incorporated as pseudo functions or megascopic dispersivity into individual blocks.

However, a representative elementary volume (REV) --the volume below which permeable media properties uniformly approach single-value limits--is not evident in the data. To justify this statement imagine that the individual data in each grid are close to the mean for a large number of measurements of several entire grids. As the size of these grids shrinks we see that the variability of their means goes through a minimum. But if there exists a REV, then the variability should approach zero at the size of the REV. Of course there still might exist a REV of size smaller than the fifth grid, but this size is not too different from the pore dimensions where grain-to-grain fluctuations begin to be important.

We mentioned the possibility of permeability being distributed as a fractal in conjunction with Fig. 10. A fractal distribution would show C_v decreasing monotonically at a rate prescribed by its factal dimension. This is not the case with the eolian outcrop which clearly shows two scales of heterogeneity. Given its correspondence with the geologic features, the two-scale interpretation is the only possible consistent interpretation.

In discussing the behavior of dispersivity we emphasized the role of the autocorrelation structure of permeability. The concentric grid sampling allows a very efficient investigation of both the magnitude and orientation of the correlation. Figure 17 shows the correlation ellipses for the correlation lengths (expressed as variogram ranges) of the five grids. To compress all of the data on to one plot shows, Fig. 17 is on log-polar (units: log-ft.) coordinates.

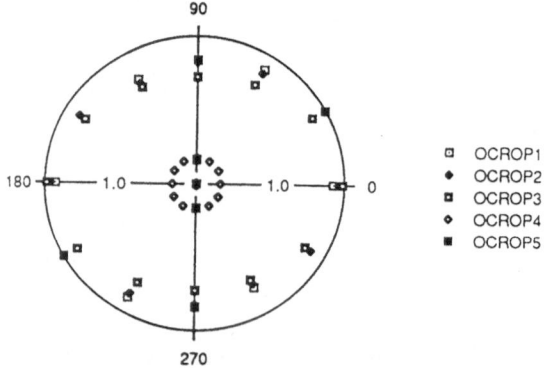

Figure 17. Ellipses of variogram range (correlation length) for the grids in Fig. 13. Three largest ellipses are essentially the same, fourth outcrop is uncorrelated and smallest outcrop has a correlation orientation parallel to local laminae.

Once again grid four is pathologic; it and grid five have different properties from the largest grids. The major correlation length for the largest grids is about 50 feet and orientated at an angle of about 30° above horizontal. The major correlation length for the small scales is about 2 feet and oriented nearly vertically. The latter scale does not appear in grid four (which appears entirely uncorrelated) because its sampling interval is about the same length as this.

Such complexity requires considerable sophistication in generating stochastic fields for conditional simulation. What is required is a generator which can deal with multiple and imbedded correlation scales each with a different variability in three dimensions. The fields must also be anisotropic and have orientation which is also scale-dependent. Generation of fields with this complexity being beyond any currently available generator, this provides another avenue for future research.

The above observations require several qualifications. Several of the variograms used to infer Fig. 17 do not show strict spherical behavior. On the largest scales they are somewhat better described by hole models, but this is entirely consistent with the observation of sampling of multiple beds. More importantly, the data in both Fig. 16 and 17 represent only a single realization on the outcrop. We have no way of knowning if a repetitive experiment on the same outcrop at a different location would give the same results.

With regards to the last point, the confidence of an interpretation is greatly increased (and the number of data required greatly diminished) when the statistics is backed up with the geologic observations. For example, we would be much more hesitant to conclude about the existance of two scales of heterogeneity in the eolian outcrop if such were not consistent with the classifications of Fig. 15. Generally speaking, the geology is consistent with the statistics (the bed orientations on the outcrop do match the orientation angle in Fig. 17) and once this is verified for several diverse outcrops we can have confidence in generating random fields directly from the geologic descriptions. Indeed, it seems reasonable that a taxonomy of reservoir types based on statistical descriptions is possible.

Concluding Remarks

Conditional simulation provides a potential means for efficiently merging geologic observation and engineering. What is required to bring this to common practice are (1) highly time-efficient computer models which can effect multiple realizations without excessive burden, (2) statistical generation procedures that can handle the complexities of geologically realistic media, (3) geologic descriptions which extract quantitative data , and (4) representations or models which can bring about a translation between geological and statistical model with accuracy and confidence.

APPENDIX

Bias and Precision of the Dykstra-Parsons Coefficient

The Dykstra-Parsons coefficient is a normalized measure of the spread of a permeability distribution that is bounded between zero and one. The formal definition is

$$V_{DP} = 1 - \frac{k_\sigma}{k}$$

where k and k_σ are the medium permeability and the permeability at one standard deviation below the median, respectively. Both quantities are taken from best-fit straight lines to the stochastic data plotted on log-probability paper. Inaccuracies caused by this practice are small when the permeability data are log-normally distributed as assumed is the case in the work here. In what follows we discuss the bias and precision in estimating V_{DP}. For log normal distributions C_v and V_{DP} are related as

$$C_v^2 = \exp[\ln(1 - V_{DP})]^2 - 1 \ .$$

Imagine that there exists a population of permeabilities consisting of a very large number of datum. The population has a known V_{DP}. Let us draw a certain number N from this population and calculate an estimated $\hat{V}_{DP}$ from this sample. Repeating the procedure several times will generate a probability distribution function (p.d.f.) for the estimate $\hat{V}_{DP}$. This function that illustrates the difficulties in estimating V_{DP}.

Suppose each repetition gives exactly the same value for $\hat{V}_{DP}$, but this value is different from V_{DP}. The estimator is biased and the magnitude of the bias b_v is the expected value of $\hat{V}_{DP}$ minus V_{DP} divided by the true value. The estimator is precise, however.

On the other hand, suppose the p.d.f. has a finite spread, but the mean of the distribution is V_{DP}. The estimator is unbiased, but imprecise and the measure of the imprecision is the standard error s_v of the p.d.f. In actuality, the estimate for V_{DP} is neither unbiased nor precise, but both can be estimated.

Figure 18 plots the bias as a function of the number of datum N in the samples and the true value of V_{DP}. As expected, the bias approaches zero with increasing sample size, and is largest for small sample sizes. The bias is also largest when V_{DP} is about 0.7; however, in no case is the bias large. We conclude that V_{DP} is essentially unbiased except perhaps at very small sample sizes.

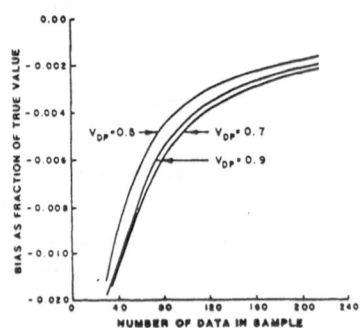

Figure 18. Bias b_v of Dykstra-Parsons coefficient $\hat{V}_{DP}$ estimator of heterogeneity as functions of number of samples and numerical value of V_{DP}. Curves were theoretically derived from a single-population, log-normal distribution.

Unfortunately, the estimator is fairly imprecise, Fig. 19. Here we are plotting N versus V_{DP} for a given standard error. Once again, precision improves as the sample size becomes larger, but even for fairly large sample sizes the error is not negligible.

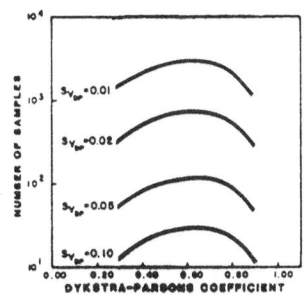

Figure 19. Number of samples required for a given level of precision for the Dykstra-Parsons coefficient V_{DP}. s_V is the standard error of the estimate at the given value of V_{DP} and sample number. Curves were derived from theoretical considerations simulations based a on single-population, log-normal distribution.

NOMENCLATURE

b_v	Bias
C	Concentration, analytic solution
C_s	Concentration, numerical solution
C_v	Coefficient of variation
D	Dimensionless diffusion coefficient
D_o	Molecular diffusion coefficient, L^2/t
$E(\cdot)$	Expectation operator
erf	Error function
f_s	Shale fraction
F	Formation resistivity factor
k	Permeability, L^2
K	Dispersion coefficient, L^2/t
L	Medium length, L
N	Number of datum in a sample
N_{Pe}	Peclet number
N_{HP}	Number of Heller points
N_{GP}	Number of grid points or blocks
s_v	Standard error in Dykstra-Parsons coefficient
t	Time, volume injected as fraction of pore volume
u	Superficial velocity, L/t

v   Interstitial velocity, L/t

V_{DP} Dykstra-Parsons coefficient

x   Distance parallel to flow, fraction of medium length

y   Distance transverse to flow

Greek:

α   Dispersivity, L

β   Exponent on velocity

Δ   Discrete change

λ   Correlation length

φ   Porosity, fraction

Subscripts:

D   Dimensionless

ℓ   Longitudinal

MA   Macroscopic

ME   Megascopic

σ   Standard deviation

Superscript

^   Estimate

Acknowledgements

This work was supported by the U. S. Department of Energy through grants, DE-AS19-82BC10744 and DE-AC19-85BC10849, and the Center for Enhanced Oil Recovery Research at The University of Texas at Austin.

REFERENCES

Arya, Atul, "Dispersion and Reservoir Heterogeneity," Ph.D. Dissertation, The University of Texas at Austin, 1986.

Haldorson, Helge H., "Reservoir Characterization Procedures for Numerical Simulation," Ph.D. Dissertation, The University of Texas at Austin, 1983.

Jensen, Jerry Lee, "A Statistical Study of Reservoir Permeability Distributions," Ph.D. Dissertation, The University of Texas at Austin, 1986.

Lake, Larry W., Alan J. Scott and Gary Kocurek, "Reservoir Characterization for Numerical Simulation," Final Report, sponsored by the U.S. Department of Energy/Bartlesville Project Office, June 1986.

Lake, Larry W., Mark A. Miller and Gary Kocurek, "A Systematic Procedure for Reservoir Characterization," Annual Report, sponsored by the U.S. Department of Energy/Bartlesville Project Office, December 1986.

Vortex Methods for Porous Media Flows

E. Meiburg   and   G.M. Homsy

Department of Chemical Engineering
Stanford University
Stanford, CA 94305

1. Introduction

Displacement processes involving two or more phases in a porous medium present the basis of certain enhanced oil recovery schemes. In most practical applications, these flows exhibit an unstable character, which manifests itself in the generation of a highly distorted displacement front (Homsy 1987). The evolving fingers of the injected fluid significantly reduce the sweep efficiency of the displacement process, thus raising the question as to how the flow might be effectively influenced. A first step towards control of the flow is the investigation of the basic instability mechanisms involving the interaction of viscous and gravitational forces as well as, in the case of immiscible displacement, surface tension forces. In addition, permeability inhomogeneities, diffusion, and dispersion can substantially alter the flow.

Our goal is to gain understanding of the principal fluid-mechanical phenomena in the nonlinear regime by means of computer simulations employing a highly accurate numerical technique. The approach to be described in the following achieves this feature by tracking the displacement front in a purely Lagrangian reference frame instead of a grid-based Eulerian one. In this way, numerical diffusion introduced by conventional finite-difference or finite-element techniques is practically eliminated. Furthermore, a locally refined resolution of the active parts of the displacement front allows to capture all physically relevant length scales for flow parameters well in the nonlinear regime.

2. Governing Equations

In the following, we limit ourselves to the case of two-dimensional flow in order to keep the equations simple, although the concept can easily be extended to three dimensions. We want to consider a porous medium of permeability $k(x,y)$, which initially is saturated by a fluid of density ρ_2 and dynamic viscosity μ_2 (Figure 1). This fluid is to be displaced in the x-direction by injecting a fluid of density ρ_1 and viscosity μ_1. u_0 is the average velocity with which fluid 1 is injected at upstream infinity. c measures the concentration of a solute in the fluid phases and is initially 0 in fluid 2 and c_1 in fluid 1. Between the two fluid phases either a sharp interface exists (immiscible displacement) or a diffusive region will form in which they mix (miscible displacement). The flowfield has a characteristic global dimension L, and we want to include the effect of gravity in our analysis.

2.1 Conservation Laws for Mass, Momentum, and Concentration

Incompressible flow in a porous medium is described by the continuity equation for the velocity vector

$$\nabla \cdot \vec{u} = 0$$

and the equation for the conservation of momentum

$$\nabla p = -\frac{\mu(c)}{k} \vec{u} + \rho(c) g \vec{e}_g - \vec{n} \delta(\vec{x} - \vec{x}_t(s)) \cdot \frac{\sigma}{R(s)}$$

This equation links the pressure gradient to the local velocity (Darcy's law) and also takes into account the effect of the gravity field g acting in the direction $\vec{e}_g$. k represents the permeability, and both the dynamic viscosity μ and the density ρ are assumed to depend on the concentration c.

For the case of immiscible displacement μ and ρ are piecewise constant, and the two fluid phases are separated by a sharp interface at $\vec{x} = \vec{x}_t(s)$ of local radius of curvature R(s), and the interfacial tension σ will affect the pressure field as well. Here s denotes the arclength parameter along the interface. The interface has to satisfy the kinematic boundary condition that its normal velocity is equal to the normal velocity of the fluid at the interface, and it can be considered infinitely thin. As a consequence, the pressure gradient caused by the interface becomes infinite at its location and is aligned with the direction of the outer normal $\vec{n}$ to the interface (Figure 2); hence the δ-function, which vanishes away from the interface.

For the case of miscible displacement, there is no interface between the fluid phases. While in immiscible displacement processes the concentration field is convected along with the fluid, miscible displacement processes give rise to a convective-diffusive equation for the concentration field

$$\frac{\partial c}{\partial t} + \vec{u} \cdot \nabla c = \nabla \cdot (D \cdot \nabla c)$$

where D is the dispersion tensor.

2.2 Scaling

The relevant dimensionless parameters governing the flow can now be derived by scaling the above equations. We introduce the following characteristic quantities for velocity, length, time, pressure, dynamic viscosity, permeability, density, concentration, and diffusion, respectively:

$$u' = u_0$$

$$l' = L$$

$$t' = L/u_0$$

$$p' = \mu_1 L u_0 / k_0$$

$$\mu' = \mu_1$$

$$k' = k_0$$

$$\rho' = \rho_1$$

$$c' = c_1$$

$$D' = D_0$$

Here k_0 represents an average permeability and D_0 denotes a characteristic diffusion coefficient for the two fluid phases.

By scaling the above equations with these quantities we obtain the following dimensionless equations for velocity, pressure, and concentration

$$\nabla \cdot \vec{u} = 0$$

$$\nabla p = -\frac{\mu(c)}{k}\vec{u} + \rho(c)\vec{e}_g \frac{g\rho_1 k_0}{\mu_1 u_0} - \vec{n}\,\delta(\vec{x}-\vec{x}_t)\cdot\frac{1}{R(s)}\frac{\sigma k_0}{\mu_1 u_0 L^2}$$

$$\frac{\partial c}{\partial t} + \vec{u} \cdot \nabla c = \frac{1}{Pe} \nabla \cdot (D \cdot \nabla c)$$

$\mu(c)$ and $\rho(c)$ are now dimensionless functions which go between 1 and R_μ and R_ρ, respectively, these being the ratios of the fluid phase properties

$$R_\rho = \frac{\rho_1}{\rho_2}$$

$$R_\mu = \frac{\mu_1}{\mu_2}$$

$D(\vec{u})$ is a dimensionless function also. We can identify three dimensionless parameters

$$B = \frac{\rho_1 g k_0}{\mu_1 u_0}$$

$$Ca = \frac{\mu_1 u_0 L^2}{\sigma k_0}$$

$$Pe = \frac{u_0 L}{D_0}$$

The parameter B describes the relative importance of gravitational and

viscous forces and becomes important when gravity plays a role. The ratio
of viscous and surface tension forces is given by the Capillary number Ca,
while the Peclet number Pe relates convective and diffusive transport of
the fluid phases.

2.3 Hele-Shaw Equations

Since it is difficult to visualize the flow in a porous medium, most of
the experimental work on unstable displacement processes has been
performed in Hele-Shaw cells (Figure 3). For recent reviews see Saffman
(1986) and Homsy (1987). These consist of two flat glass plates of
spanwise extent 2L separated by a narrow gap of width b. In the limit
b/L<<1 the creeping flow of a single phase averaged over this gap to a first
approximation is governed by the same equations for the conservation of
mass and momentum as those given in 2.2 if the permeability k is assumed
to be of the constant value

$$k = \frac{b^2}{12}$$

This form of the governing equations is called the Hele-Shaw equations
(e.g. Lamb 1932, §330). The parameter B and the Capillary number for
Hele-Shaw flows have the form

$$B = \frac{\rho_1 g b^2}{12 \mu_1 u_0}$$

$$Ca = \frac{12\mu_1 u_0}{\sigma} \cdot \frac{L^2}{b^2}$$

For flows involving more than one phase, the analogy between porous media flows and Hele-Shaw flows is of limited value, see Homsy (1987) and references therein. Nevertheless, both flows exhibit very similar features on a large scale, and an investigation based on the Hele-Shaw equations is expected to prove useful for the understanding of porous media flows as well. Saffman and Taylor (1958) used the Hele-Shaw equations as a basis for the linear stability analysis of the experimentally observed viscous fingering instability and showed that, without surface tension, the linear growth rate of a wavy perturbation of the initially flat interface is proportional to the wavenumber. Both Saffman and Taylor (1958) and Chouke, van Meurs, and van der Poel (1959) demonstrated the stabilizing effect of surface tension, resulting in a cutoff wavelength below which all waves are linearly stable.

2.4 Vorticity-Streamfunction Formulation

By rewriting the governing equations in terms of the vorticity and streamfunction variables we will be able to integrate the full initial value problem by employing numerical techniques similar to certain ones known from the field of inviscid potential flows. This recasting of the porous media flow equations into the vorticity and streamfunction variables was first suggested by Josselin de Jong (1960). We introduce the vorticity

$$\omega = \nabla \times \vec{u} = \frac{\partial v}{\partial x} - \frac{\partial u}{\partial y}$$

and the streamfunction ψ

$$\nabla \psi = \begin{pmatrix} -v \\ u \end{pmatrix}$$

By using the vorticity-streamfunction formulation we satisfy the continuity equation identically. Upon taking the curl of the momentum equation and applying the chain rule we obtain

$$\frac{\mu}{k}\omega = \mu \nabla \psi \cdot \nabla(k^{-1}) + \frac{\nabla \psi}{k}\nabla \mu + B \nabla \times (\rho \vec{e}_g) - \frac{1}{Ca}\delta(\vec{x}-\vec{x}_t(s)) \cdot \frac{\partial}{\partial s}\left(\frac{1}{R(s)}\right)$$

The vorticity-streamfunction formulation is completed by deriving an expression that yields the velocity field as a function of the vorticity field. This can either be achieved by combining the definitions of vorticity and streamfunction to yield the Poisson equation

$$-\omega = \frac{\partial^2 \psi}{\partial x^2} + \frac{\partial^2 \psi}{\partial y^2}$$

or by explicitly using the continuity equation and the definition of vorticity to yield the Biot-Savart law (e.g. Batchelor 1967, p. 87)

$$\vec{u}(\vec{x},t) = -\frac{1}{2\pi} \int \frac{(\vec{x}-\vec{x}') \times \vec{e}_z \omega(\vec{x}',t)}{|\vec{x}-\vec{x}'|^2} \, d\vec{x}' + \vec{u}_{pot}(\vec{x},t)$$

In general, the Poisson equation is more convenient to solve on a fixed grid, especially when fast Poisson-equation solving routines can be applied. The Biot-Savart law, on the other hand, offers advantages when the evolution of a relatively small vorticity field of simple configuration is to be tracked with high accuracy in a Lagrangian reference frame. An overview over two- and three-dimensional vortex methods for incompressible inviscid and nearly inviscid flows is given by Leonard (1980), (1985).

The momentum equation in its vorticity form presents the key to a numerical simulation of porous media flows based on the vorticity variable, since it tells us that the flowfield becomes rotational as a result of

- permeability inhomogeneities

- mobility variations

- density variations in a gravity field

- surface tension effects

It furthermore has the remarkable feature of being an algebraic equation, which means that local information about the flow and the porous medium suffices for the determination of the value of the vorticity variable. Consequently, we do not have to solve a large system of equations by matrix inversion as would be the case if the vorticity were governed by a differential equation.

3. Numerical Technique

As a first step, we have developed a numerical technique for the simulation of two-dimensional immiscible displacement processes in a porous medium of constant permeability, i.e. for the solution of the Hele-Shaw equations. We assume the flow to be periodic in the spanwise y-direction. The momentum equation in its vorticity form states that for this kind of flow the bulk of the two fluid phases is irrotational, since μ and ρ are piecewise constant. Vorticity is present only at the interface between the fluid phases where jumps in the mobility and the density occur in addition to surface tension forces. As a result, we can effectively simulate the full two-dimensional flow by tracking the evolution of the vorticity distribution at the fluid interface, i.e. the dimension of the problem can be reduced by one.

For choosing the optimum way of discretizing the vorticity sheet it is important to notice that the only physical length scale in the problem is set by the surface tension. It has the effect of damping small scales that otherwise would exhibit linear growth rates inversely proportional to their wavelength. Consequently, for a study of the physically most interesting limit of low surface tension, i.e. high Capillary number, extreme care has to be taken in the process of discretizing the vorticity field. The goal is to minimize the effect of introducing a numerical length scale through the discretization procedure. The crudest approach would be to represent the circulation by many point vortices, but it is obvious that this highly singular distribution would set a numerical length scale in the form of the distance between the vortices. Vortex-in-cell methods represent an improvement, as shown in the work by Tryggvason and Aref (1983, 1985), but the mesh size still provides a numerical cutoff length. In the current work, we try to gain an even smoother and more accurate representation of the vorticity sheet by discretizing it into n circular arcs as shown in Figure 4. Mangler and Smith (1959) were the first ones to represent a vortex sheet by a single circular arc. The continuous distribution of the vortex sheet strength along arc i is assumed to be of the form

$$\gamma_i(\phi) = \alpha_i + \beta_i \sin\phi$$

Mangler and Smith showed that this distribution has the advantage that the velocity field associated with it can be obtained by integrating the Biot-Savart law analytically. Here α_i is the average vortex sheet strength of the two endpoints of the circular arc and β_i provides a continuous transition along the arc between these endvalues. Recently, this form of discretization has been successfully employed by Higdon and Pozrikidis

(1985) in their investigation of the evolution of a vortex sheet in inviscid flow.

After setting an initial interface shape, the calculation proceeds in discrete time steps as follows. At every time step, the velocities of n marker points $\vec{x}_i$ (one per circular arc) on the interface are determined by integrating over the vorticity distribution of all other circular arcs using the Biot-Savart law, which for an infinitely thin vortex sheet yields

$$\vec{u}(\vec{x}_t(s),t) = -\frac{1}{2\pi}\int \frac{(\vec{x}_t(s)-\vec{x}'_t(s)) \times \vec{e}_z \omega(\vec{x}'_t(s),t)}{|\vec{x}_t(s)-\vec{x}'_t(s)|^2} d\vec{x}'_t(s) + \vec{u}_{pot}(\vec{x}_t(s),t)$$

where the Cauchy principal value is taken. An analytical expression taking into account the periodic images of the interface in the spanwise y-direction can be derived by analytical summation (Higdon and Pozrikidis 1985). We evaluate this integral using a trapezoidal finite difference scheme, which is sufficient since the overall error in evaluating the velocity of $\vec{x}_i$ is determined by the accuracy of the integration over the interfacial section next to $\vec{x}_i$. For the same reason, the velocity induced on a marker particle by a vortex sheet arc is evaluated by means of the point vortex approximation if the distance between arc and marker particle exceeds the width of the Hele-Shaw cell.

Once the velocities of the marker particles are known, these are advanced over the time step by a fourth order Runge-Kutta method. As a result, we have the position of the interface at the new time at n discrete points $\vec{x}'_i$. We now fit a new set of n new circular arcs through these points in the following way. First a cubic spline is fitted through the updated $\vec{x}'_i$, with the straight-line distance between the points serving as the splining parameter. With the help of this cubic spline we then find the midpoints $\vec{xc}'_i$ between the $\vec{x}'_i$. Subsequently, a new circular arc is fitted through each set $\vec{xc}'_i$, $\vec{x}'_i$, $\vec{xc}'_{i+1}$.

As a next step, the vorticity distribution has to be updated. For this purpose, we first determine the values of the vortex sheet strength at the $\vec{x}'_i$ using the momentum equation in its vorticity form. As an initial estimate, $\nabla\psi$ is approximated by the velocity values of the previous time step. By applying a cubic spline we find the vortex sheet strength at $\vec{xc}'_i$ and $\vec{xc}'_{i+1}$, upon which the coefficients α'_i and β'_i can be determined in the way described above. An iterative procedure is subsequently applied until the new values of the velocities have converged.

At fixed time intervals, the marker points $\vec{x}_i$ along the interface are redistributed and new ones are added in order to avoid deteriorating resolution as a result of local interface generation. This remeshing procedure takes into account the local Capillary number at $\vec{x} = \vec{x}_t(s)$ formed with the local velocity component $\vec{u}(\vec{x}_t(s)) \cdot \vec{n}(\vec{x}_t(s))$ normal to the interface

$$Ca(\vec{x}_t(s)) = \frac{12\mu_1\vec{u}(\vec{x}_t(s))\vec{n}(\vec{x}_t(s))}{\sigma} \cdot \frac{L^2}{b^2}$$

and requires that even the shortest wavelength with a positive linear growth rate for this Capillary number

$$l_c = \frac{2\pi}{Ca^{1/2}}$$

is resolved by a certain number of points, typically four or eight in the test calculations. In order to provide adequate resolution, the spacing Δs between the marker points in addition is reduced in regions of high interface curvature by requiring

$$\Delta s + \frac{c}{R(s)}$$

instead of Δs to be less than this fraction of l_c, where c is a constant. Similar remeshing criteria were employed by DeGregoria and Schwartz (1986) in their boundary integral technique.

As the initial interface becomes more and more convoluted, the rate of change of the flow variables increases rapidly, so that the time step has to be reduced. It is adjusted in such a manner that the difference between the final corrector velocity and the initial predictor velocity multiplied by the actual time step nowhere in the flow exceeds a certain numerical value established in test calculations. Acceleration effects are most pronounced in the 'active' sections of the interface, i.e. the finger tips. Those sections left behind in the displacement process, on the other hand, move at small and almost constant velocities. Our numerical algorithm takes advantage of this fact by employing a local time-stepping procedure which updates the velocities of the most active interface sections five times as often as those of the rest of the interface. This results in significant savings especially in the advanced stages of finger growth since a large fraction of the interface becomes passive.

In summary, we apply a numerical procedure which achieves high spatial accuracy while minimizing the effect of setting a numerical length scale by discretizing the vorticity sheet along the interface into circular arcs with a continuous vorticity distribution. The method is of fourth order in time and partly compensates the $O(N^2)$ operation count associated with the direct interaction algorithm by using a locally refined discretization as well as local time-stepping.

4. Validation of the Numerical Algorithm

In the following, we compare numerical results for immiscible flows without gravity obtained with our algorithm with analytical as well as experimental results of other authors in order to validate our numerical procedure and investigate its limitations. The calculations to be presented were carried out on the CRAY X-MP of the San Diego Supercomputer Center and typically required less than one minute of CPU time.

4.1 Flows with vanishing surface tension

Aitchison and Howison (1985) report analytical results for the evolution of the interface in a linear Hele-Shaw cell of width 2π in the case of vanishing surface tension. They find that if at t=0 the boundary between the two fluid phases consists of a slightly perturbed planar interface of the shape

$$x = \varepsilon\cos\varphi \ , \quad y = \varphi - \varepsilon\sin\varphi \ , \quad 0 \leq \varphi \leq 2\pi$$

the perturbation will grow until at a time

$$t_{crit} = \frac{1}{U_0}\left(\frac{\varepsilon^2}{2} - \frac{1}{2} + \ln\left(\frac{1}{\varepsilon}\right)\right)$$

a cusp will form at the interface. U_0 is the velocity at which fluid is injected into the cell. For this flow there exists no natural length scale, which makes its simulation a difficult test case for numerical methods. Since the problem does not provide for a damping mechanism, any computational technique that does not introduce some numerical damping will break down after a finite time $t < t_{crit}$ due to the fact that discretization and roundoff errors trigger the growth of short wavelength oscillations. Consequently, the time up to which a numerical scheme without artificial damping is able to yield a smooth solution for different levels of discretization presents an indication of the numerical stability characteristics of the computational algorithm. We have attempted to study the evolution of the interface for the case $\varepsilon = 0.2$ and $U_0 = 1$, in which

the cusp should appear at $t \approx 1.129$. Figure 5 shows the minimum radius of curvature of the interface as a function of time for successively finer discretizations of the interface. The number of marker points given in Figure 5 indicates the initial number; remeshing took place after time intervals of 0.05. The time step was reduced according to the criterion that the product of the time step and the difference between the initial predictor velocity and the final corrector velocity should for none of the marker points exceed a constant $c_{\Delta t}$.

As the resolution of the vortex sheet is refined and the time step is reduced, the correct time for the formation of the cusp is approached, thus demonstrating convergence of our numerical scheme. Figure 6 shows the evolution of the interface with time, and the emerging singularity at the tip is clearly visible. The numerical algorithm is able to follow the decrease of the radius of curvature at the tip over more than an order of magnitude, well into a regime where the interface exhibits a small region of very high curvature. This set of test calculations demonstrates the ability of the numerical scheme to correctly track a variety of length scales. Numerical instability effects do not noticably influence the solution until very late in time, when the minimum radius of curvature has become very small.

4.2 Steady Fingers with Surface Tension

Finite values of surface tension introduce a physical length scale into the problem. They damp very short waves and prevent the evolution of a singularity in the interface shape. Instead, for relatively low values of the surface tension above the threshold value for the onset of linear instability, one observes the evolution of fingers of constant width. Their shape is steady in the reference frame moving with the finger tip. McLean and Saffman (1981) calculated the width of steady Hele-Shaw fingers as a function of the Capillary number. These results have been used by other authors (Tryggvason and Aref 1985, DeGregoria and Schwartz 1986) to check their time-dependent numerical procedures. Figure 7 compares our numerically obtained results to the exact ones of McLean and Saffman. For clarity, we present our results in the same form as Tryggvason and Aref and DeGregoria and Schwartz, respectively, expressing the surface tension parameter as the ratio of the most amplified wavelength and the cell width $l_{max}/2L$. For the calculation of the width of a steady finger, we start our calculation with a sinusoidal perturbation of amplitude 0.1 of the flat interface. The ratio λ of the width of the steady finger to the width of the Hele-Shaw cell was calculated from the condition of mass conservation

$$u_{tip} \lambda = 1$$

where u_{tip} is the converged value of the numerically obtained tip velocity. We have carried out two sets of calculations in which the discretization is determined by the condition that the shortest wavelength linearly unstable under local conditions should be resolved by at least four or eight points, respectively. The time step is empirically determined by the criterion that a reduction by 50 per cent of the time step should not change the value of u_{tip} by more than 0.1 per cent. The marker points at the interface were remeshed after time intervals of 0.05. The numerical algorithm demonstrates that the results converge as the discretization is improved. For the finer discretization it reproduces the exact results by McLean and Saffman with high accuracy for the whole range of dimensionless surface tension from the onset of linear instability of the displacement process to the appearance of a tip splitting instability at around $l_{max}/2L\approx0.3$. We do not confirm findings by Tryggvason and Aref (1985) of fingers considerably narrower than predicted by the results of McLean and Saffman.

4.3 The Experimentally Observed Shape of the Tip of Steady Fingers

Pitts (1980) studied the shape of the tip of steady viscous fingers as a function of the dimensionless finger width, λ, which is in turn determined by the Capillary number. For the coordinate system $\bar{x}$, $\bar{y}$ moving with the finger tip, he found an analytical expression that matched the experimentally observed tip shape for a range of finger widths

$$e^{(\pi\bar{x}/2\lambda)} \cos(\pi\bar{y}/2\lambda) = 1$$

However, he did not give a rigorous derivation of this expression. Figure 8 presents a comparison between our result and the finding of Pitts for $\lambda=0.54$, a value at which Pitts' result reproduces experimental observations quite accurately. Again, the numerical results agree well with the analytical expression, which establishes confidence in the numerical procedure that it can reproduce the dynamics of the finger tip with a high degree of accuracy.

5. Conclusion

We have described and tested a new numerical procedure for calculations of porous media flows. The technique is based on the vorticity formulation of the momentum conservation equation and tracks the interface in a purely Lagrangian reference frame. In this way, it avoids the dissipative effects introduced by conventional grid-based methods and allows a highly accurate discretization of the interface. Test calculations for Hele-Shaw flows show good quantitative agreement with exact results. As a next step, we plan to carry out simulations for higher Capillary numbers at which the fingers develop a tip splitting instability. Subsequently, the numerical code is to be extended in order to account for diffusion effects. Thus, we will be able to calculate miscible displacement processes as well.

6. Acknowledgements

Acknowledgement is made to the Donors of The Petroleum Research Fund, administered by the American Chemical Society, for support of this research, and to the San Diego Supercomputer Center for providing computer time on the CRAY X-MP.

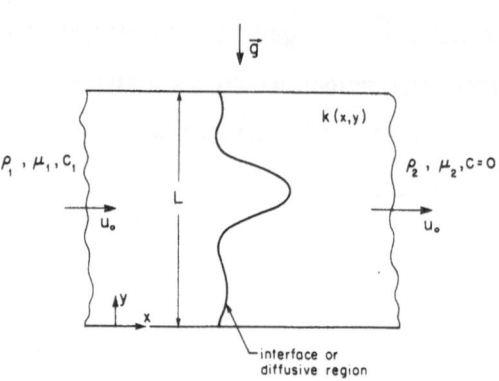

Figure 1:   Sketch of a two-dimensional displacement process in a porous medium.

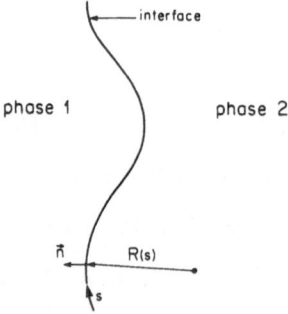

Figure 2:   We characterize the interface in an immiscible displacement process by its arclength parameter s, the local radius of curvature R(s), and the outer normal $\vec{n}$.

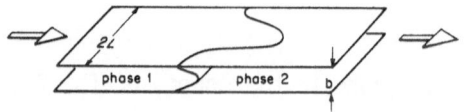

Figure 3:   The Hele-Shaw configuration.

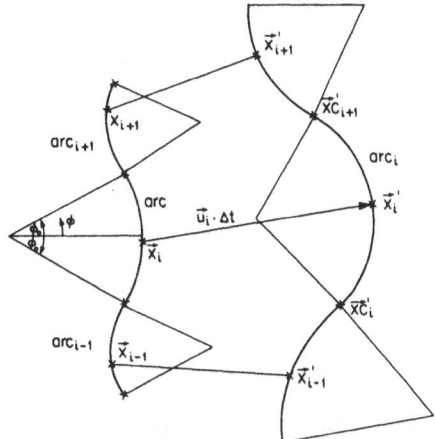

Figure 4: The interface is discretized into circular arcs. Each arc is represented by a marker point $\vec{x}$ which is advanced over a time step Δt. Subsequently, the end points $\vec{xc}$ of the new circular arcs at the updated time level are found as the centers between the $\vec{x}$ on a spline representation of the interface based on all marker points.

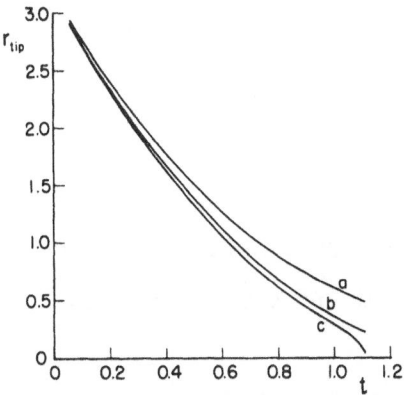

Figure 5: The temporal decrease of the minimum radius of curvature r_{tip} along the interface for various discretizations.

a: initial discretization of the interface into 9 segments

b: initial discretization of the interface into 15 segments

c: initial discretization of the interface into 19 segments

The analytical solution of predicts the appearance of a cusp singularity at a time $t_{crit} \approx 1.129$.

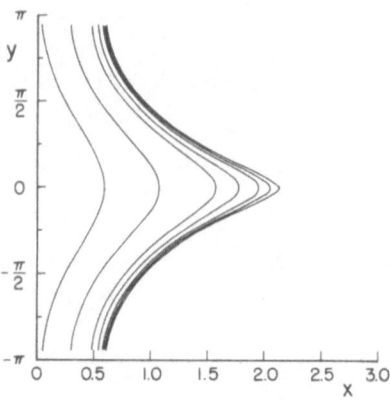

Figure 6: The evolution of the interface in time. At the tip of the interface a cusp forms.

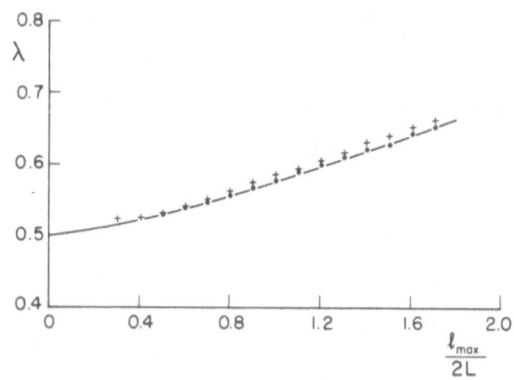

Figure 7: Comparison of the numerically obtained ratio of finger width to cell width as a function of the dimensionless surface tension with the exact solution by McLean and Saffman (1981).

+   discretization into four segments per shortest linearly unstable wavelength

*   discretization into eight segments per shortest linearly unstable wavelength

--   exact solution

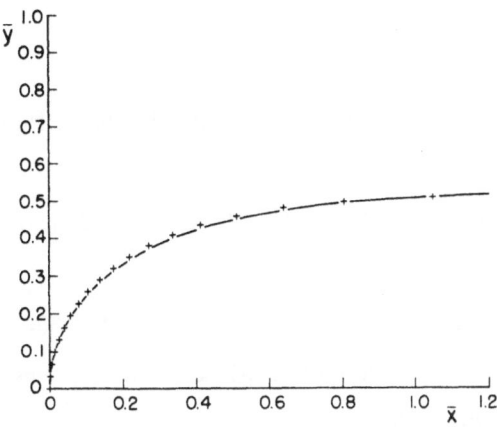

Figure 8:   Comparison of the numerically obtained tip shape with the relation given by Pitts (1980) for the ratio of finger width to cell width of 0.54.

7. References

Aitchison, J.M. and Howison, S.D. 1985 Computation of Hele-Shaw Flows with Free Boundaries. J. Comp. Phys. 60, 3, p.376-390.

Batchelor, G.K. 1967 An Introduction to Fluid Dynamics. Cambridge University Press.

Chouke, R.L., van Meurs, P., van der Poel, C. 1959 The Instability of Slow, Immiscible, Viscous Liquid-Liquid Displacements in Permeable Media. Trans. AIME 216, p.188-194.

DeGregoria, A.J. and Schwartz, L.W. 1986 A Boundary-Integral Method for Two-Phase Displacement in Hele-Shaw Cells. J. Fluid Mech. 164, p. 383-400.

Higdon, J.J.L. and Pozrikidis, C. 1985 The Self-Induced Motion of Vortex Sheets. J. Fluid Mech. 150, p. 203-231.

Homsy, G.M. 1987 Viscous Fingering in Porous Media. Ann. Rev. Fl. Mech. 19, p. 271.

de Josselin de Jong, G. 1960 Singularity Distributions for the Analysis of Multiple-Fluid Flow through Porous Media. J. Geophys. Res. 65, 11, p. 3739-3758.

Lamb, H. 1932 Hydrodynamics. 6th edn. Cambridge University Press.

Leonard, A. 1980 Vortex Methods for Flow Simulation. J. Comp. Phys. 37, p. 289-335.

Leonard, A. 1985 Computing Three-Dimensional Incompressible Flows with Vortex Elements. Ann. Rev. Fluid Mech. 17, p. 523-559.

Mangler, K.W. and Smith, J.H.B. 1959 A Theory of the Flow past a Slender Delta Wing with Leading Edge Separation. Proc. Roy. Soc. Lond. A251, p. 200-217.

McLean, J. and Saffman, P.G. 1981 The Effect of Surface Tension on the Shape of Fingers in a Hele-Shaw Cell. J. Fluid Mech. 102, p. 455-469.

Pitts, E. 1980 Penetration of a Fluid into a Hele-Shaw Cell. J. Fluid Mech. 97, p. 53-64.

Saffman, P.G. 1986 Viscous Fingering in Hele-Shaw Cells. To appear in J. Fluid Mech.

Saffman, P.G. and Taylor, G.I. 1958 The Penetration of a Fluid into a Porous Medium or Hele-Shaw Cell Containig a More Viscous Liquid. Proc. Roy. Soc. Lond. A245, p. 312-329.

Tryggvason, G. and Aref, H. 1983 Numerical Experiments on Hele-Shaw Flow with a Sharp Interface. J. Fluid Mech. 136, p.1-30.

Tryggvason, G. and Aref, H. 1985 Finger-Interaction Mechanisms in Stratified Hele-Shaw Flow. J. Fluid Mech. 154, p. 287-301.

A PARAMETRIC STUDY OF VISCOUS FINGERING IN MISCIBLE DISPLACEMENT BY NUMERICAL SIMULATION

D.E. Moissis, C.A. Miller and M.F. Wheeler
Rice University, Houston, TX 77251

Numerical simulation is used to study the effects of several parameters on miscible viscous fingering. A miscible flood of a rectangular slab is simulated in two spatial dimensions. The parameters, obtained by the dedimensionalization of the governing equations, are the viscosity ratio, Peclet numbers associated with molecular diffusion, longitudinal dispersion and transverse dispersion and the aspect ratio of the slab. The effects of local permeability variations and the overall heterogeneity of the porous medium are also considered. A finite element modified method of characteristics is used for the solution of the concentration equation, combined with a mixed finite element method for the solution of the pressure equation. This scheme is essentially free of numerical dispersion.

The results suggest that the local permeability distribution near the entrance of the porous medium plays an important role in finger generation, while the permeability distribution downstream does not significantly affect fingering. The number of developing fingers and their growth rates depend strongly on the mobility ratio. The aspect ratio of the slab also influences significantly the number of fingers.

Introduction

Viscous fingering is a phenomenon that is important in miscible EOR processes. Its understanding and accurate quantitative prediction may be critical in decisions on the applicability of certain recovery processes, especially where expensive solvents are to be used.

Viscous fingers are initiated by heterogeneities in both the porous medium (permeability variations) and the fluid flow; the way they grow or are damped is a result of the combined effect of several factors, such as the mobility ratio, dispersion, gravity and the heterogeneity of the porous medium.

Several simplified models of miscible viscous fingering have been developed, in which the factors affecting viscous fingering were

introduced as a number of parameters. Thus, Koval[2] proposed a one-parameter model; Todd and Longstaff[3] proposed a 2-parameter model. More
recently Fayers[4] introduced a model with physically interpretable
parameters. In all these models the heterogeneity of the porous medium
was treated indirectly, by adjusting the value of a single parameter.
However, heterogeneities can exist at different length scales, with
different qualitative and quantitative effects on the displacement[5,6].

In this paper we proceed with a dedimensionalization of the equations of miscible displacement to identify the dimensionless parameters that determine the solution. We then use numerical simulation to
study the effect of each of these parameters on the dynamics of viscous fingering and on the displacement in general. The effects are
expressed in terms of the number of fingers, their width, their locations and the root mean square growth rate.

We also study the effect of the permeability distribution and the
heterogeneity of the porous medium on the generation and development
of fingers. In the present paper we are interested in small scale heterogeneities, i.e. within a slab suitable for laboratory experiments.

We use a finite element modified method of characteristics[7] for the
approximation of the concentration of the invading fluid. The method
is essentially free of numerical dispersion. It is combined with a
mixed finite element method, used for the calculation of an accurate
velocity field.[9]

The Model

We assume two-dimensional, incompressible, horizontal miscible displacement in a rectangular slab of a porous medium. The pore space is
initially filled with oil (C=0) and flooded at the side x=0 with pure
solvent (C=1). So x is the principal flow direction and y is the
direction transverse to the flow. Under these assumptions the displacement can be modeled by the following set of equations :

$$u = - \frac{k}{\mu} \frac{\partial p}{\partial x} , \tag{1a}$$

$$v = - \frac{k}{\mu} \frac{\partial p}{\partial y} , \tag{2a}$$

$$\frac{\partial u}{\partial x} + \frac{\partial v}{\partial y} = 0 , \tag{3a}$$

$$\phi\frac{\partial C}{\partial t} + u\frac{\partial C}{\partial x} + v\frac{\partial C}{\partial y} = \nabla\cdot(D\nabla C) , \tag{4a}$$

where D is the dispersion tensor $D=D(u,v,\mathcal{D},a_\ell,a_t)$, $\mu=\mu(C,\mu_o,\mu_s)$ and $k=k(x,y)$. $\mathcal{D}$ is the molecular diffusion and a_ℓ and a_t are the longitudinal and transverse mixing lengths. μ_o and μ_s are the oil and solvent viscosities respectively.

By scaling lengths in the x direction by the length of the slab L_x, lengths in the y direction by the width of the slab L_y, velocities by the average flowrate per unit area q, viscosity by the oil viscosity μ_o, permeability by the arithmetic mean permeability $\bar{k}$, pressure by $q\mu_o L_x/\bar{k}$, and by expressing time in Pore Volumes Injected (PVI), the above equations are reduced to the following dimensionless form :

$$u = -\frac{k}{\mu}\frac{\partial p}{\partial x} , \tag{1}$$

$$v = -a\frac{k}{\mu}\frac{\partial p}{\partial y} , \tag{2}$$

$$\frac{\partial u}{\partial x} + a\frac{\partial v}{\partial y} = 0 , \tag{3}$$

$$\frac{\partial C}{\partial t} + u\frac{\partial C}{\partial x} + av\frac{\partial C}{\partial y} = \nabla\cdot(D\nabla C) , \tag{4}$$

where : $D=D(u,v,Pe_m,Pe_\ell,Pe_t)$, $\mu=\mu(C,M)$ and $k=k(x,y)$, and the various dimensionless parameters are as defined below.

The dimensionless dispersion tensor has the following form :

$$D = Pe_m^{-1}\begin{bmatrix} 1 & 0 \\ 0 & a^2 \end{bmatrix} + \frac{Pe_\ell^{-1}}{\|u\|}\begin{bmatrix} u^2 & auv \\ auv & a^2v^2 \end{bmatrix} + \frac{a^{-1}Pe_t^{-1}}{\|u\|}\begin{bmatrix} v^2 & -auv \\ -auv & a^2u^2 \end{bmatrix} \tag{5}$$

The boundary conditions are :

$v=0$, $\frac{\partial C}{\partial y}=0$ at y=0 and y=1 (nonflow boundary),
$p=p_0(t)$, $C=1$ at x=0 (inflow end) and
$p=0$, $\frac{\partial C}{\partial t}+u\frac{\partial C}{\partial x}=0$ at x=1 (outflow end).

The value of p_0, which is a function of time, is adjusted at each time step so that the average flowrate per unit area q remains constant.

All variables and parameters contained in the above equations are dimensionless. The parameters are the following :

Viscosity ratio : $M = \mu_0 / \mu_s$, Aspect ratio of the slab : $a = L_x / L_y$
and Peclet numbers : $Pe_m = q L_x / \phi \, \mathcal{D}$, $Pe_\ell = L_x / a_\ell$ and $Pe_t = L_y / a_t$.
The three Peclet numbers give the ratios of convective effects to those of molecular diffusion, longitudinal dispersion and transverse dispersion respectively.

It is clear from the above formulation that the solution depends on the values of these five parameters, as well as on the spatial dependence of permeability. In this paper numerical simulation was used to study the effect of each of these parameters on the displacement and more specifically on the number of fingers, their locations, growth rates, and local behaviour.

As a quantitative measure of the instability of the displacement we use the root mean square (RMS) finger length for the concentration contour $C = C_0 = 0.5$, which is defined as follows :

$$RMS = \left\{ \int_0^1 [x(y, 0.5) - \bar{x}]^2 \, dy \right\}^{1/2} \tag{6}$$

where $x(y, C_0)$ are the x coordinates of all points (x, y) for which $C(x, y) = C_0$ and

$$\bar{x} = \int_0^1 x(y, 0.5) \, dy \tag{7}$$

The RMS finger size is simply the standard deviation of the advance of the iso-concentration contour $C = C_0$; here C_0 was chosen equal to 0.5. This quantity is a reasonable measure of the instability of the displacement.

The porous medium considered as a base for this study was a 2ft X 2ft X 0.5in. Berea sandstone slab, which was used in experiments by Giordano, Salter and Mohanty.[ ] They measured the permeability in each of the 1600 (40 X 40) 0.6in. X 0.6in. X 0.5in. parallelepipeds that comprise the slab. The resulting permeability map is shown schematically in Figure 1; we assumed this permeability distribution in our simulations. The permeability was assumed isotropic, with a constant value in each block. The permeability characteristics of the slab are given in Giordano et al.[ ] The arithmetic mean permeability was found to be 430md and the coefficient of variation (ratio of the standard deviation to the arithmetic mean) was 0.133. The latter is a measure of the heterogeneity of the porous medium^2 and was changed in some of our simulation runs.

The following values of the physical parameters were used in the simulation of the base case :

$$\phi=0.195 \ , \ M=75 \ , \ \mathcal{D} =7.5 \ \text{X}10^{-6} \ cm^2/sec \ , \ a_\ell=2.76 \ \text{X}10^{-3} \ ft \ ,$$
$$a_t=7.73 \ \text{X}10^{-5} \ ft \ , \ q=2ft/day.$$

The values of the dimensionless parameters for the various runs are given in Table 1. The value of Pe_m was the same (29409) for all runs except No.7, for which $Pe_m=5882$. In any event it was found that, for the range of the values considered, the role of molecular diffusion in determining longitudinal dispersion was negligible and in determining transverse dispersion rather small.

A quarter power mixing law was used for viscosity (dimensionless):

$$\mu^{-1/4} = 1+C(M^{1/4}-1) \tag{8}$$

The Numerical Method

The coupled system of the pressure equation (Equations 1-3) and the concentration equation (4) is solved sequentially.

The solution of the concentration equation is approximated using a finite element modified method of characteristics. This scheme was first introduced by Douglas and Russell[7] and then analysed by Russell[8] for miscible displacement. In this algorithm the time derivative and the convective terms are combined as a directional τ derivative; time stepping is along the characteristics. The characteristics are tracked backward, using small time steps. The dispersion terms are treated implicitly using large time steps. Continuous bilinears are employed for the concentration approximation space. The truncation error is $O(\Delta t)C_{\tau\tau} + O(\Delta x)^2$, where Δt is the time step and Δx the spatial grid spacing. Since the derivative $C_{\tau\tau}$ that multiplies the $O(\Delta t)$ term is along the characteristic, it is very small and consequently it makes the first term of the above error estimate very small. The remaining term is of order $(\Delta x)^2$.

This method is not conservative and in practice the time step is limited by mass balance requirements. The numerical dispersion generated by this scheme is very small; this feature is particularly important in convection-dominated problems such as the one considered here.

Because the method uses extrapolated values of velocity, high accuracy is required for the flow calculations. Therefore a mixed finite element scheme is used to solve the pressure equation (Wheeler et al[9]). The method does not combine Equations (1)-(3) but solves them separately seeking a finite element approximation for both pressure and velocity. Next lowest order Raviart-Thomas spaces are used as approximating spaces. The benefit of this method is that it provides velocities, which are globally 2nd order accurate; velocities are 3rd order accurate at certain superconvergence points.[10] This high order of accuracy allows the use of a coarser grid than needed for standard methods. In addition, this method conserves mass in each gridblock. This method, in conjunction with the above mentioned scheme for the solution of the concentration equation, has been applied in miscible displacement problems by Ewing, Russell and Wheeler[11] and Russell and Wheeler.[12,13]

In all the simulations presented in this paper an 80 X 80 uniform grid was used for both pressure and concentration. Thus our simulation grid was twice as fine as the permeability grid. After some experimentation with time-stepping, a time step of 0.01 Pore Volumes Injected (PVI) was selected. All simulations were run on a CRAY-XMP at Mendota Heights, MN.

Effect of Permeability Distribution

The concentration profiles (contours of constant concentration, C=0.25, 0.50 and 0.75) at different times are plotted in Fig. 2 for the simulation of the base case. It can be seen that at early times the profile is quite irregular with a fairly large number of small fingers (Fig. 2a). At later times (Fig. 2b) some of the initial fingers (in this case five) grow faster than the rest of the fingers and they eventually dominate the displacement. The growth of the smaller fingers is suppressed and later they merge with larger ones (Fig. 2c,d), leaving unswept areas of oil that can be fairly extensive. All these processes result in a number of large fingers, which grow quite independently of each other, at least until breakthrough. These fingers will be referred to as "active fingers". Although the number of active fingers varies with mobility ratio, dispersion, aspect ratio and the heterogeneity of the medium, the qualitative behaviour presented here is present in all our runs.

The first question that was addressed is to what extent the permeability distribution affects the displacement at both early and later times. To relate the fingering pattern at early times to the permeability distribution near the entrance of the medium, the average (harmonic mean) permeability in the first 4 blocks in the x direction was calculated for each of the 40 divisions in the y direction. This permeability map was then correlated to the fingering pattern at t=0.1 PVI (Fig. 2a). It was observed that a finger appeared almost consistently at each block where this average permeability exhibited a (local) maximum. The largest finger appears at the block where the average permeability has its highest value. The length of the rest of the fingers however does not always correspond to the permeability value; this result is not surprising since the permeability averaging was done in a somewhat arbitrary fashion. (The idea is that the injected fluid in 0.1 PVI has covered on the average 4 permeability blocks in the x direction). Furthermore this result is independent of the viscosity ratio, i.e. the number of initial fingers, their locations and their relative lengths at t=0.1 PVI are the same for all viscosity ratios studied, though the lengths of the fingers are, of course, different. This result strongly suggests that the permeability distribution near the entrance of the medium determines the fingering pattern at early times. This conclusion was corroborated by simulations where the slab was flooded from two of its other three sides. In all cases poor correlation was obtained when the overall average (harmonic mean) permeability was considered, i.e. when all 40 blocks in the x direction were averaged.

To study the importance of the permeability distribution in the rest of the medium, a simulation was run for a slab that was homogeneous, except for the first 8 X 40 blocks near the entrance, i.e. the permeabilities of the remaining 32 X 40 blocks were replaced by a constant value, namely their arithmetic average (Run No.2). The concentration profile at t=0.3 PVI is shown in Fig. 3. By comparing this to Fig. 2c, it follows that modest heterogeneity downstream has little effect on viscous fingering, other than to cause local (small wavelength) irregularities in the shape of the fingers. The effect on the number of active fingers, their locations, their lengths and the RMS finger growth rate is negligible. The same qualitative results were obtained in simulations with M=7.5. (The concentration profiles obtained in Runs No. 3 and No. 6 were approximately the same). Of course the model of the porous medium used is quite homogeneous, in the sense that the permeability variations that exist are rather small. We expect large permeability variation downstream to have a significant effect on viscous fingering.

In summary, viscous fingering is very sensitive to the permeability distribution near the entrance of the porous medium but rather insensitive to permeability variations downstream. Consequently, if viscous fingering is to be modeled correctly, two different time periods have to be considered: an initial period when permeability distribution is important and when there exists a large number of fingers and the later stage of the displacement, when permeability plays little or no role and there exists a number of active fingers, fairly independent of each other. In the former period the displacement can be modeled as a fingered mixing zone; in the latter period it can be modeled as a set of large individual fingers growing with little or no interaction.

It also follows that in unstable EOR processes the values of permeability must be known with a good degree of accuracy near the entrance of the medium (e.g. near wells and especially injection wells), if the process is to be simulated accurately; such a high degree of accuracy is not as important for the permeability of the rest of the medium.

Effect of Heterogeneity of the Porous Medium

In Run No. 4 the permeabilities of all 1600 blocks were altered according to the formula :

$$k_{new} = k_{original} - 0.5 (k_{original} - \bar{k}) \qquad (9)$$

This modification made the medium more homogeneous, by reducing the standard deviation σ of the permeability distribution by half, while the value $\bar{k}$ of the average (arithmetic mean) permeability remained the same ($\bar{k}=1.00$) and the permeability map was not qualitatively altered. Similarly, in Run No. 5 the permeabilities were altered as follows :

$$k_{new} = k_{original} + 0.5 (k_{original} - \bar{k}) \qquad (10)$$

The medium was now made more heterogeneous, by increasing the standard deviation by 50%, while preserving the value of $\bar{k}$ and the permeability map.

The concentration profiles for t=0.4 PVI are shown in Figs. 4 and 5 for runs No. 4 and No. 5 respectively. By comparing Figs. 4,5 and 2d it follows that heterogeneity of the medium affects the number of active fingers, the length of the fingers and the local behaviour, i.e. the smoothness or irregularity of the fingers. Thus a high heterogeneity enhances merging of fingers resulting in a smaller number of active fingers (2 active fingers in Fig. 5 as opposed to 5 fingers in Figs. 4 and 2d). It also yields more irregularly shaped fingers. The difference between Figs. 4 and 2d is rather small, except for some more splitting and merging that occurs for higher SD; however between these two and Fig. 5 there is a significant difference.

Finally the RMS finger growth rate increases with heterogeneity as shown in Fig. 6. It can be observed here that the largest difference in the growth rates for the three cases occurs in the initial 0.1 PVI. This result suggests that it is the permeability distribution near the entrance that mainly causes the difference in growth rates and demonstrates once more the importance of the permeability variations in this region.

Effect of Viscosity Ratio

The concentration profiles at t=0.3 PVI for Runs No. 6 and No. 7 (Viscosity ratios 7.5 and 750 respectively) are shown in Figs. 7 and 8. Comparing these to Fig. 2c (M=75) it follows that the behaviour of the displacement is quite different at low unfavourable viscosity ratios (Fig. 7) and at high and very high viscosity ratios (Figs. 2c and 8).

The behaviour at M=75 was discussed in a previous section. The main characteristic is that the large initial number of fingers is reduced by merging to a smaller number of "active fingers". The effect of the viscosity ratio on this process is very important : At low viscosity ratios (7.5) there are no fingers that grow dramatically faster than the rest; consequently merging and growth suppression occur to a much lesser extent, resulting in a larger number of active fingers, whose growth rates do not vary much. So at low viscosity ratios the displacement can be modeled as a fingered mixing zone even at later times. On the other hand at very high viscosity ratios (750) the active fingers start to outgrow the others earlier than for M=75; more merging and suppression of growth occur and this results in a smaller number of long active fingers. It can also be noticed that as the viscosity ratio increases the fingers are more unstable. Thus, for M=7.5 the fingers are quite smooth, for M=75 they are irregularly shaped and exhibit some splitting and for M=750 they are very irregular and exhibit large scale splitting, growing secondary fingers.

There are two causes which account for this increasing irregularity of the fingers with increasing viscosity ratio. First, the few fingers present at high viscosity ratios grow wider (in the y-direction). Hence they are more subject to small wavelength instabilities than the relatively narrow fingers that appear at low viscosity ratios. Second, the thickness of the displacement front (represented in the plots as the distance between different concentration contours) decreases as viscosity ratio increases. The relatively diffuse front which appears at low viscosity ratios is less susceptible to instabilities than the sharp front that exists at high viscosity ratios.

Fig. 9 shows how the viscosity ratio affects the RMS finger growth rate. The effect seems to be more pronounced at low values of M than at high ones. As viscosity ratio tends to infinity, the RMS growth rate seems to approach an asymptotic value; this could however be a numerical effect, because a finer grid is needed for the accurate simulation of displacements with very high viscosity ratio. It can also be noticed from Fig. 9 that the RMS finger length grows almost linearly with time, except in the initial stage of the displacement. It was also found that growth of active individual fingers is also fairly linear in most cases, except again at early times. This fact is yet another indication of the different phenomena that dominate the displacement at early and at later times.

Effect of Longitudinal Dispersion

Fig. 10 shows the concentration profiles at t=0.3 for Run No. 8. This, compared to Fig. 2c, illustrates the effect of longitudinal dispersion (Although both longitudinal and transverse dispersion in Run No. 8 are 5 times larger than in the base case, the transverse dispersion is still relatively small and increasing the longitudinal dispersion is the dominant effect).

As expected, large longitudinal dispersion yields a thick (more diffuse) displacement front, which makes the fingers less prone to flow disturbances arising from small scale heterogeneities. In Run No. 8 longitudinal dispersion is large enough to make the fingers virtually free of small wavelength disturbances. In addition, since the front is more diffuse, the waves corresponding to small values of concentration grow faster and breakthrough occurs earlier. This may give the impression that fingers grow faster when longitudinal dispersion is larger; in fact, it is hard to define finger lengths or growth rates in this case, because the displacement front is very diffuse; what we can do is to measure RMS growth rates for different concentration contours. In our case we find that the RMS growth rate for $C_0=0.5$ is virtually unaffected by longitudinal dispersion. It is clear that the RMS growth rate increases with longitudinal dispersion for $C_0<0.5$ and decreases for $C_0>0.5$.

Effect of Transverse Dispersion

Figs. 11 and 12 show the concentration profiles at t=0.3 for Runs Nos. 9 and 10 to illustrate the effect of transverse dispersion at different viscosity ratios (Compare to Figs. 7 and 2c respectively).

It should be noted that, although in these two runs transverse dispersion was increased by a factor of 10, its effect is still not very significant. In this range larger transverse dispersion for both low and high viscosity ratios results in shorter and wider fingers and in a thicker displacement front. The fingers are also smoother (the effect of the diffuse front more than compensates for the effect of wider fingers). The number of active fingers has not been affected by transverse dispersion in the cases studied. However merging of small fingers with bigger ones occurs faster when transverse dispersion is larger. We expect even larger transverse dispersion to cause significantly more merging, resulting in a smaller number of active fingers.

RMS finger growth rate decreases as transverse dispersion increases, as is shown in Fig. 13. The effect seems to be more pronounced at low values of M. Also the suppression of finger growth becomes increasingly important as transverse dispersion increases. We expect that sufficient transverse dispersion will completely wipe out fingering at very low viscosity ratios.

Effect of Aspect Ratio

In Run No. 11 the aspect ratio of the slab was changed to 2. The permeability distribution remained the same but the 40 X 40 permeability grid blocks were changed to rectangles of aspect ratio 2. Likewise in Run No. 12 the aspect ratio was changed to 4. The concentration profiles at t=0.3 for these two Runs are shown in Figs. 14 and 15 respectively. The effect of the aspect ratio is made quite clear by comparing these figures to Fig. 2c (aspect ratio 1). When the slab is made narrower, the initial fingers are closer to one another and consequently their interaction is stronger. The result is that growth suppression of smaller fingers and merging occur in the initial stages of the displacement, yielding, even at early times, a small number of active fingers. In our case, for a viscosity ratio of 75, we get five active fingers for a=1, two active fingers for a=2 and one active finger for a=4. The results suggest that, when the aspect ratio is large, the displacement has to be modeled as a number of individual fingers even at early times. This fact causes the RMS growth rate to increase significantly with aspect ratio, as shown in Fig. 16.

Conclusions

The factors that affect viscous fingering can be expressed as dimensionless parameters, most of which arise during the dedimensionalization of the miscible displacement equations. The main parameters are the viscosity ratio, longitudinal and transverse dispersion, the aspect ratio and the permeability characteristics of the porous medium. The effect of each parameter can be summarized as follows :

The permeability distribution near the entrance of the porous medium determines the initial number, locations and relative growth rates of the fingers; none of these depends on the viscosity ratio. Subsequent merging reduces the initial number of fingers to a smaller number of "active fingers", which continue growing through the porous medium. The permeability distribution downstream (away from the entrance) has little effect on the merging process, on the number of active fingers and on the displacement in general, provided permeability variations are not very large.

The root mean square finger growth rate increases with the heterogeneity of the porous medium. Also the number of active fingers decreases with increasing heterogeneity, because larger heterogeneity facilitates merging. The effect of heterogeneity is more prominent in the initial stages of the displacement.

The RMS finger growth rate increases with viscosity ratio; the effect is more prominent at low viscosity ratios than at high viscosity ratios. At low viscosity ratios little merging occurs and this results in a large number of thin active fingers with comparable growth rates. At high viscosity ratios a smaller number of wider active fingers forms. The thickness of the displacement front decreases with increasing viscosity ratio. The higher the viscosity ratio the more irregular and unstable the fingers that form.

Longitudinal dispersion causes the displacement front to be thicker and the fingers smoother. It does not affect the RMS growth rate (for the C=0.5 concentration contour) nor the number of active fingers.

Transverse dispersion causes earlier merging. The resulting fingers are wider and somewhat smoother. RMS finger growth rate decreases with increasing transverse dispersion. The effect is stronger at low viscosity ratios.

A large aspect ratio (narrower slab) yields a smaller number of active fingers. Merging occurs early in the displacement. At very small aspect ratios a single active finger forms. RMS finger growth rate increases with aspect ratio.

References

1. R.M. Giordano, S.J. Salter and K.K. Mohanty, *The Effects of Permeability Variations on Flow in Porous Media,* paper SPE 14365, presented at the 60th Annual Technical Conference and Exhibition of SPE, Las Vegas, NV, September 1985.
2. E.J. Koval, *A Method for Predicting the Performance of Unstable*

Miscible Displacement in Heterogeneous Media, Trans., AIME, **228** , pp. 145-154 (1963).

3. M.R.Todd and W.J.Longstaff, *The Development, Testing and Application of a Numerical Simulator for Predicting Miscible Flood Performance,* Trans., AIME, **253** , pp. 874-882 (1975).

4. F.J. Fayers, *An Approximate Model with Physically Interpretable Parameters for Representing Miscible Viscous Fingering,* paper SPE 13166 presented at the 59th Annual Technical Conference and Exhibition of SPE, Houston, TX, September 1984.

5. K.J. Weber, *How Heterogeneity affects Oil Recovery,* Reservoir Characterization, edited by L.W. Lake and H.B. Carroll Jr., Academic, Orlando, FL, 1986, pp. 487-544.

6. T.J. Lasseter, J.R. Waggoner and L.W. Lake, *Reservoir Heterogeneities and their Influence on Ultimate Recovery,* Reservoir Characterization, edited by L.W. Lake and H.B. Carroll Jr., Academic, Orlando, FL, 1986, pp. 545-559.

7. J. Douglas, Jr. and T.F. Russell, *Numerical Methods for Convection-dominated diffusion Problems Based on Combining the Method of Characteristics with Finite Difference or Finite Element Procedures,* SIAM J. Numer. Anal., Vol. 19, No. 5, October 1982, pp. 871-884.

8. T.F. Russell, *Finite Elements with Characteristics for Two-Component Incompressible Miscible Displacement,* paper SPE 10500, Proc. Sixth SPE Symp. on Reservoir Simulation, New Orleans, LA, Jan-Feb 1982.

9. M.F. Wheeler and R. Gonzalez, *Mixed Finite Element Methods for Petroleum Reservoir Engineering Problems,* Computing Methods in Applied Sciences and Engineering VI, edited by R. Glowinski and J.L. Lions, North Holland, New York, 1984, pp. 639-658.

10. M. Nakata, A. Weiser and M.F. Wheeler, *Some Superconvergence Results for Mixed Finite Element Methods for Elliptic Problems on Rectangular Domains,* The Mathematics of Finite Elements and Applications V, MAFELAP 1984, edited by J.R. Whiteman, Academic, London, 1985, pp. 367-389.

11. R.E. Ewing, T.F. Russell and M.F. Wheeler, *Simulation of Miscible Displacement Using Mixed Methods and a Modified Method of Characteristics,* paper SPE 12241, presented at the Reservoir Simulation Symposium, San Francisco, CA, November 1983.

12. T.F. Russell and M.F. Wheeler, *Finite Element and Finite Difference Methods for Continuous Flows in Porous Media* The Mathematics of Reservoir Simulation, edited by R.E. Ewing, SIAM, Philadelphia, PA, 1983, pp.35-105.

13. T.F. Russell, M.F. Wheeler and C. Chiang, *Large-Scale Simulation of Miscible Displacement by Mixed Characteristic Finite Element Methods,* Mathematical Computational Methods in Seismic Exploration and Reservoir Modeling, edited by W.E. Fitzgibbons, SIAM, Philadelphia, 1986, pp. 85-107.

Table 1. Simulation Runs

Run	M	a	Pe_l	Pe_t	σ
1	75	1	725	25873	0.133
2	75	1	725	25873	0.063*
3	7.5	1	725	25873	0.063*
4	75	1	725	25873	0.067
5	75	1	725	25873	0.201
6	7.5	1	725	25873	0.133
7	750	1	725	25873	0.133
8	75	1	145	5175	0.133
9	7.5	1	725	2587	0.133
10	75	1	725	2587	0.133
11	75	2	725	25873	0.133
12	75	4	725	25873	0.133

(*) Uniform permeability in 32 X 40 downstream blocks

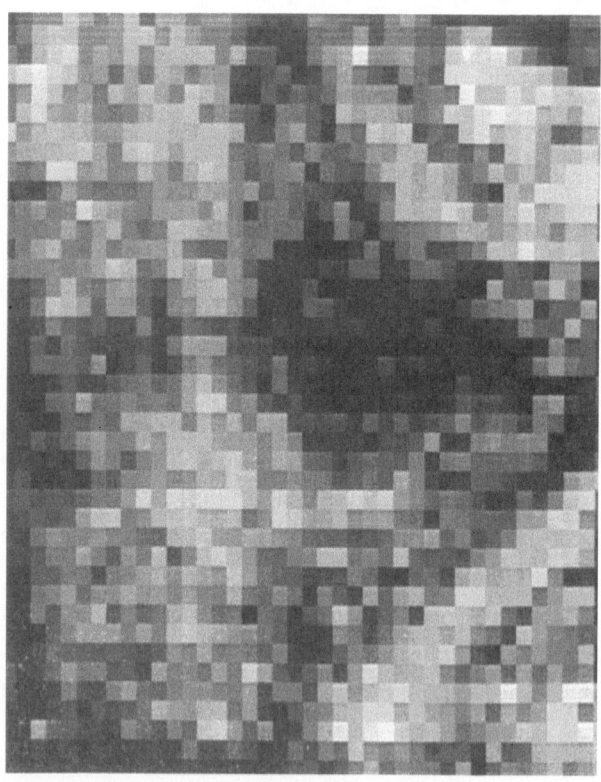

Fig. 1. Permeability map of the slab used in the simulations
(Bright areas correspond to high permeability values)

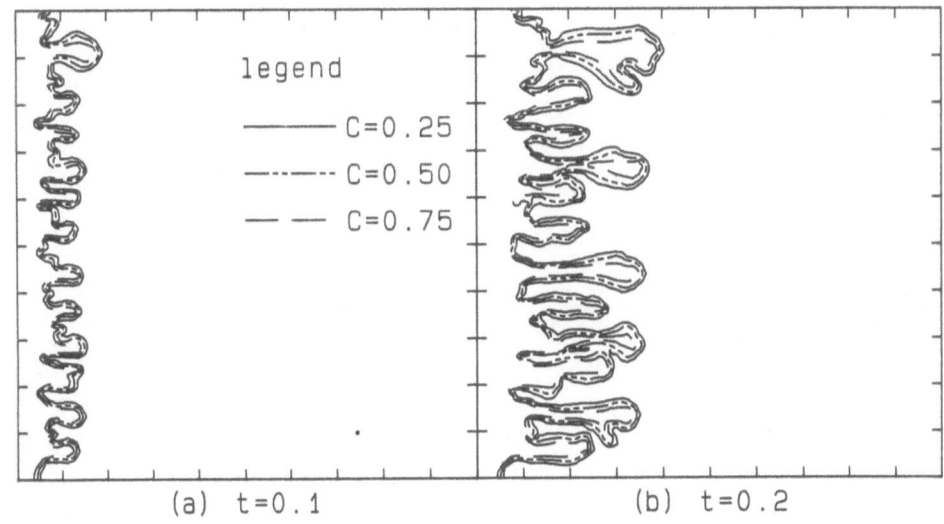

(a) t=0.1   (b) t=0.2

legend

—— C=0.25

—·—·· C=0.50

— — C=0.75

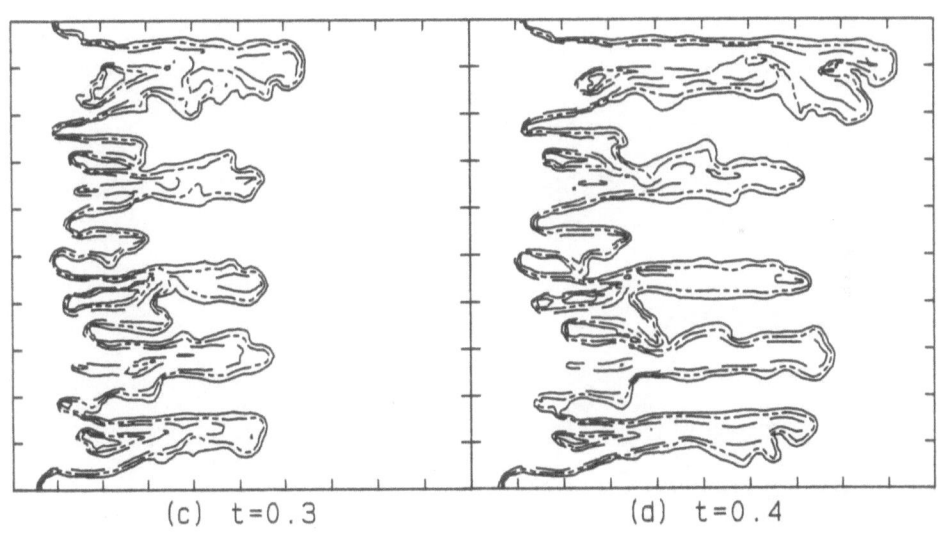

(c) t=0.3   (d) t=0.4

Fig 2. Simulation of base case, M=75 (Run 1)

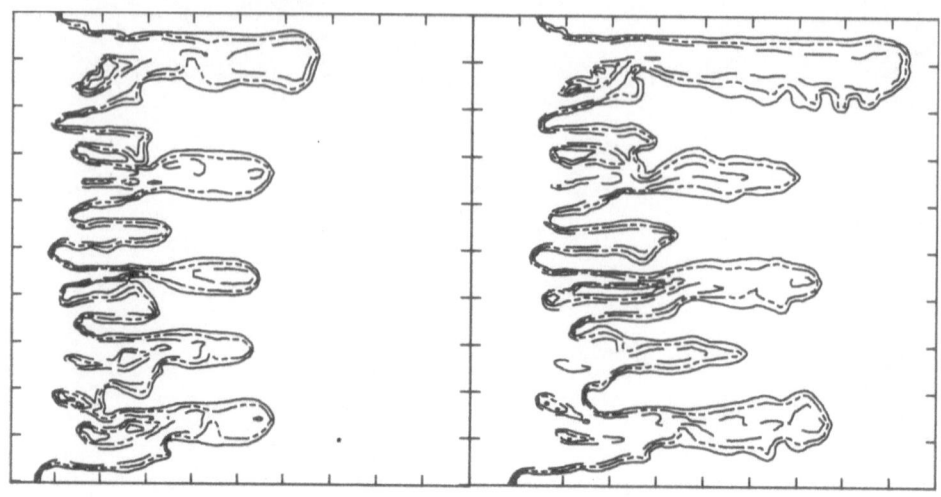

Fig 3. Uniform permeability
downstream (Run 2), t=0.3

Fig 4. Heterogeneity reduced
by 50% (Run 4), t=0.4

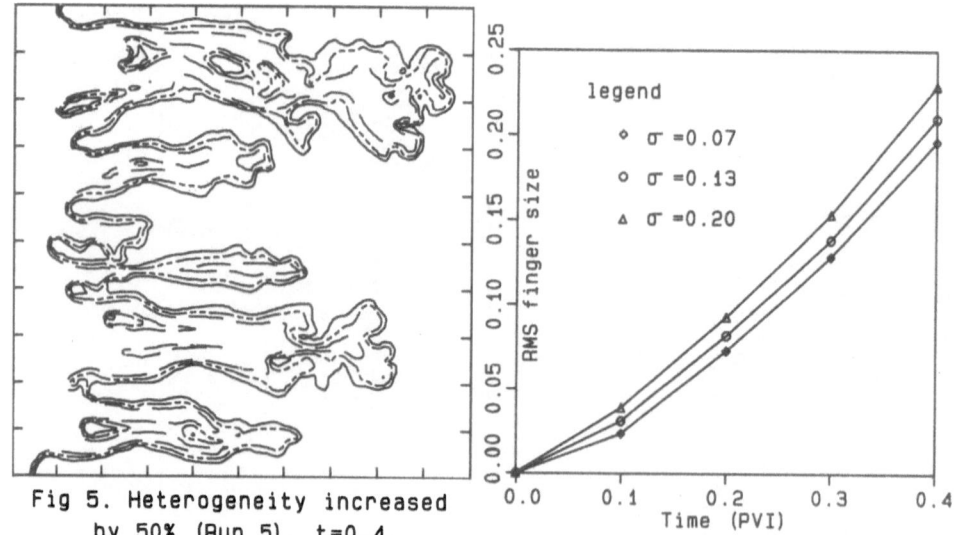

Fig 5. Heterogeneity increased
by 50% (Run 5), t=0.4

Fig 6. Effect of heterogeneity
on RMS finger growth rate

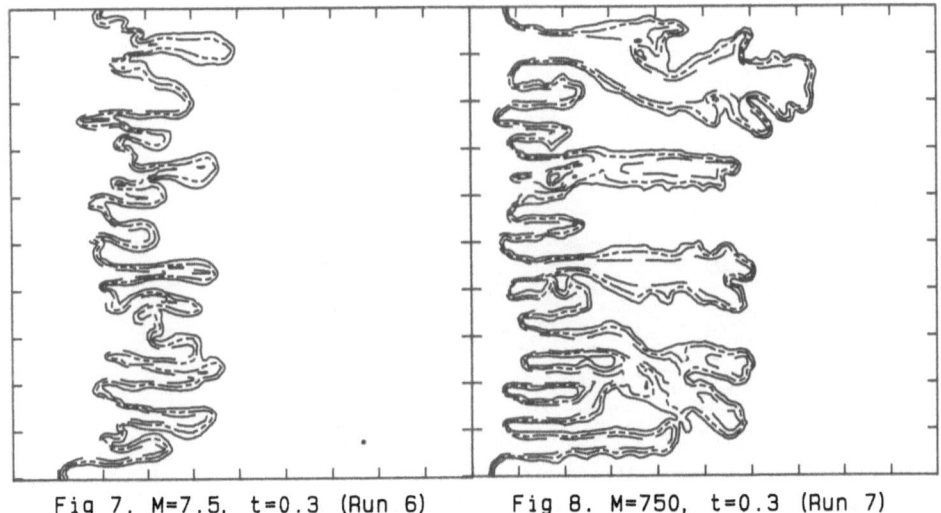

Fig 7. M=7.5, t=0.3 (Run 6)   Fig 8. M=750, t=0.3 (Run 7)

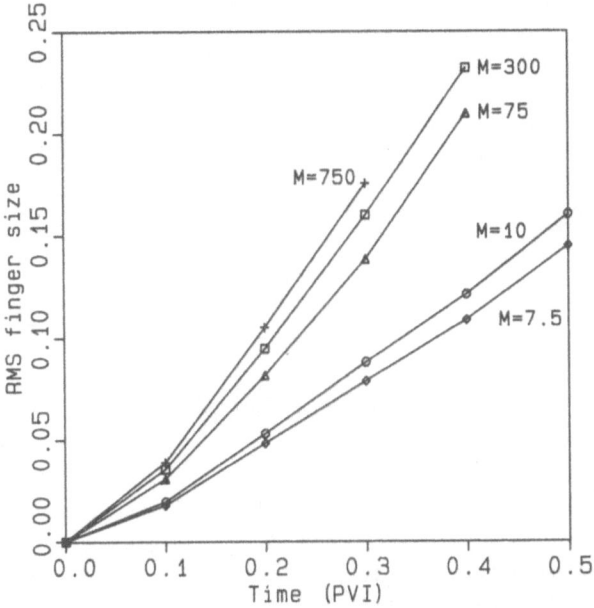

Fig. 9. Effect of mobility ratio on RMS finger growth rate

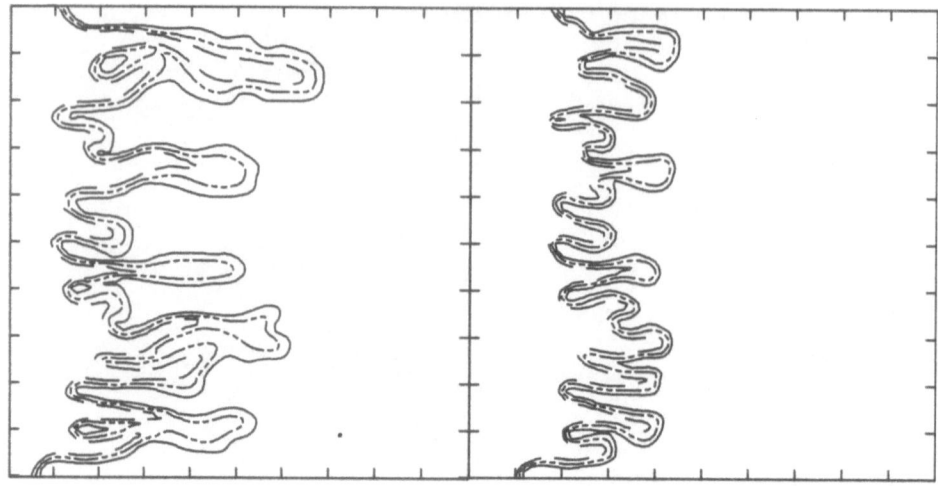

Fig 10. Dispersion 5 times
higher, t=0.3 (Run 8)

Fig 11. Transverse dispersion 10
times higher, M=7.5, t=0.3 (Run 9)

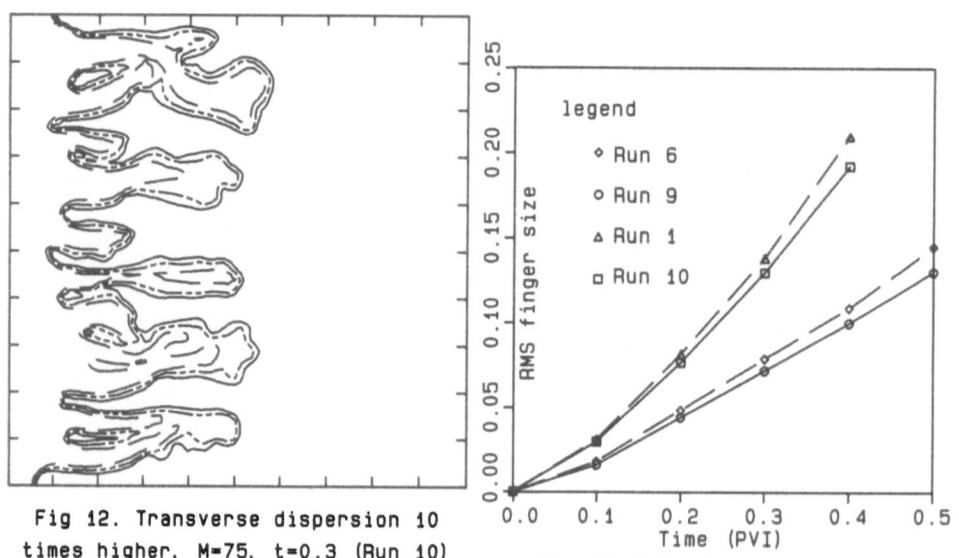

Fig 12. Transverse dispersion 10
times higher, M=75, t=0.3 (Run 10)

Fig 13. Effect of transverse
dispersion on RMS growth rate

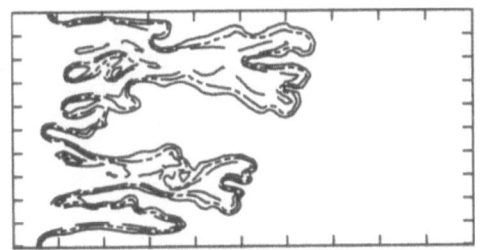

Fig 14. a=2, M=75, t=0.3 (Run 11)

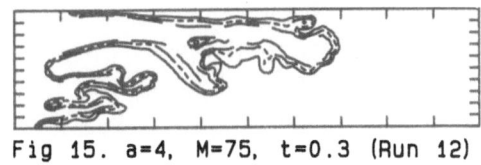

Fig 15. a=4, M=75, t=0.3 (Run 12)

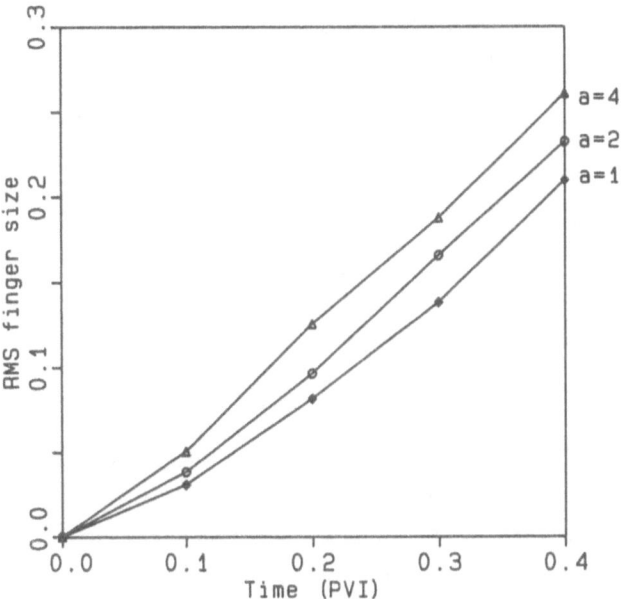

Fig. 16. Effect of aspect ratio on RMS finger growth rate

RECENT DEVELOPMENTS IN HELE-SHAW FLOW MODELING

Leonard W. Schwartz

Department of Mechanical & Aerospace Engineering
Rutgers University
New Brunswick, New Jersey

The unsteady Hele-Shaw problem is a model nonlinear system that, for a certain parameter range, exhibits the phenomenon known as viscous fingering. While not directly applicable to multiphase porous-media flow, it does prove to be an adequate mathematical model for unstable displacement in laboratory parallel-plate devices. We seek here to determine, by use of an accurate boundary-integral front-tracking scheme, the extent to which the simplified system captures the canonical nonlinear behavior of displacement flows and, in particular, to ascertain the role of noise in such systems. Flow patterns in both rectilinear channels and a five-spot pattern will be discussed when the driving fluid is considered to be inviscid. Finite driver viscosity cases will be included for the five-spot geometry. We also indicate that the model system can be derived from an energy minimization principle.

1. Introduction

A Hele-Shaw cell is a simple laboratory device consisting of two plates of

glass separated by a small constant distance. A displacement process where, for

example, a more viscous fluid is pushed out by a less viscous one may thus be

visualized. Some time ago it was shown[1,2] that this process is inherently un-

stable and, provided the cell is of sufficient size, an initially straight

interface between the fluids will develop undulations. When these undulations

grow to large amplitude, the less viscous fluid may be observed to form

"fingers" into the more viscous one; this is the celebrated viscous fingering

instability. A variant of the Hele-Shaw cell is the sand-pack or bead pack

where the space between the plates is filled with a packing of small particles.

Viscous fingering has also been observed in these devices when the "pusher"

fluid is less viscous[3]. Unstable displacements can be divided into two classes:

they are called immiscible when the fluid interface is sharp and possesses a

finite value of interfacial tension; else they are termed miscible.

Hele-Shaw cells have found application in the oil industry[3] because they are (grossly) simplified models that illustrate the flow behavior when viscous oil is pushed out by a less viscous fluid such as water. The analogy to porous media flows is, of course, quite incomplete; the displacement process in a parallel plate device using immiscible fluids is almost perfect. That is, in the model device, the displaced fluid is almost completely removed from that portion of the cell past which the interface has moved. In waterflooding of oil reservoirs, there is usually connate water present in the nominally oil-filled region; similarly residual oil is left behind after the passage of the front. The amount of residual oil is, apparently, dependent on the wetting properties of the liquids. On the other hand, in bead pack experiments when the beads are water-wet and the displacing fluid is water, virtually complete evacuation has been observed.[4] Undoubtedly the viscous fingering instability, and flows resulting from it, are pertinent to actual waterflooding. Thus, Hele-Shaw cells, and the mathematical equations that model these flows, are of more than passing interest.

The following section will give an outline of the problem to be solved and introduce the boundary-integral methods used to obtain the time-dependent frontal histories. Section 3 contains results of simulations in two flow geometries; flow patterns produced by the computations are similar to experimentally observed flows. The effect of "noise" associated with, for example, variations in permeability is illustrated in section 4. There we demonstrate that the flow patterns shown previously are subject to nonlinear instability; the critical disturbance amplitude to trigger this instability decreases as the capillary number of the displacement is increased. For the five-spot pattern, this results in significant reductions in sweep efficiency. Two other aspects are discussed in section 5. These include an extension of the model to include wetting-layer effects, and the development of an energy principle for Hele-Shaw flows.

2. Mathematical Formulation

Starting with the Stokes equations for slow viscous motion specialized to flow between closely-spaced parallel plates and requiring that the capillary number be small, a simplified two-dimensional problem can be formulated. These Hele-Shaw equations, for finite viscosity ratio, can be written as

$$\nabla^2 p_i = 0, \qquad (x,y) \in D_i, \quad i = 1,2 \qquad (1)$$

where D_1 is the domain occupied by the displaced fluid and D_2 corresponds to the displacing phase. The velocity potentials are related to the pressures p_i by

$$\phi_i = - \frac{k}{\mu_i} p_i . \qquad (2)$$

The permeability k for a Hele-Shaw cell is $b^2/12$ where b is the plate spacing. At the interface ∂D, whose equation is of the form $F(x,y,t)=0$, the jump conditions are

$$p_1 - p_2 = \sigma \kappa \qquad (3a)$$

and

$$\frac{\partial \phi_1}{\partial n} = \frac{\partial \phi_2}{\partial n} . \qquad (3b)$$

Here κ is the curvature of the interface in the plane of the cell and σ is the interfacial tension. The kinematic condition determining the interface motion is

$$(x_t, y_t) \cdot \vec{n} = \frac{\partial \phi}{\partial n} . \qquad (3c)$$

Lengths may be nondimensionalized using a macroscopic length L as reference. This is the width of the cell for a rectilinear channel or the interwell distance for the five spot. The areal flux Q in either geometry is used as reference potential. The viscosity ratio

$$m = \frac{\mu_2}{\mu_1}$$

is less than one for unstable floods. The mobility ratio M, as usually defined, is the reciprocal of m. The geometry is shown in figure 1.

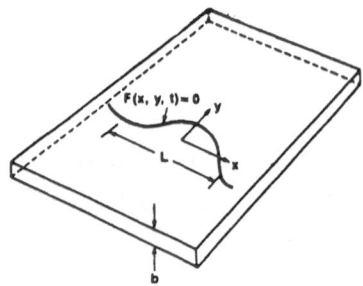

Figure 1. Displacement in a Hele-Shaw cell.

At an instant of time the interface ∂D is given. Because of (2) and (3a), the pressure and velocity potential discontinuities are then known. Subject to the condition of constant areal flow rate, we seek a vortex distribution $\omega(s)$ on ∂D whose strength is determined as part of the solution. With this singularity distribution, condition (3b) is satisfied identically, while the tangential velocity discontinuity on ∂D is simply equal to the local value of ω. The vorticity distribution satisfying (1), (2), and (3) is the solution of the singular second-kind Fredholm equation

$$\frac{(1+m)}{2}\omega(z_0) + (1-m)\ \text{Re}\left[\frac{ie^{i\alpha_0}}{2\pi}\int\frac{\omega(s)ds}{z-z_0}\right] = \tau\frac{dx}{ds_0} - (1-m)\ \text{Re}\left[\frac{df}{dz_0}e^{i\alpha_0}\right] \tag{4}$$

where the integral is to be interpreted as a Cauchy principal value and Re
denotes the real part of a complex function. The dimensionless surface tension
parameter τ is referenced to the viscosity of the displaced phase, i.e.

$$\tau = (\sigma b^2)/(12\mu_1 QL) \ . \tag{5}$$

Equation (4) is obtained by differentiating (3a) in the tangential direction
and replacing the velocities in favor of the vorticity distribution. α_0 is the
angle subtended by a tangent vector to the interface and a reference axis. The
function $f(z_0)$ is the complex potential for the simple single-phase flow in
either geometry. Once the vorticity distribution has been determined, the normal
velocity of points on the boundary is given by

$$\frac{\partial\phi}{\partial n} = -\text{Im}\left[\frac{df}{dz}e^{i\alpha_0}_{\ 0} + \frac{i}{2\pi}e^{i\alpha_0}\int\frac{\omega(s)}{z_0-z}ds\right] \ . \tag{6}$$

Numerically, the algorithm includes the following features:

(a) The time integration in condition (3c) is done implicitly using the LSODE[5]
package.

(b) The discretization of equations (4) and (6) employs a method that is
equivalent to separating the singular and nonsingular parts of the Cauchy
principal value integral. The singular part is integrated analytically. The
remainder is treated by assuming the vorticity to be piecewise linear. Certain
other details are similar to the procedures discussed by Moore[6] however the
point spacing along the interface is determined by local stability considerations and simplifications possible with equally-spaced points are inapplicable
here.

(c) Points are periodically redistributed along a cubic spline representation of the interface so as to provide resolution of all locally unstable perturbations, using the linear stability result given by Saffman & Taylor[1]. That is, for the unstable cases, corresponding to m < 1, the point spacing is no greater than

$$\Delta s = B \left[\frac{\tau}{v_n (1-m)} \right]^{1/2} \tag{7}$$

where B is a constant and v_n is the normal component of velocity locally. Closer point spacing is used when necessary in regions of large interface curvature.

(d) The linear algebraic system corresponding to equation (4) is solved by Gauss elimination.

For the special case of m=0, it is possible to use a computationally superior boundary method that employs a source, rather than a vortex, distribution on ∂D. The superiority stems from the fact that near neighbor interactions using sources do not introduce possible numerical instabilities which can become important for vortex methods. Details of both schemes can be found in references 7 and 8.

3. Simulation Results in Linear Channels and Isolated Five-Spots

The above algorithms have been used to generate flood histories, initially in linear channels. In the experiments of Saffman & Taylor[1] air was used to displace a glycerin-water solution from between plates separated by about 1 mm in a straight channel that was several cm wide. Instability of an initially plane interface led, for a range of displacement speeds, to the ultimate formation of a single finger which continued to propagate down the channel without futher change of form. Some time later, McLean & Saffman[9] used a steady-state boundary integral technique and Newton iteration to calculate a family of solutions for various values of r and m=0. Qualitative similarity was obtained; however the theoretical model underpredicted the width of the experimentally-observed fingers. Somewhat more disconcerting was a stability analysis for such fingers, once formed, that identified growing modes and concluded that such fingers should not be stable, apparently contradicting the experimental result.

Our simulation of this process, starting with an initial sine-wave is shown in figure 2a for a particular value of r. Succesive positions of the interface are shown at equally-spaced time increments. Thus we may conclude, operationally, that the ultimate steadily-propagating finger is stable. To numerically investigate the question of stability, we then introduced tiny bumps near the tip of the developed finger and monitored the subsequent time history of these perturbations. In general the disturbance grew initially; however it was convected with the local particle velocity and ultimately arrived at the side of the finger where it became essentially stagnant in a laboratory frame of reference. The amplitude then slowly decreased until it no longer was detectible. To the extent that the disturbance did grow initially, this may explain the previous finding of instability.

For the five-spot geometry, similar time histories could be constructed.[8] Such an example in given in figure 2b for m=0. Using the numerical algorithm, qualitatively similar patterns could be produced for values of $r \geq O(10^{-4})$. The trend was clear; reductions in r produced monotonically decreasing values of η, the breakthrough sweep efficiency, which is equivalent to the area swept out prior to the front reaching the sink in the upper right hand corner of the figure. When extended to finite values viscosity ratio, it is possible to assess the combined effect of varying the two parameters r and m. The result of a number of such runs is shown in figure 3. The horizontal axis is the mobility ratio $M = m^{-1}$ in conformance with standard practice in the petroleum literature. In his monograph on waterflooding, Craig[3] shows a similar figure which is a compilation of laboratory results using a variety of techniques. He offers no explanation for the observed divergence of results when the mobility ratio is unfavorable (M > 1). Using the Hele-Shaw mathematical model, we see a similar divergence, due in our case to varying the speed of flooding, say, via changes in the parameter r.

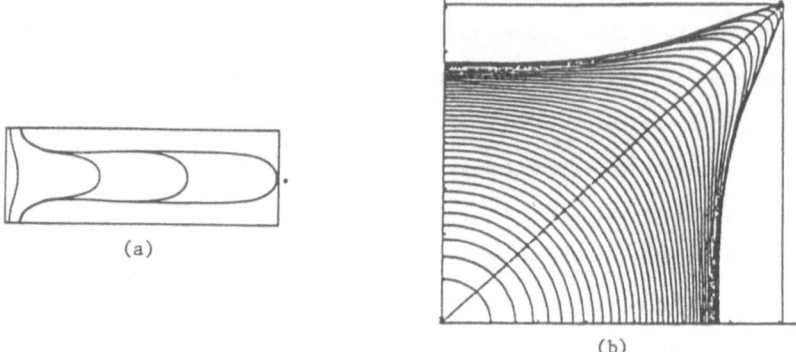

Figure 2. Calculated flood histories without noise for m=0; (a) linear channel, r=0.0004 ; (b) 5-spot geometry, r=0.001 .

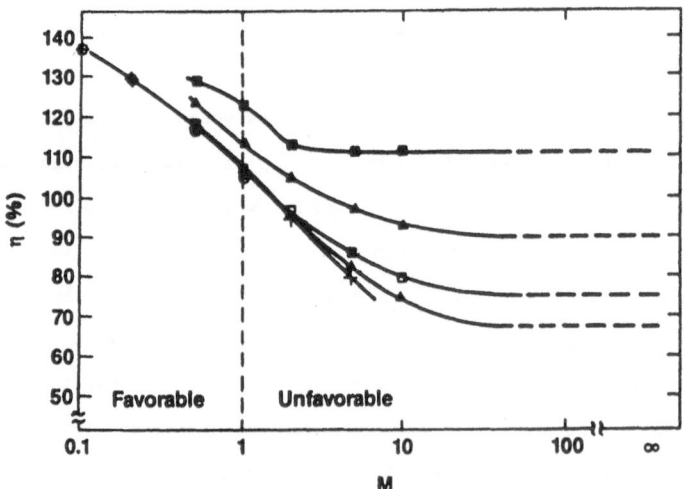

Figure 3. Variation of sweep efficiency η with mobility ratio M for several values of r. The curves form a monotonic family with larger r values corresponding to larger η.

4. Simulations with Noise

In discussing the question of the stability of fingers once formed, we noted that very small disturbances produced only a transient alteration of the profile of the finger shape. When large disturbances are introduced, it is possible to induce an effect which has become known as tip-splitting. There is a critical value of the disturbance amplitude for which the finger can no longer recover and splits into two branches. These branches move together until one of them gets ahead of the other. The winning branch blocks or shields the other which ultimately comes to rest, remaining as a residual nub. The winner grows to fill the channel and if disturbances are continually present, the process will occur repeatedly. The disturbance amplitude or "noise" can be related to the varia- tion in permeability[8]. Such a computation[10], for a linear channel with noise exceeding the critical value, is shown in figure (4a). It may be compared, qualitatively, with an experimental Hele-Shaw flood at high speed[11]and with a recent immiscible displacement in a bead pack[4]. The two experimental "fingers" are shown in figures (4b) and (4c) respectively. The bead pack was preferen- tially wetted by the displacing phase (i. e. imbibition); quite different patterns, breaking up on the bead level were observed if the non-wetting phase was used as the driver.

The sensitivity to noise has been demonstrated numerically to increase dramatically with decreasing r. In fact, for $r < O(10^{-4})$ no steady fingers without splitting could be obtained for either test geometry when m=0. This we attribute to the numerical approximation of continuous functions inherent in this, or any other, computational scheme. For the five-spot geometry, a flood with repeated tip splitting is shown in figure 5. The particular pattern was produced by a tiny disturbance placed near the asterisk on the starting profile. Full-scale conditions, i.e. a "Hele-Shaw cell" with 1 Darcy permeability and characteristic dimension of perhaps 100 m, would have a r value as small as

258

10^{-10} or less. At this value, only the noisy results have any physical significance. The anticipated sweep efficiency can be expected to be
correspondingly low.

(a)
(b)
(c)

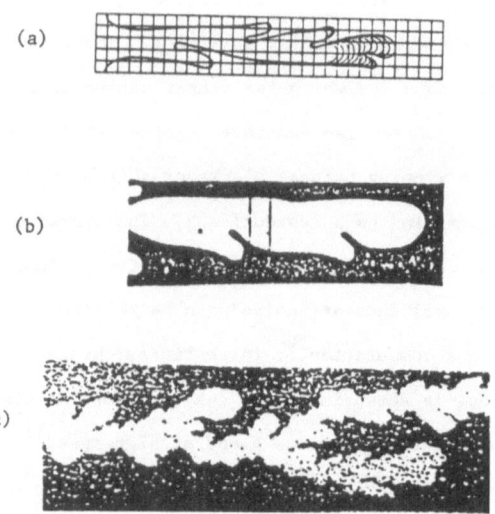

Figure 4. Noise-induced tip splitting in linear channels; (a) simulation, (b) experimental result in a Hele-Shaw cell[11] (c) experiment in a bead pack[4].

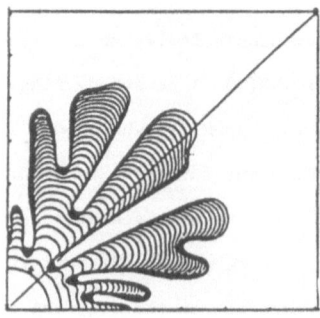 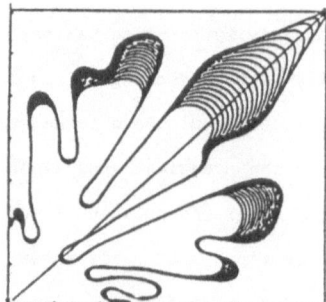

Figure 5. Simulation in a 5-spot pattern for $r=0.00005$, $m=0$. A small disturbance is introduced on starting profile. Early and late parts of the flood.

5. Other Aspects

We have also resolved the discrepancy between theory and experiment concerning ultimate finger widths in the stable regime. At relatively low displacement rates, where the greatest disagreement between observed and measured finger widths occurs, it has been shown that the simplified model needs to be augmented by a velocity-dependent pressure correction inserted in the boundary condition (2).[12,13] For an inviscid driver, the fundamental instability of a plane interface is modified to become

$$\omega = \frac{kV_0 - (\pi/4)\sigma Mk^3}{1 + (4/3)(J\sigma MkCa^{2/3}/bV_0)} \tag{8}$$

where $Ca = \mu V_0/\sigma$ is the capillary number and J is a numerical constant determined to be 3.90. Equation (8) gives the rate of growth ω as a function of wavenumber k. M is the single-phase mobility which is equal to $b^2/(12\mu)$ for a Hele-Shaw cell. If the constant J is set to zero, the basic stability result of Saffman & Taylor is recovered. The origin of the correction is the additional pressure drop associated with the presence of a thin wetting film left behind on the walls of the cell after passage of the displacement front. The critical wavenumber for the onset of instability is unchanged; however the wavenumber of maximum growth and its growth rate are both reduced. When this correction is included in the numerical calculation of finger evolution, we find that the discrepancy between theory and experiment is largely removed. All of the qualitative features of the basic model remain unchanged however. That is, the system retains its fundamental properties: that linear instability leads to finger formation and that a nonlinear balance of effects arises that allows the finger to ultimately propagate without further change of form, in a manner that is stable to small disturbances. Thus the basic Hele-Shaw model remains a valid candidate for studies of pattern formation in nonlinear systems, the experimental discrepancies in linear channels notwithstanding.

An alternative form of the basic moving-boundary problem has also been formulated. We have recently developed an energy-minimization principle that is equivalent to the differential statement of the problem.[14] Let $\dot{E}_{tot}$ represent the instantaneous rate of energy dissipation functional, given by

$$\dot{E}_{tot} = (b^3/12\mu)\iint_R (p_x^2+p_y^2)dxdy - (\pi\sigma b^3/24\mu)\int_C (\kappa + \varsigma)(\partial p/\partial n) \ ds. \qquad (9)$$

The principle states that the motion of the interface will be such as to minimize $\dot{E}_{tot}$ at each instant of time. The first integral on the right of (9) represents viscous dissipation in the fluid region R, while the second is the rate of work done against surface tension in increasing the length of the interface C. ς is an arbitrary Lagrange multiplier associated with the rate of injection of the displacing fluid. More simply, the principle is equivalent to the statement that the motion takes the "path of least resistance." Implementation of the principle should may allow the development of approximate solutions techniques using Ritz or Galerkin approaches, for example.

6. Cited References:

1. Saffman, P. & Taylor, G., Proc. R. Soc.(Lond.)A245, 312, 1958.
2. Chouke, R., van Meurs, P. & van der Poel, C., Trans. AIME 216,188, 1959.
3. Craig, F.,The Reservoir Engineering Aspects of Waterflooding, Soc. Petr. Engrs., Dallas, 1971.

4. Stokes, J., Weitz, D., Gollub, J., Dougherty, A., Robbins, M., Chaikin, P., & Lindsay, H., Phys. Rev. Lett., Oct. 16, 1986.

5. Hindmarsh, A., **Livermore Solver for Ordinary Differential Equations**, Lawrence Livermore Laboratory, 1980.

6. Moore, D., "Numerical and analytical aspects of Helmholtz instability," Proc. IUTAM, 1985, 263.

7. DeGregoria, A. & Schwartz, L., J. Fluid Mech. **164**, 383, 1986.

8. Schwartz, L. & DeGregoria, A., to appear J. Austral. Math. Soc., Ser. B, 1987.

9. McLean, J. & Saffman, P., J. Fluid Mech. **102**, 455, 1981.

10. DeGregoria, A. & Schwartz, L. Phys. Fluids **28**, 2313, 1985.

11. Park, C. & Homsy, G., Phys. Fluids **28**, 1583, 1985.

12. Schwartz, L., Phys. Fluids **29**, 3086, 1986.

13. Schwartz, L. & DeGregroria, A., Phys. Rev. A, **34**, Jan. 1987.

14. Schwartz, L. & DeGregoria, A. sub. to Phys. Rev. Lett., Dec. 1986.

Loss of Strict Hyperbolicity of the Buckley-Leverett Equations for Three Phase Flow in a Porous Medium.

Michael Shearer[1]

Department of Mathematics

North Carolina State University

Raleigh, North Carolina

1. **Introduction**. In this paper, I examine the loss of strict
hyperbolicity of 2×2 systems of conservation laws modelling
one-dimensional flow of three immiscible incompressible fluids (e.g.
oil, water, gas) in a porous medium, taking capillary effects to be
negligible. In §2, I review the classical scalar Buckley-Leverett
equation for two-phase flow, writing down appropriate hypotheses
(constitutive assumptions) for the relative permeabilities. The
corresponding Buckley-Leverett equations for three phase flow are
required to satisfy these assumptions when one of the phases is
absent.

In §3, I prove the main result, which says that if certain
interaction conditions hold between the two-phase relative
permeabilities and the third (absent) phase, then the equations fail
to be strictly hyperbolic everywhere in the three-phase flow regime

[1]Supported in part by U.S. Army Research Office grant
DAAL03-86-K0004.

(where all three fluids are present). The proof suggests a graphical method for investigating the loss of strict hyperbolicity of any given model, even those not satisfying the interaction conditions.

When strict hyperbolicity fails for the Buckley-Leverett equations, there is typically a small region in the ternary diagram of normalized saturations in which the equations are elliptic. As a first step in studying equations with small elliptic regions, it is reasonable to consider the elliptic region shrunk to a point, called an umbilic point. Indeed, this is exactly the situation for the Marchesin model [8]. In §4, I summarize recent work with David Schaeffer on equations with umbilic points. These equations have a rich mathematical structure which differs markedly from that of strictly hyperbolic equations.

§2. Equations of motion.

The equations of motion for the slow flow of incompressible immiscible fluids in a porous medium are expressed by Darcy's law, which takes the place of conservation of momentum, and the conservation of mass. The classical Buckley-Leverett equation [2], for two phase flow in one dimension, is obtained by eliminating the pressure from Darcy's law, on the assumption that capillary forces are negligible.

Remark. Taking the capillary pressure to be zero is a mathematical idealization that makes sense in the context of understanding and tracking sharp interfaces or fronts. The capillary pressure would smooth these shock wave solutions of the Buckley Leverett equation, just as viscous effects smooth shock waves in gas dynamics.

Let $s_1 = s$, $s_2 = 1-s$ be the volume fractions of the two fluids (e.g. oil and water), with $k_j(s_j)$, $j = 1$ or 2, denoting their relative permeabilities (divided by viscosity). The Buckley-Leverett equation is then

$$s_t + \left[\frac{k_1(s)}{k_1(s) + k_2(1-s)} \right]_x = 0 \qquad (2.1)$$

Typical properties of k_1 and k_2 are reflected in the following assumptions:

k_j is C^2 (twice continuously differentiable) on $[0,1]$, $j = 1,2$;

$$k_j(0) = 0 = k_j'(0), \quad k_j''(0) > 0, \quad 0 \le s \le 1, \quad j = 1,2. \qquad (2.2)$$

Note that (2.2) implies that the flux
$$F(u) = k_1(u)/(k_1(u) + k_2(1-u)), \quad 0 \le u \le 1,$$
has the characteristic S shape with a single inflection point, where $F''(u) = 0$.

The generalization of (2.1) to three phase flow is straightforward. We let u, v, w = 1-u-v be the three volume fractions and let f, g, h be the corresponding relative permeabilities, divided by viscosity. Then conservation of mass leads to two independent equations

$$u_t + [f/D]_x = 0$$

$$v_t + [g/D]_x = 0 \qquad (2.3)$$

where $D = f + g + h$. The functions f, g, h depend on (u,v) in the <u>saturation triangle</u>

$$0 \leq u, \; v, \; 1-u-v \leq 1 \qquad (2.4)$$

(In the Engineering literature, this triangle is transformed into an equilateral triangle known as the <u>ternary diagram</u>.)

Two phase flow corresponds to one of the volume fractions u, v, $1-u-v$ being zero. System (2.3) should then reduce to the Buckley-Leverett equation (2.1) with the appropriate definitions of k_j and s, satisfying (2.2). For example, on $v = 0$, we require

$$f(0,0) = f_u(0,0) = 0, \; f_{uu}(u,0) > 0$$

$$g(u,0) \equiv 0$$

$$h(1,0) = h_u(1,0) = 0, \; h_{uu}(u,0) < 0 \; .$$

We shall impose corresponding conditions on f, g, h on the remaining two sides $u = 0$ and $1-u-v = 0$ of the saturation triangle (2.4). I refer to these consistency conditions as the (B-L) conditions.

Equations (2.3) are an example of a 2×2 system of conservation laws, which we write in general in the compact form

$$U_t + F(U)_x = 0, \tag{2.5}$$

where $U = (u,v)$ and $F = \mathbb{R}^2 \to \mathbb{R}^2$. The type of system (2.5) of differential equations depends on the eigenvalues $\lambda_1(U)$, $\lambda_2(U)$ of $dF(U)$. If $\lambda_1(U)$ and $\lambda_2(U)$ are real and distinct, then (2.5) is strictly hyperbolic at U. If $\lambda_1(U)$ and $\lambda_2(U)$ are complex conjugates, then (2.5) is elliptic at U.

To understand the loss of strict hyperbolicity, it is convenient to project $dF(U)$ onto the space of trace-free matrices. This projection, or deviator of dF is defined by

$$\text{dev } dF(U) = dF(U) - \frac{1}{2}(\text{tr } dF(U))I$$

As in [8], we choose coordinates so that

$$\text{dev } dF(U) = \begin{bmatrix} X(U) & Y(U)+Z(U) \\ Y(U)-Z(U) & -X(U) \end{bmatrix}.$$

Note that $dF(U)$ has distinct real eigenvalues if and only if $\text{dev } dF(U)$ lies outside the cone

$$x^2 + y^2 = z^2 \tag{2.6}$$

On the surface of the cone, $dF(U)$ has coincident real eigenvalues, while if $\text{dev } dF(U)$ lies inside the cone, $dF(U)$ has complex eigenvalues.

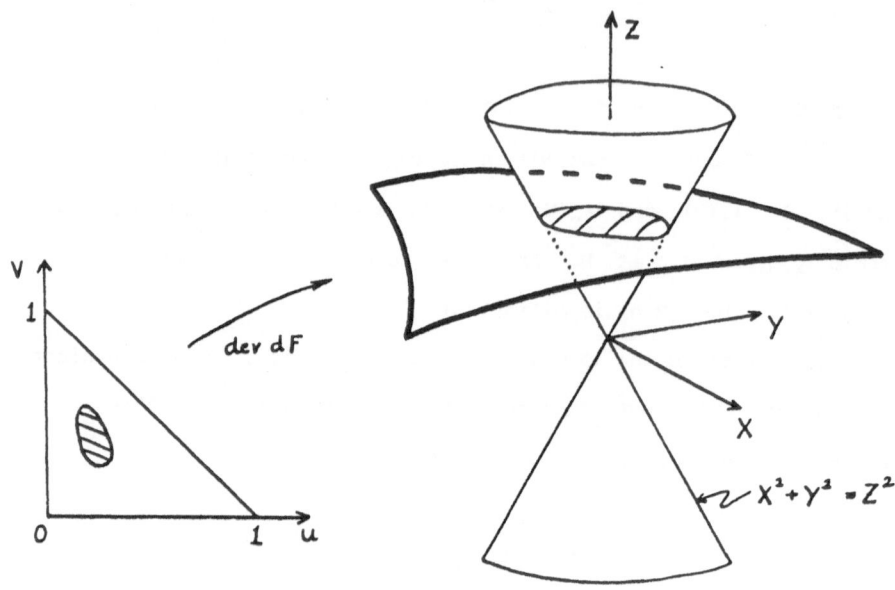

Figure 1.

The mapping $U \longrightarrow$ dev dF(U) defines a surface in R^3. In Figure 1, I illustrate a typical intersection of this surface with the cone (2.6), showing how equations (2.5) can have a bounded elliptic region (shaded in Fig. 1) in U-space. It is clear from Figure 1 that with a continuous perturbation of F(U) we can arrange for the surface $\{$dev dF(U)$\}$ to intersect the cone only at the origin, thus shrinking the elliptic region to a point U^*. Then dev dF(U^*) = 0, and dF(U^*) = λI, for some $\lambda \in R$. Such a point U^* is called an umbilic point. Hyperbolic equations with isolated umbilic points were introduced in [8]. Related nonstrictly hyperbolic equations are discussed in [4,6].

In Figure 1, consider a simple closed path Γ in U-space that encloses the (shaded) elliptic region. The image Γ' of Γ under the map $U \rightarrow \text{dev } dF(U)$ wraps around the cone exactly once. The orientation of Γ' with respect to that of Γ depends on whether the map $U \rightarrow \text{dev } dF(U)$ is orientation preserving or reversing. This distinction has an important consequence for the eigenvectors of $dF(U)$, $U \in \Gamma$. Let Γ have the counterclockwise orientation. If dev dF is orientation preserving, then the right eigenvectors of $dF(U)$ rotate counterclockwise through 180° as U traverses Γ. If dev dF is orientation reversing, then the eigenvectors of $dF(U)$ rotate through 180° in the clockwise direction. The way to see this is to note that without loss of generality, we may take Γ' to be the circle $\left\{ (X,Y,0) : X^2 + Y^2 = 1 \right\}$. The right eigenvectors of dev $dF(U)$ (and hence of $dF(U)$) are then

$$\begin{bmatrix} \cos \frac{\theta}{2} \\ \sin \frac{\theta}{2} \end{bmatrix} \quad , \quad \begin{bmatrix} \sin \frac{\theta}{2} \\ -\cos \frac{\theta}{2} \end{bmatrix} \quad , \text{ where } \quad \begin{array}{l} X = \cos \theta \\ Y = \sin \theta \end{array} \quad .$$

We now have a division of equations (2.5) with bounded elliptic regions into two classes, those equations for which dev dF is orientation preserving and those for which the mapping is orientation reversing.

§3. Loss of strict hyperbolicity

I now return to the Buckley-Leverett system (2.3) and discuss constitutive assumptions in addition to (B-L) of §2. First observe that (B-L) implies

$$dF(U_o) = \begin{bmatrix} 0 & 0 \\ 0 & 0 \end{bmatrix}$$

at each vertex U_o of the saturation triangle. Thus each vertex is an umbilic point. Secondly, (B-L) implies that equations (2.3) are hyperbolic along each edge of the saturation triangle. In fact, the edges are invariant lines for the equations. (E.g. if $v(x,0) = 0$, then $v(x,t) \equiv 0$ for all $t \geq 0$.) To put it another way, if one of the fluids is absent everywhere at time zero, then it is also absent everywhere at all future time.

These two observations about the boundary of the saturation triangle are not enough to characterize the type of the differential equation in the interior. As a first step in understanding the interior, consider the Marchesin model, in which f, g and h are functions only of their respective volume fractions:

$$f = f(u), \; g = g(v), \; h = h(w). \tag{3.1}$$

Then (B-L) becomes

$$f(0) = f'(0) = 0 \;\;, \; f''(u) > 0, \tag{3.2}$$

and similarly for g, h.

Theorem 3.1 If f, g, h satisfy (3.1), (3.2), then the Marchesin model is hyperbolic everywhere in the saturation triangle, and strictly hyperbolic except for umbilic points at the vertices, and a single umbilic point in the interior.

Proof Everything in this theorem is proved in [8, Appendix], except for showing the model is hyperbolic. To see this, we note that the lines

$$f'(u) = g'(v)$$

$$f'(u) = h'(1-u-v) \qquad (3.3)$$

$$g'(v) = h'(1-u-v)$$

intersect only at the interior umbilic point. In the notation of §2, it is easy to calculate that the discriminant $X^2 + Y^2 - Z^2$ is a positive multiple of the quantity

$$\Delta = f^2(h'-g')^2 + g^2(f'-h')^2 + h^2(g'-f')^2 - 2fg(f'-h')(h'-g')$$

$$(3.4)$$

$$-2gh(f'-g')(g'-f') - 2fh(h'-g')(g'-f').$$

Since exactly two of the quantities

$$A = f'-h' \ , \ B = h'-g' \ , \ C = g' - f' \qquad (3.5)$$

have the same sign, except on the lines (3.3), let us suppose without loss of generality, that $A > 0$, $B < 0$, $C < 0$. I.e., $f' > g' > h'$. (Corresponding values of (u,v) lie in a sector adjoining $1-u-v = 0$ and containing $u = 1$, $v = 0$.) Then

$$\Delta = (fB - gA + hC)^2 + 4fhBC > 0 \qquad (3.6)$$

The symmetry in the expression (3.4) for Δ means it can always be written as the sum of a square and a positive term. If one of A, B or C is zero, then the other two have opposite signs, which again leads to $\Delta > 0$ (e.g., let $B = 0$ in (3.5), (3.6)). This completes the proof.

The Marchesin model has no interaction between the different fluids. Other models, such as Stone's model [11], take two of the relative permeabilities (say, f,g) to depend solely on their respective volume fractions u,v, but have h(u,v) in some form that interpolates between f and g in a manner that allows some interaction. In the analysis here, I will allow a limited amount of interaction in all of the relative permeabilities. The following conditions (I1) and (I2), which I call interaction conditions, limit the interaction between fluids near the boundary of the saturation triangle. Condition (I1) is to hold on each edge of the triangle, away from the vertices:

(I1) On $v = 0$, $0 < u < 1$, $g_v < (hf_u - fh_u)/(f + h)$,   (3.7)
    and similarly on the edges $u = 0$, $1-u-v = 0$.

We can read (3.7) as saying that the effect of the absent phase (v in (3.7)) is weak compared to that of the two present phases. Note that f, h, f_u and $-h_u$ are all strictly positive, so that $g_v(u,0) \equiv 0$ satisfies (3.7). A weaker sufficient condition is clearly

$$g_v(u,0) < \min \{ f_u(u,0), -h_u(u,0) \}.   (3.8)$$

That is, (3.8) implies (3.7). Condition (I1), together with (B-L), ensure that equations (2.3) are strictly hyperbolic on each edge, and that the right eigenvector of dF(U) that is parallel to the edge when U lies on the edge, is associated with the faster characteristic speed.

The second interaction condition concerns the direction of the right eigenvectors near the vertices of the triangle. Let $r_k(U)$ ($k = 1,2$) denote the right eigenvector (of length one) of $dF(U)$ corresponding to the eigenvalue $\lambda_k(U)$, with $\lambda_1(U) \leq \lambda_2(U)$. Condition (I1) implies that $r_2(U)$ is parallel to each edge of the saturation triangle, when U lies on that edge. At the vertices, $r_2(U)$ has to rotate through the angle between the edges. (In the coordinates of this paper, $r_2(U)$ rotates through $90°$ at the origin and $45°$ or $135°$ at each of the other two vertices.) The second interaction condition ensures that at each point U near a vertex, $r_2(U)$ points into the saturation triangle. Of course, $r_2(U)$ is not uniquely defined at the vertices (which are umbilic points), so the condition (I2) is stated initially in global terms.

Let $\theta(u,v)$ be the angle that $r_2(u,v)$ makes with the horizontal.

(I2): $\tan \theta(u,v) \geq 0$ for all $u > 0$, $v > 0$ near $u = v = 0$,
and similarly near the other vertices of the saturation triangle.

In terms of the relative permeabilities, (I2) is guaranteed by the simple sufficient condition

(I2)' $f_v(u,v) \geq 0$, $g_u(u,v) \geq 0$ for all $u > 0$, $v > 0$ near $u = v = 0$,
and similarly at the other vertices.

(I2)' is simple, but also easily violated. (E.g., $f(u,v) = u^2(2-v)$, $g(u,v) = v^2$). In fact, a useful general condition guaranteeing (I2)

is not yet known. Nor is it known if (I2) can be justified on physical grounds. However, (I2)' is obviously an interaction condition; a weaker interaction condition that implies (I2) is

(I2)" $\quad f_v(u,v) > (g_v + h_v)f/(g + h)$;

$\qquad g_u(u,v) > (f_u + h_u)g/(f + h)$, near $u = v = 0$,

and similarly at the other vertices.

Condition (I2)" ensures that the off-diagonal entries in $dF(U)$ are positive for $U \neq 0$ near the origin, which in turn implies (I2). Note however, that (I2)" is not a necessary condition for (I2). Note also that near the origin, the right hand sides of (I2)" are negative, since $h_u(0,0) < 0$, $h_v(0,0) < 0$, and $g_v(0,0) = f_u(0,0) = 0$. Consequently, (I2)' implies (I2)".

We are now ready to state and prove the main result.

Theorem 3.2 Assume (B-L) and (I1), (I2). Then equations (2.3) cannot be strictly hyperbolic in the interior

$$0 < u, v, 1-u-v < 1$$

of the saturation triangle.

$\cdot$

Proof Let Γ be a simple convex closed curve in the saturation triangle $0 \leq u, v, 1-u-v \leq 1$ such that (2.3) is strictly hyperbolic at each point of Γ. Then the map R_Γ: $(u,v) \longrightarrow r_2(u,v)$ takes Γ continuously into real projective space P_1. (I.e., ignoring the orientation of the unit vector $r_2(u,v)$). If (2.3) is strictly hyperbolic within Γ, then R_Γ is homotopic to a constant. (The homotopy is obtained by shrinking Γ continuously onto a point.)

Now let Γ be obtained from the three edges of the saturation triangle by cutting off the corners of the triangle in a C^1 manner close to the corners. Hypotheses (I1), (I2) then imply that $r_2(u,v)$ rotates 180° in a clockwise fashion as (u,v) traverses Γ in an anticlockwise manner (see Fig. 2). Thus R_Γ is not homotopic to a constant map. This completes the proof.

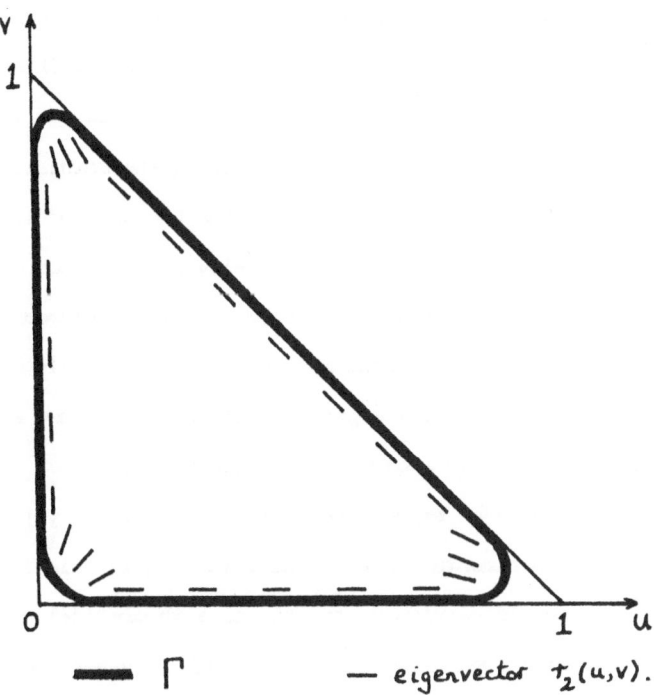

Figure 2.

Remarks. 1. The proof of Theorem 3.2 yields the additional information that the map $(u,v) \longrightarrow \det dF(u,v)$ is orientation reversing in a neighborhood of the boundary of the saturation triangle. Moreover, the image of Γ (as in the proof) encircles the cone (2.6) exactly once. If we reverse the inequalities in condition (I2) at each vertex, then the conclusion of the Theorem still holds, and dev $dF(\Gamma)$ encircles the cone twice, since the eigenvectors experience a 360° rotation. However, reversing the inequalities in (I2) does not seem to be physically reasonable.

2. The proof of Theorem 3.2 suggests a simple graphical test of any given model equations of the form (2.3), whether or not they satisfy (B-L) and (I1), (I2). It is easy to display $r_2(u,v)$ at each hyperbolic point on a selected grid covering the saturation triangle. It is important to have some points on the boundary of the triangle, since orientations can change dramatically near the boundary. It is then easy to see any rotation of $r_2(u,v)$ around closed curves. Preliminary tests with various models, including models with moderate linear terms in the relative permeabilities, have shown a remarkable robustness of elliptic regions or umbilic points.

3. It should be emphasized that no attempt has yet been made to incorporate experimental three phase data. Until this is done, Theorem 3.2 should be viewed strictly as a theoretical consequence of consistency with two phase flow and of weak interactions in nearly two phase regimes.

Examples

1. The Marchesin model (3.1), (3.2) automatically satisfies conditions (I1), (I2). The loss of strict hyperbolicity predicted by Theorem 3.2 occurs at the interior umbilic point of Theorem 3.1.

2. Stone's model. Here, I discuss a simplified version of

Stone's model for which it is easy to examine (I1), (I2). Let u, v, $w = 1 - u - v$ be the volume fractions of water, oil and gas, respectively. Let

$$f = au^2, \quad h = bw^2, \quad g = v(1 - u)^m (1 - w)^n ,$$

where a,b are positive constants (viscosity ratios), and $m,n > 1$ are not necessarily integers. Condition (I1) is trivial except on the edge $v = 0$, where it becomes

$$(1 - u)^{m-1} u^{n-1} (au^2 + b(1-u)^2) < 2ab \ , \ 0 < u < 1. \qquad (3.9)$$

Since oil is more viscous than gas or water, we have $a,b > 1$. If $m \geq 2$ and $n \geq 2$, (3.9) is easily seen to be satisfied.

Condition (I2) is trivial except in the corner $u = v = 0$. Here, we have

$$(g + h)f_v - (g_v + h_v)f \ \sim \ 2au^2 > 0, \quad \text{and}$$
$$(f + h)g_u - (f_u + h_u)g \ \sim \ nv(u + v)^{n-1} > 0$$

as $u,v \rightarrow 0$, thus verifying (I2)" and hence (I2).

These calculations for Stone's model suggest that condition (I2) is more intrinsic than (I1). Indeed, it is possible to violate (I1) by simply adjusting viscosity ratios (a,b in (3.9)), providing there is some interaction in the model. However, numerical trials indicate that violating (I1) falls far short of removing an elliptic region.

§4 Equations with umbilic points.

Rather little is known about equations with bounded elliptic
regions. The Cauchy problem is not well-posed for the equations
linearized about an elliptic point. However, growth in oscillations
associated with the elliptic region can be expected to be limited by
nonlinear effects in the surrounding hyperbolic region. This appears
to be happening in published numerical results of Bell, Trangenstein
and Shubin [1]. There is no theory describing this mechanism.

In this section, I consider equations with isolated umbilic
points. This class of hyperbolic equations is relevant to three phase
flow for two reasons. First, the Marchesin model has an isolated
umbilic point. Second, as discussed in §2, an isolated umbilic point
is an elliptic region shrunk to a point. Thus, new results and
phenomena discovered in connection with equations with umbilic points
will be reflected in equations with bounded elliptic regions. As
suggested by Theorem 3.2 and preliminary numerical tests discussed in
§3, equations (2.3) typically have a small bounded elliptic region.

The objective of this section is to show how the rotation of the
eigenvectors of $dF(U)$ is resolved at the umbilic point. Recall that
rarefaction wave solutions $U = U(x/t)$ of (2.5) are continuous
piecewise smooth solutions whose values lie on an integral curve of
$r_k(U)$, $k = 1$ or 2, with $x/t = \lambda_k(U)$. These integral curves are
called rarefaction curves; they are oriented by the direction of
increasing λ_k, since $\lambda_k(U)$ must increase with x/t, i.e., from left
to right through the wave. If $\lambda_k(U)$ has a turning point at U^* on
a rarefaction curve (more precisely, if $d\lambda_k(U^*).r_k(U^*) = 0$), we say
U^* is an inflection point. Curves of inflection points are called
inflection loci. The inflection points are the analogues for 2×2

systems of the inflection point in the flux for the Buckley-Leverett equation (2.1). In both cases, an inflection point corresponds to the loss of genuine nonlinearity [7].

For strictly hyperbolic 2×2 systems, the rarefaction curves form everywhere a local curvilinear coordinate system, with a corresponding pair of Riemann invariants that flatten the coordinate lines. For nonstrictly hyperbolic 2×2 systems however, the integral curves are not so well behaved.

In [8], it is shown that to study properties of 2×2 systems (2.5) near an isolated umbilic point, it is enough to consider equations with quadratic nonlinearities. Thus, consider the system

$$U_t + Q(U)_x = 0 \qquad\qquad (4.1)$$

where $Q : \mathbb{R}^2 \to \mathbb{R}^2$ is a homogeneous quadratic function such that $dQ(U)$ has real eigenvalues for all U. Since basic properties of equation (4.1) are unchanged by constant linear changes of coordinates in U, we need only consider equivalence classes of Q. This leads to a normal form for (4.1). Specifically, up to a linear change of coordinates,

$$Q(u,v) = d(au^3/3 + bu^2v + uv^2), \qquad\qquad (4.2)$$

for some a,b, apart from obvious degenerate cases. Thus, basic properties of all 2×2 hyperbolic conservation laws with an isolated umbilic point are described by studying a two parameter family of equations.

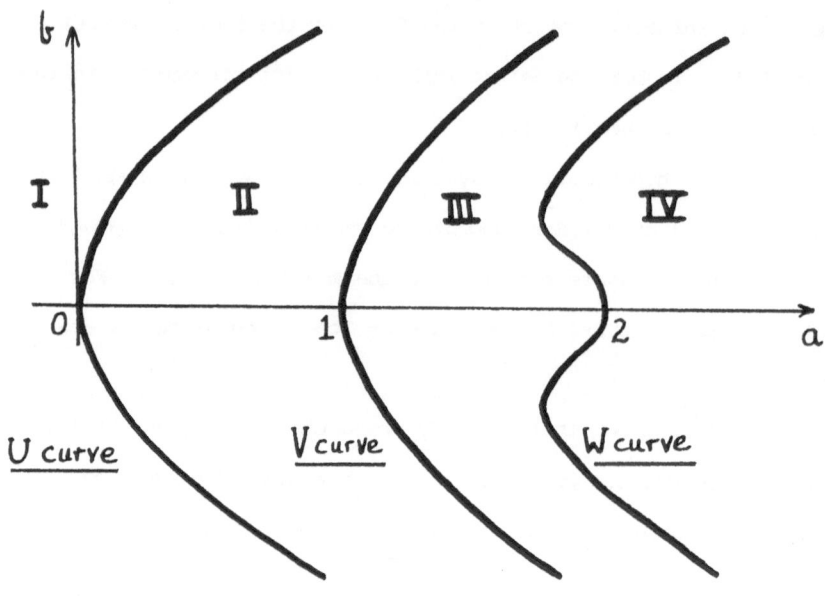

Figure 3.

    I now describe the classification of equations (4.1), (4.2) in terms of the rarefaction curves and inflection loci. There are four patterns of rarefaction curves, depending on the location of (a,b), as shown in Figures 3 and 4. In Figure 3, the U,V and W curves separating the four cases have the following interpretation. The V curve $a = 1 + b^2$ separates the cases according to whether the mapping dev dQ is orientation preserving (Cases III, IV) or orientation reversing (Cases I, II). In Figure 4, you can see the rotation of the eigenvectors. The arrows on the rarefaction curves indicate the direction of increasing characteristic speed. The W curve, given by a quartic equation, separates Case IV, in which one straight rarefaction curve passes through the origin, from Cases I-III, in which there are three straight rarefaction curves. Finally, the U curve $a = 3b^2/4$ separates case I, in which there are three inflection loci (all the curved rarefaction curves have inflection points), from Cases II-IV, in each of which there is only a single inflection locus.

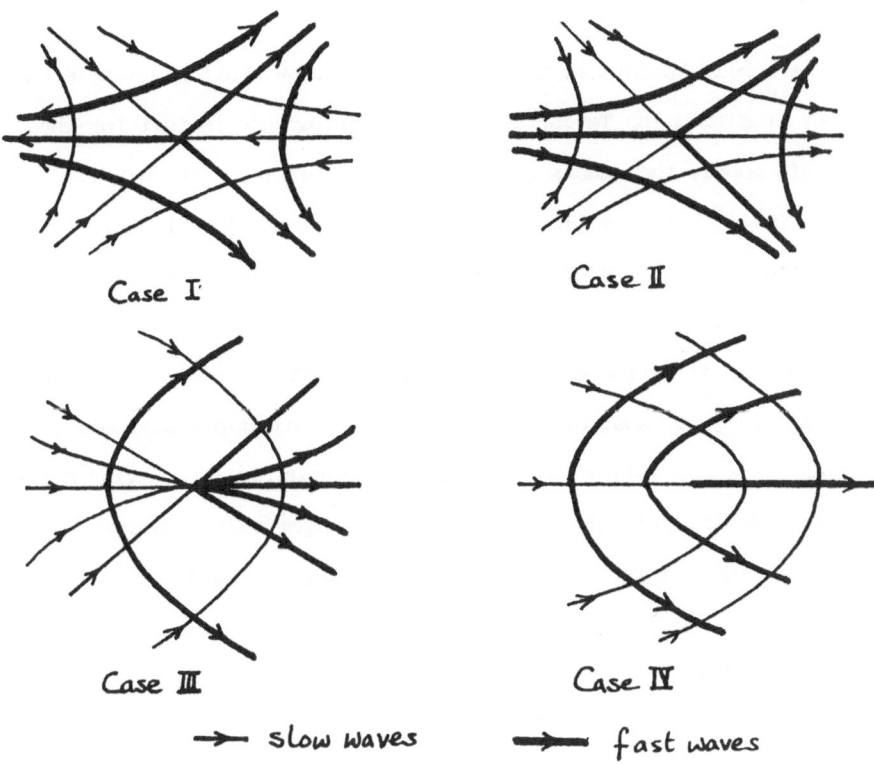

Case I   Case II

Case III   Case IV

→ slow waves   ⟹ fast waves

Figure 4.

Since the argument in Theorem 3.2 depends upon showing that dev dF is orientation reversing, one expects Cases I and II to be especially important in oil recovery. Here is a result concerning the Marchesin model:

Theorem 4.1. [8] The Marchesin model corresponds to cases I or II near the umbilic point. That is, the quadratic terms in the Taylor series expansion of the Marchesin model are equivalent to $Q(u,v)$ given by (4.2) with a,b in region I or II of Figure 3.

The umbilic points at the vertices of the saturation triangle can also be studied using Taylor series. The difference is that for the Marchesin model, the quadratic terms are degenerate, with $a = 1$, $b = 0$. Preliminary results indicate that the inclusion of cubic terms gives rise to a pattern of rarefaction curves similar to those of Case III in Fig. 4.

The classification of equations with umbilic points is described in detail in [8]. Most of the effort on these equations has gone into solving Riemann problems [3,5,9,10]. A particularly interesting paper by Tang and Ting [12] discusses nonstrictly hyperbolic equations arising in the context of nonlinearly elastic plane waves.

References.

1. J.B. Bell, J.A. Trangenstein and G.R. Shubin, Conservation Laws of mixed type describing three phase flow in porous media. SIAM J. Appl. Math., to appear.

2. S.E. Buckley and M.C. Leverett, Mechanism of fluid displacement in sands. Trans. AIME, 146 (1942), 107-116.

3. H. Holden, On the Riemann problem for a prototype of a mixed type conservation law. New York University preprint, 1986.

4. E. Isaacson, Global solution of a Riemann problem for a nonstrictly hyperbolic system of conservation laws arising in enhanced oil recovery. J. Comp. Phys., to appear.

5. E. Isaacson, D. Marchesin, B. Plohr, and B. Temple, The classification of solutions of quadratic Riemann problems, I,II,III. MRC, University of Wisconsin Technical Reports, 1985,1986.

6. B.L. Keyfitz and H.C. Kranzer, A system of hyperbolic conservation laws arising in elasticity theory. Arch. Rat. Mech. Anal. 72 (1980), 219-241.

7. P.D. Lax, Hyperbolic systems of conservation laws II. Comm. Pure Appl. Math. 10 (1957), 537-566.

8. D.G. Schaeffer and M. Shearer, The classification of 2×2 systems of nonstrictly hyperbolic conservation laws, with application to oil recovery; Appendix with D. Marchesin, P.J. Paes-Leme. Comm. Pure Appl. Math., to appear.

9. D.G. Schaeffer and M. Shearer, Riemann problems for nonstrictly hyperbolic 2×2 systems of conservation laws. Trans. A.M.S., to appear.

10. M. Shearer and D.G. Schaeffer, D. Marchesin and P.J. Paes-Leme, Solution of the Riemann problem for a prototype 2×2 system of nonstrictly hyperbolic conservation laws. Arch. Rat. Mech. Anal., to appear.

11. H.L. Stone, Probability model for estimating three-phase relative permeability. J.P.T. (1970), 214-218.

12. Z. Tang and T.C.T. Ting, Wave curves for the Riemann problem of plane waves in simple isotropic elastic solids. Preprint, Univ. of Ill., Chicago, 1985.